Routledge Reviva.

The Life Prince Henry of Portugal, Surnamed the Navigator; and its Results:

The Life Prince Henry of Portugal, Surnamed the Navigator; and its Results:

Comprising the Discovery, Within One Century, of Half the World.

by
Richard Henry Major

Routledge
Taylor & Francis Group

First published in 1967 by Frank Cass & Co. Ltd.

This edition first published in 2019 by Routledge
2 Park Square, Milton Park, Abingdon, Oxon, OX14 4RN
and by Routledge
52 Vanderbilt Avenue, New York, NY 10017

Routledge is an imprint of the Taylor & Francis Group, an informa business

© 1967 by Taylor & Francis

Publisher's Note
The publisher has gone to great lengths to ensure the quality of this reprint but
points out that some imperfections in the original copies may be apparent.

Disclaimer
The publisher has made every effort to trace copyright holders and welcomes
correspondence from those they have been unable to contact.
A Library of Congress record exists under ISBN:

ISBN 13: 978-0-367-26417-8 (hbk)
ISBN 13: 978-0-367-26420-8 (pbk)
ISBN 13: 978-0-429-29314-6 (ebk)

CASS LIBRARY OF AFRICAN STUDIES

TRAVELS AND NARRATIVES

No. 24

Editorial Adviser: JOHN RALPH WILLIS

THE LIFE OF
PRINCE HENRY OF PORTUGAL

SURNAMED

THE NAVIGATOR

AND ITS RESULTS

FROM AUTHENTIC CONTEMPORARY DOCUMENTS

BY

RICHARD HENRY MAJOR

FRANK CASS & CO. LTD.

1967

Published by
FRANK CASS AND COMPANY LIMITED
67 Great Russell Street, London WC1

First edition   1868
New impression   1967

Printed in Great Britain by
Thomas Nelson (Printers) Ltd., London and Edinburgh

ualentie bretaine

THE LIFE

OF

PRINCE HENRY OF PORTUGAL,

SURNAMED

THE NAVIGATOR;

AND ITS RESULTS:

COMPRISING

THE DISCOVERY, WITHIN ONE CENTURY, OF HALF THE WORLD.

WITH

NEW FACTS IN THE DISCOVERY OF THE ATLANTIC ISLANDS;
A REFUTATION OF FRENCH CLAIMS TO PRIORITY IN DISCOVERY;
PORTUGUESE KNOWLEDGE (SUBSEQUENTLY LOST) OF THE NILE LAKES;

AND

THE HISTORY OF THE NAMING OF AMERICA.

From Authentic Cotemporary Documents.

BY

RICHARD HENRY MAJOR, F.S.A., F.R.S.L., ETC.,

KEEPER OF THE DEPARTMENT OF MAPS AND CHARTS IN THE BRITISH MUSEUM;
AND HON. SEC. OF THE ROYAL GEOGRAPHICAL SOCIETY.

ILLUSTRATED WITH PORTRAITS, MAPS, ETC.

LONDON:
A. ASHER & CO., 13, BEDFORD STREET, COVENT GARDEN;
AND BERLIN, 20, UNTER DEN LINDEN.
1868.

TO

HIS EXCELLENCY

DOM FRANCISCO D'ALMEIDA,

COUNT DE LAVRADIO,

ENVOY EXTRAORDINARY AND MINISTER PLENIPOTENTIARY OF

HIS MOST FAITHFUL MAJESTY

AT THE COURT OF ST. JAMES'S,

ETC. ETC. ETC.,

THE LINEAL DESCENDANT OF

FRANCISCO D'ALMEIDA,

FIRST VICEROY OF THAT INDIAN EMPIRE

FOR WHICH PORTUGAL WAS PRIMARILY INDEBTED

TO THE GENIUS AND PERSEVERANCE OF

PRINCE HENRY THE NAVIGATOR,

THIS WORK IS INSCRIBED,

WITH EVERY SENTIMENT OF

GRATEFUL AFFECTION AND RESPECT,

BY

THE AUTHOR.

CONTENTS.

CHAPTER XX.

CHAPTER XXI.

CORRIGENDA.

Page 57, line 5, *for* " Utensilitus" *read* " Utensilibus."
" 64, " 1, *for* " 1320" *read* " 1306."
" 89, " 12, *for* " coast" *read* " east."
" 105, " 12, *for* " navigation" *read* " migration."
" 121, " 3 from bottom, *for* " Deschelier" *read* " Desceliers."
" 139, " 20, *for* " Alphonso fourth" *read* " Alphonso IV."
" 144, " 7, *for* " summits" *read* " summit."
" 156, " 9, " and" omitted at beginning.
" 159, " 9, *for* " Ferdinand" *read* " Fernando."
" 165, " 32, *for* " prince" *read* " Prince."
" 186, " 30, *for* IX. *read* XI.
" 197, last line but two, *for* " consists" *read* " consisted ;" last line, *for* " are" *read* " were."
" 211, " 9, *for* " Gilianez Datayde" *read* " Gonsalvez de Atayde."
" 256, " 6, " a" omitted.
" 260, " 28, *for* " beside" *read* " besides."
" 305, " 2, in note, *for* " Neile" *read* " Neale."
" 324, " 27 } *for* " John" *read* " João."
" 325, " 36 }
" 350, last line, *for* " writen" *read* " written."
" 368, " 2, *for* " Alonza" *read* " Alonzo."
" 391, " 20, *for* " preceding chapter" *read* " last chapter but one."
" 393, " 1, in note, *for* " Neile" *read* " Neale."
" 401, in pagination, *for* " 410" *read* " 401."
" 408, last line, *for* " impreter" *read* " interpreter."
" 414, " 16, *for* " do" *read* " de."
" 417, in running title, *for* " Westward" *read* " Eastward."
" 424, " 8, *for* " order" *read* " orders."
" 441, line 7, *for* " 1572" *read* " 1571."

PREFACE.

It may perhaps be fairly regarded as a matter of surprise
that no Englishman has hitherto attempted to prepare a
monograph of the life of Prince Henry the Navigator. If a
phenomenon without example in the world's history, resulting from the thought and perseverance of one man, might
be supposed of interest enough to tempt the pen of the
biographer, assuredly that inducement was not wanting.
When we see the small population of a narrow strip of the
Spanish Peninsula, limited both in means and men, become,
in an incredibly short space of time, a mighty maritime
nation, not only conquering the islands and Western Coasts
of Africa and rounding its Southern Cape, but creating
empires and founding capital cities at a distance of two thousand leagues from their own homesteads, we are tempted to
suppose that such results must have been brought about by
some freak of fortune, some happy stroke of luck. Not so:
they were the effects of the patience, wisdom, intellectual
labour, and example of one man, backed by the pluck of a race
of sailors who, when we consider the means at their disposal,
have been unsurpassed as adventurers in any country or in
any age. Doubtless, the geographical position of Portugal,
at the extremity of the European continent, had much to
do with the suggestion of its glorious mission ; but what else

besides danger and death could the formidable waves of the
Atlantic have suggested to her mariners, had it not been for
the courageous conception and unflinching zeal of one who,
during forty long years of even limited success, knew how
to blend patience with enthusiasm, and conquer disappointment by devoted persistence in what he had prescribed to
himself as a duty. The story of the life of such an one can
surely not be deemed an uninteresting or unimportant
matter.

Till a comparatively recent date, it is true, the materials
for anything approaching to a satisfactory biography had
not been brought within the reach of the historical student.
The Livy of Portugal, as he has been called, João de Barros,
had handed down to us some incomplete details from the
scattered papers of a contemporaneous historian whose
collected work has, as we shall presently see, recently been
made public property.

Two centuries after Barros, an elegant Portuguese writer,
José Freire, better known by the name of "Candido Lusitano"
(his pseudonym as a member of the Academy of the
Arcades), produced in 1758, at Lisbon, in 4to., a life of
the Prince, which was more to be commended for the graces
of its style than for the abundance or the accuracy of the
information it supplied. A translation of this work into
French, with a preliminary discourse, by the Abbé Cour-
nand, appeared in Lisbon and Paris in 1781, small 8vo.,
but afforded no additional knowledge respecting its subject.

After that time, the glorious little kingdom which, by the
mere energy of its children, had in old times maintained
itself for more than a century in the first rank of European
Powers, until condemned by the disaster of Alcaçer-quivir
in 1580 to a paralysing condition of silence and inaction,

began anew to do itself honour in the field of literature, as of old it had not failed to do in the same career as well as on the perilous surface of the ocean. In 1779 the Royal Academy of Sciences at Lisbon, which had been founded in the beginning of the century, was reorganised by the care of the Duke of Lafoens, and since then we have had brought to light, from time to time, a series of valuable materials for history, which the Archives of Portugal could alone be expected to supply.

Amongst these, some of the most important were embodied in the Collecção de livros ineditos de Historia Portugueza dos Reinados de João I., D. Duarte, D. Affonso V., and D. João II., edited by Don José Corrêa da Serra, in three volumes folio. Lisbon, 1790—1793.

But the most important of all the precious relics of the fifteenth century thus exhumed, was discovered in 1837 in the Royal Library at Paris by M. Ferdinand Denis, the distinguished *conservateur* of the Library of St. Geneviève; and one cannot but feel that Fortune was unusually just in allotting the glory of so noble a *trouvaille* to one who, out of Portugal, stands second to none. in intimate acquaintance with the details of Portuguese history and literature. This beautiful manuscript, which was drawn up in 1448, and fairly completed in 1453, was published in Paris in 184], 8vo., with a title of which the following is a translation* :—
" Chronicle of the discovery and conquest of Guinea, written by command of King Affonso V., under the scientific direc-

* Chronica do descobrimento e conquista de Guiné, escripta por mandado de el Rey D. Affonso V. sob a direcção scientifica e secundo as instrucções do illustre infante D. Henrique, pelo chronista Gomes Eannes de Azurara, fielmente trasladada do manuscrito original contemporaneo que se conserva na Bibliotheca Real de Pariz e dada pela primeira vez a luz per diligencia do Visconde da Carreira, precedida de uma introducçao e illustrada com algunas notas pelo Visconde de Santarem. Pariz, 8vo.

tion, and in conformity with the instructions, of the illustrious Infant Don Henrique [the Prince Henry of the present work], by the Chronicler Gomes Eannes de Azurara, faithfully copied from the original contemporaneous manuscript in the Royal Library at Paris, and now first edited by the Visconde da Carreira, Envoy Extraordinary and Minister Plenipotentiary of His Most Faithful Majesty at the Court of France; preceded by an Introduction, and illustrated with notes by the Visconde de Santarem," &c.

The title and period of this narrative will themselves convey some idea of its importance. The original manuscript is a magnificent specimen of the calligraphic skill required by King Affonso V., the first King of Portugal who endowed his country with a library. It is necessary, however, to state that in that condition it was not an original emanation from the hand of its recognised author Azurara. It was compiled from the rough narrative of one of Prince Henry's sailors, Affonso de Cerveira, who had himself been engaged in those great discoveries which we have now to narrate, and had given a description of them under the title of " History of the Conquest of the Portuguese along the coast of Africa."*

This precious monument of the glory of Portugal sets forth from cotemporary testimony the attempts of the brave men who first penetrated the Sea of Darkness (as the Arabs called the Atlantic beyond the Canaries) which till then had baffled the efforts of the most experienced navigators of Europe. Although in the composition of the original chronicle there is a display of pedantry common to the period, a little reflection will show that such pedantry is highly excusable in works produced before the invention of

* Historia da Conquista dos Portuguezes pela Costa d'Africa.

printing, when erudition could be acquired only through the medium of manuscripts which were naturally at the command of only a very few. Nay, it was more than excusable, it was valuable, for it informed the reader of the sources from which the author's information was derived. At the same time it will be obvious that the pedantry and prolixity which may be so justly excused in a writer of the first half of the fifteenth century, would prove but a wearisome repast except to the palate of an antiquary especially interested in the subject. To such the original is within as easy access as are most of the works which offer to an antiquary the pabulum in which he delights; but for the present purpose it has been deemed sufficient to extract the *facts* with which the author supplies us. In this respect his work is invaluable, for he not only lived with Prince Henry, but was personally acquainted with most of his intrepid explorers, more than fifty of whom were attached to the Prince's household and received their nautical instruction under his auspices. We may, therefore, feel sure that he supplied much that was wanting in the original manuscript of Cerveira which formed the basis of his chronicle. So much of the subject matter of the work as was subsequently brought together by Barros in his "Asia", published at Lisbon in 1552-53, fol., was derived by him from scattered, torn, and mixed fragments of Azurara's original rough drafts.

In the " Paléographie Universelle " of Silvestre, Paris, 1841, fol., is a facimile of the first page of the fair copy of Azurara's manuscript, which is described as a small folio volume written on parchment and consisting of three hundred and nine pages with two columns in each page. At the end it is stated to have been written by João Gonsalvez, calligrapher to Affonso V., and completed on the 18th of February, 1453; and in a letter of the same date the author dedicates his work to the same sovereign by whose orders it had been composed.

Very soon after this date the work disappeared from Portugal. Damião Goes, the chronicler of the Life of King João I., Prince Henry's father, knew nothing of the book beyond the name of its author. It would seem to have been presented by Prince Henry to a King of Naples, inasmuch as it appears to tally with a book to which the celebrated Fr. Luis de Souza makes the following allusion in his Historia de S. Domingos, P. I., Liv. vi. cap. 15, p. 629, edition of 1767. Referring to the Prince's motto, "Talant de bien faire," and the oak leaves and acorns and pyramids which formed his device (see frontispiece), he says, that they occur "in a book of the Prince's discoveries which Prince Henry himself sent to a King of Naples, and which I saw in Valencia amongst some choice curiosities belonging to the Duke of Calabria, the last male descendant of those princes, and who was there as Viceroy." The Vicomte de Santarem propounds a very reasonable conjecture that King Affonso V., for whom, and not, as Souza states, for Prince Henry, the work was compiled, presented it to his uncle Alfonso King of Naples, surnamed the Magnanimous, between the years 1453 and 1457, for in the year 1457 (see Santarem's "Quadro Elementar," tom. i., p. 358) Martim Mendes de Barredo was sent as ambassador from Portugal to Naples, and Alfonso was a man who took great interest in literature, especially in accounts of voyages of discovery, and was well acquainted with the language in which this book was written. However this may have been, the manuscript was still in Spain at the beginning of the

eighteenth century, for on one of the blank leaves at the
end is a note to the effect that it belonged to the library of
the late Don Juan Lucas Cortes, Member of the Royal
Council of Castile, anno 1702. No one knows how or when
it became the property of the Imperial Library in Paris.
The Vicomte de Santarem had reason to suspect that it was
long after the revolution, and that the acquisition had been
a rather recent one. Immediately after its discovery by
M. Ferdinand Denis, the Vicomte de Carreira, Envoy Extraordinary and Minister Plenipotentiary of Portugal in France,
obtained permission to print it, and to secure accuracy,
copied the text with scrupulous fidelity with his own hand.
A learned Portuguese philologist, Senhor José Ignacio
Roquete, revised the proofs and made a glossary of old and
obsolete words and phrases, which would have been otherwise absolutely unintelligible to the general reader.

The miniature which forms the frontispiece of the present
volume is a chromolithograph from an exact facsimile of
that which is in the original manuscript, for procuring which
I have to acknowledge my indebtedness to the kind intervention of M. Prosper Merimée. No greater proof could be
adduced of the excellence of the copyist's handiwork than
the perfection of the picture when exposed to the test of
photography, for no modification of tint in the faithfully
coloured copy produced the slightest deviation from perfect
drawing in the monochrome of the photograph, a result
greatly to the honour of M. Avril, the artist to whom this
copy is due. The original miniature in the manuscript
is the only authentic portrait of Prince Henry which the
Portuguese possess. The Prince is represented dressed in
mourning, his head covered with the large barret cap without
any insignia, and his hair cut short according to the custom

of the time on such occasions. As the chronicle was finished
in 1448, and as the Prince's brother Dom Pedro lost his life
at Alfarrobeira on the 20th of May, 1449, it is most probable that the portrait was taken while the Prince was in
mourning for his illustrious brother, for the fair copy of the
chronicle was not completed till 1453.

Azurara had intended, as he himself states in his last
chapter but one, to write a second volume, containing further discoveries made during the life of Prince Henry; but
this volume, if written, has not yet been discovered. Azurara was also the author of the " Chronicle of the Conquest
of Ceuta " which has supplied the material of the chapter
headed " Ceuta " in this volume, and also of the " Chronicles of Dom Pedro, and Dom Duarte de Meneses," the first
governors of that place, which describe the warfare carried
on in Africa, and may be considered as a continuation of his
" Conquest of Ceuta."

The dates of the birth and death of Azurara are entirely
unknown; but according to Mattheus de Pisano, the preceptor of Affonso V. and translator into Latin of Azurara's
" Conquest of Ceuta", it was only in middle life that he
applied himself to study, having till then been entirely
ignorant of literature and solely occupied with warlike pur-
suits. This is the more remarkable as we find him held in
such high estimation by Affonso V., as to be nominated by
him (on the 6th June, 1454) keeper of the archives of the
Torre do Tombo, in succession to no less a person than the
venerable Fernam Lopez, the father of Portuguese history,
and, beyond all question, the best chronicler of any age or
any nation. Azurara himself went to Africa and remained
there a long time to make himself acquainted with the
scenes and circumstances of the deeds which he had to de-

scribe. While at Alcaçer Çeguer he received the celebrated
letter from the King as to his merit as a chronicler, which is
printed at the beginning of his "Chronicle of Dom Duarte
de Meneses", a letter which does even greater honour to the
King who wrote it than to the subject who received it, for
its affable and even affectionate expressions clearly show the
sovereign's consciousness that he was doing himself an
honour when he honoured the intellect of his subject with
the familiarity of friendship.

It is greatly to be regretted that a valuable MS. work in
the library at Evora entitled "Esmeraldo de Situ Orbis," by
Duarte Pacheco, a knight of the household of King João II.,
should still remain unprinted. It is a sort of historical and
geographical description of the discoveries of the Portuguese,
which, if we may judge by the titles of the chapter supplied
to us by Joaquim da Rivera, the librarian at Evora, in his
excellent catalogue of the MSS. in that library, would throw
much light on the geographical details of these early discoveries. Fortunately, however, we possess some important
extracts therefrom, which have been given us by Albano da
Silveira in his "Memoria chronologica ácerca do descobrimento das terras do Preste João das Indias e Embaixadas
que a elle enviaram os Portugueses."

Another manuscript, but recently given to the world in
print, is valuable more for the quality than the quantity of
the material which it supplies for the illustration of our
subject. This is the "Leal Conselheiro" or "Faithful
Adviser," from the pen of King Duarte, Prince Henry's
elder brother. So simple, dignified, and loveable a picture
of the home affections as existing among the members of a
regal family, is perhaps not to be found elsewhere. It is
the unaffected, nay almost unconscious, exposition of every

manly and gentle virtue that could dignify the character of
a prince as a Christian, a patriot, and a soldier. But more
conspicuous than all the other qualities which are therein
exhibited as characteristic of the members of this family, is
the strong and loving affection existing between all of them,
tempered by a lofty tone of mutual honour and respect which
finds its culmination in the profound reverence of all of
them for the sacred persons of the King and Queen. No
higher eulogium could be unconsciously paid to the training
bestowed upon their children by King João and Queen Philippa than the *tone* as well as the words of this noble production. When Affonso V. first established a library in his
palace at Lisbon, one of his first cares was to exhibit this
beautiful and richly ornamented manuscript which had been
left him by the King his father. It is now in the Imperial
Library in Paris. It was not till the year 1842 that it was
published in Paris by the Reverend J. I. Roquete.

In the absence of the second volume promised by Azurara,
it is not of little moment that we possess the accounts of the
Venetian Cadamosto's voyages occupying the interval between
the completion of Azurara's first volume and the death of
Prince Henry. But although these Venetian narratives,
which have been reprinted several times, have been highly
commended for their minuteness of detailed description, I
cannot very cordially join in the eulogium; for on points
where Cadamosto's accuracy can be tested, as for example in
the matter of dates, I have almost invariably found him
wrong, and in such a case minuteness of detailed description
only enlarges the field for misgiving and distrust. Indeed,
as will be seen in the chapter on the Cape Verde Islands,
I have shown that Cadamosto's description of his second
voyage abounds in inaccuracies and inconsistencies, and, what

is worse, of mal-appropriation of credit which did not appertain to him.

A remarkable instance of his jumbling two accounts together will be seen on pages 284 and 319, where the discovery and naming of Cape Roxo and parts adjacent are at first claimed by himself, as occurring in his second voyage, and afterwards ascribed to Pedro de Cintra, of whose voyage he is also the narrator.

But, happily, another document, never hitherto translated into English, has been brought within my reach by the recent researches of our learned fellow-antiquaries in Ba : varia. In the year 1847, the Academy of Sciences of Munich printed a memoir by Dr. Schmeller on a most interesting collection of manuscript documents formed by a German in Lisbon, in the year 1507. Although bearing the Portuguese-sounding name of Valentim Fernandez, he was a Moravian by birth, but, being of German descent, he styles himself occasionally Valentim Aleman and sometimes Valentim the Moravian.

Before referring to those documents in his collection which have been of especial service in this work, I will state briefly the history of the collection itself and of its collector. Valentim Fernandez was a printer. At that time the art of printing led many Germans into foreign countries, and he wandered into Portugal. We find him in 1495 at Lisbon engaged in conjunction with another German, Nicholas of Saxony, in printing the " Life of Christ," by the Carthusian Monk Ludolph of Saxony, but which had been translated into Portuguese in 1445, by Bernardo, a monk of the Cistercian Monastery of Alcobaça. On account of his knowledge of the German language, he was appointed Notary for the Germans in Lisbon, that he might draw up all the

agreements and written negociations which took place with
German merchants, and also authenticate translations from
the Latin.   Soon after Valentim Fernandez appeared not as
a printer only, but as an editor.   Dom Pedro, Prince Henry's
brother, had in 1428, brought back from Venice a valuable
manuscript of Marco Polo, which had been presented to him
as a compliment by the Signoria of the Republic.   From
this manuscript, and from the Latin text of the Dominican
Friar Pepino of Bologna, which had been sent from Rome
to King João II., Valentim made a translation of the work
into Portuguese, together with the "Travels into India" of
the Genoese Geronimo de Santo Stephano.   He also translated the travels of the Venetian Niccolo de' Conti from the
Latin text of Poggio Bracciolini.   The importance of these
works to the King Dom Manoel, by whose order they were
translated, may be judged by the fact of their containing descriptions of journeys into India at that early period.   English translations both of Conti and of Santo Stefano, the
former by John Winter Jones, Esq., the present Principal
Librarian of the British Museum, will be found in the volume
which I had the honour to edit in 1857, for the Hakluyt
Society, under the title of " India in the Fifteenth Century."

It was doubtless in connection with such studies as these
that Fernandez subsequently compiled the collective geographical work which is immediately under our notice, and
which was intended to furnish an account of the countries
discovered by the Portuguese in Africa and India.   Its contents are as follows :—

1. Azurara's " Chronicle of Prince Henry's Discoveries
of Guinea," down to 1448.

2. Diogo Gomez' Narrative, down to 1463.

3. Narrative of Gonzalo Pirez, down to 1492.

4. Narrative of João Rodriguez, down to 1493.
5. "Journal of Hans Mayr," 1505-6.
6. Fernandez' " Description of Africa," 1507.
7. Fernandez' " Account of the Islands in the Atlantic,"
with Plans.
8. " Ships' Routes, or Instructions for Pilots."

Of these the narrative of Diogo Gomez has been given by
Dr. Schmeller in full; but as the earlier portion is a recital of
voyages made under the auspices of Prince Henry, but with
which he himself had nothing to do, and which being derived
from hearsay, would not be regarded as absolutely trustworthy, while they are better narrated by Azurara, I have
extracted that portion only which describes his own adventures.

The document is the work of a half-educated man, much
more of a sailor than a student, but it throws light upon a
subject—the discovery of the Cape Verde Islands—on which
I am able to demonstrate that Cadamosto had written with
the greatest inaccuracy.

Gomez' success as an explorer was remarkable, and his
power of conciliating even hostile native chiefs by dint of
sheer courage and tact was beyond all praise. Another but
minor point of interest in his narrative is, that it is the
only document that I have met with in which the slightest
detail has been preserved of the death and burial of Prince
Henry, whereas Gomez was, by especial order of the King,
placed in immediate guardianship over the remains of his
revered master until they were consigned to the tomb.

The other documents in Valentim Fernandez' work have
not been printed by Dr. Schmeller verbatim, but simply described with a running commentary by himself. One of them,
however, entitled " Das ilhas do Mar Oceano," was of so much

importance to the early history of the discovery of Madeira,
that I procured from the library at Munich a verbatim copy,
for which I am indebted to the kindness of Dr. Halm the
distinguished chief of that important library, and also to the
most obliging care of Professor Kunstmann who afforded me
the benefit of his learned supervision of the transcription from
the quaint and difficult Portuguese of the early manuscript.
The value of this freely rendered kindness was the greater
that no one was so competent as this eminent *savant* to deal
with the difficulties of this task, inasmuch as we have
already received from the hands of Professor Kunstmann a
variety of most valuable memoirs on the various documents
comprised in the collection of Valentim Fernandez,* and which
have been of much service to myself in the present work.

By means of the document on Madeira, combined with
other evidence, it has been my good fortune to *establish* the
truth of the story, hitherto much disputed, of the accidental
discovery of Madeira in the fourteenth century by the
Englishman Machin; for this document is *earlier* than the
earliest yet produced in which that story was related, and
being entirely independent of any other, is a proof of the
derivation of all the accounts from an earlier source.
Demonstrative evidence of the former existence and
genuineness of that original source is adduced in the
chapter on " Porto Santo and Madeira."

* Afrika vor den Entdeckungen der Portugiesen. München, 1853.
   Die Handelsverbindungen der Portugiesen mit Timbuktu im xv. Jahrhun-
derte.
   Valentin Ferdinand's Beschreibung der Westküste Afrika's bis zum Senegal
mit Einleitung und Anmerkungen. München, 1856.
   Valentin Ferdinand's Beschreibung der Westküste Afrika's vom Senegal bis
zur Serra Leoa. München, 1860.
   Valentin Ferdinand's Beschreibung der Serra Leoa mit einer Einleitung über
die Seefahrten nach der Westküste Afrika's im vierzehnten Jahrhunderte.
München, 1861.

History, for its own sake, is more zealously cultivated in
Germany than in England, and it is in Germany that of late
years the name of Prince Henry the Navigator, almost unknown in England, has been found to engage the attention
of the learned. In 1842 a biography of the Prince was commenced by Professor J. E. Wappæus of Göttingen, but was
unfortunately not proceeded with beyond the first volume,
which was entirely occupied with preliminary matter of the
most erudite and laboured character. It is probable that the
author, whose industry and zeal could lead him to devote
three hundred and sixty-five pages of close octavo print to
historical events anterior to the birth of the subject of his
biography, would not be satisfied without exhausting the
contents of the Torre do Tombo itself, when he came to con-
front in reality the task which he had proposed to himself.
But "non cuivis homini contingit adire Corinthum ;" and
I, for my own part, with all the advantages already mentioned at my disposal, am quite prepared to suppose that
the biography of Prince Henry could only have full justice
done to it by one who had the opportunity, the talent, and
the industry to investigate the cotemporary treasures of the
Torre do Tombo. A labour so Herculean is more, perhaps,
than we may hope to see undertaken, unless possibly by the
greatest of modern Portuguese historians, Alexandre de
Herculano, who is perhaps more intimately acquainted than
any other with the valuable contents of that great historical
treasure-house.

So recently as 1864, a life of Prince Henry in Germany
was published in Dantzick, 8vo., by a German clergyman,
named Gustav de Veer, of Dantzick. This gentleman had
resided two years in Madeira, and took a loving interest in
the life and deeds of the noble Prince whose name is idolized

in that country.   The history of discovery was, moreover,
a subject for which Herr de Veer may be said to have an
inherited attachment, for he is the lineal descendant of that
famous Dutchman, Gerrit de Veer, who wrote the accounts
of three remarkable voyages made by the Dutch in 1594-
1596, with the view of finding the way to China by the
North-East, and in the last two of which he was himself
engaged.   In the earlier of these two, Spitzbergen was discovered and circumnavigated.

   In the pages which precede the actual life of Prince Henry
himself, Herr de Veer has given an account of the doings of
the Portuguese navy before the Prince's time.   This, in my
humble judgment, does not appear to be of very great
moment, and in the present work I have thought it prefer-
able to relate what had been done, or said to have been done,
in previous times on the face of that vast ocean which was
to be the field of the Prince's fame.   1 have but one word of
objection to make to Herr de Veer's otherwise interesting
and able publication; viz., that he has inserted a *fancy*
portrait of Prince Henry, a portrait not only based on no
authority whatever, but a slander on the masculine character
of the Prince himself.   A portrait, if faithful, will convey at
a glance more information than pages of written description:
it is manifest, therefore, that no process could more effectually neutralize the purpose of a biography, or show more
disregard for men's opinions of him who is depicted, than to
present a portrait without even the pretence of a prototype,
and which neither in face nor figure contains one single
characteristic of the original.   I feel sure that all the gravity
of this incontestable fact did not suggest itself to the mind
of Herr de Veer, when he allowed himself to put forth the
delineation of the emasculated creature, which forms the

frontispiece to his book, as a portrait of that firm and large-minded man to whose genius and perseverance we are indebted for our knowledge of one half of the world: I say it deliberately, to whom we are indebted for our knowledge of one half of the world; and it is for this reason that this work is entitled, " The Life of Prince Henry the Navigator, and its Results."

The glory of Prince Henry consists in the conception and persistent prosecution of a great idea, and in what followed therefrom. This book, then, is rather a record of the glory than of the mere life of Prince Henry. That glory is not a matter of fancy or bombast, but of mighty and momentous reality, a reality to which the Anglo-Saxon race, at least, have no excuse for indifference.

The coasts of Africa visited;—The Cape of Good Hope rounded;—The New World disclosed;—The seaway to India, the Moluccas, and China laid open;—The globe circumnavigated, and Australia discovered: within one century of continuous and connected exploration. " Such," as I have stated in my closing chapter, " were the stupendous results of a great thought, and of indomitable perseverance in spite of twelve years of costly failure and disheartening ridicule. Had that failure and that ridicule produced on Prince Henry the effect which they ordinarily produce on other men, it is impossible to say what delays would have occurred before these mighty events would have been realized, for it must be borne in mind that the ardour not only of his own sailors but of surrounding nations owed its impulse to this pertinacity of purpose in him."

In my remarks on the slave-trade I have been largely indebted to a paper on the subject in the " Revista Literaria," 13° Janeiro, 1839.

The "Ensaio sobre a statistica das Possessões Portuguezas no ultramar," by the careful José Joaquim Lopez de
Lima, has been of much service to me in the description
of discovery of the West Coast of Africa and the islands in
the Gulf of Guinea.

For the important voyage of Vasco da Gama I have
followed the "Roteiro," edited, in 1861, by A. Herculano
and the Baron de Paiva.

The voyage of Magellan has been digested from Pigafetta's account, collated with Peter Martyr, Herrera, Gomara
and Navarrete.

Elsewhere I have given a list of the principal works
which have been resorted to in the construction of the
volume.

After that I had sent to the press my refutation of the claims
of the French to priority in discovery of the coast of Guinea—
claims till now uncorroborated by any documents—a new book
reached me, entitled "Les Navigations Françaises, et la Révo-
lution Maritime du xiv⁰ au xvi⁰ siècle, d'après les documens
inédits tirés de France, d'Angleterre, d'Espagne, et d'Italie,"
par Pierre Margry, Paris, 1867, 8vo., in which those claims
were re-asserted, with the following important addition.
The author stated that in the year 1852 a friend of his,
M. Lucien de Rosny, had occasion to visit the British
Museum for the purpose of making philological researches.
While there, a Mr. William Carter, drolly described as "un
homme distingué d'Oxford Street," "seeing him search for
old French texts, placed at his disposal a volume containing a series of detached pieces, copied probably towards the
middle of the 17th century. M. Lucien de Rosny having
found in this collection some very curious things connected
with his studies in the French language, obtained per-

mission to copy them, and among them was a document of which the following was the title: ' Briev Estoire del navigaige Mounsire Jehan Prunaut Roenois en la tière des noirs homes et iles a nous incogneus avec les estranges façons de vivre des dits noirs et une colloque en lor language.' " M. Margry then, not very intelligibly, says, " This last part is wanting, as well as some lines of the narrative effaced by time or damp."

The following is the text :—

" Ou mois de Septembre, M IIIc. soixante et quatre de l'Incar[2] nacion nostre Signor, ceus de Diêpe et Roan, aparillerent deux naues et orent por amirax* Mesires Jehan li Roanois, home de grant ronom en la tière de Normendie, et singlèrent longement en mer, à la noel, au liu d'Ovideg,† où onc n'avoient esté encoire cil Normendie, et ancrèrent par de lâ pour avencier lor afaires as ung liu moult chaleureus, q'on apiele as jor cap Bugiador, qui siet au réaume de la Guinoye. Li Gilofs (ainsi sont apielés les gents cel partie, qui tot noir sont de visaige et de pel et tot nus, sinon la o côvient de mucer), onc n'avoient vu homs blancs, si que ceus qui virent la nes furent espoventez, et tost retornèrent de rechef ensemble jouste la mer, à grant plante de lor compaignons, pour veoir ceus Normans, mes point ne cuidoient entrer sor lor nes, jusques ils furent asseurtez que cil Normans ne voloient mi les le dangier o les navrer. Les boun naviors, qui tos estoient de grant cuer, lor dounnérent a fuson petits juiaus et présouns, et les firent boire boun vin vermail, com que moult les esjouirent et les affierent. Adoncques les gents noirs de céans lor douèrent morphi, piaus de bestes sauvages et autres coses de lor pais fort estranges à veoir. Quât lur not fut plein d'aveirs precios et autre belle rien que ce estoit mervelle. Mesire Jehan, soun frère Legier et les compaignons de sa navie, de joie resbaudis, firent entendre cel homes noirs qu'ils rétorneroient enkoires là l'an ensuyvant et

* Amiral veut dire ici simplement commandant.
† Dans sa première décade de l'Asie, Barros dit qu' Ovidech est le nom que les naturels donnaient dans leur langue au fleuve que les Portugais ont depuis appelé Sanaga, du nom d'un des principaux du pays.

qu'ils se approvisionassent de cel marchâdises, cô que il li asseure-
rent. Adoncqs si drecièrent les veiles et despleierent à vent et
vers Normendie retornérent et sîglerent as mois. . . . . . .

. . . .
        (Plusieurs lignes effacées par l'humidité et illisibles.)
. . . .
li dits naviors et lor chef mesire Jehan là Roenois firent lor apareil
por quere aveir. . . . .

. . . .
                    (Lignes illisibles.)
. . . .
orent iiij naus et s'en retornerent par illuec et il besoignerent avoec
ces homs noirs. Mes la lonc ten ne porent estre porce que les
naus furent molt adomagiés par les pluies et grant boraskes orrible
et ténébrosse avoecq bise qui vient d'Orient et qui lor estoient.
Adoncques messire Jehan requerit gens d'illecq permission de
prenre tiere et bastir plusor masons por i mestre cels marchandises
et eus à saveté. Co que les seingnors volontiers li otroierent et les
aidierent à fere cels masons, adonqs cels de la nave traistérent lor
nes sur la costière. Les seingnors cel partie moult desiroient
l'alliance messire Jehan, et de ce tems comença li fait de mar-
chandise avoec li naviors de Normandie et cils homs noirs. Lors
fist asambler mesire Jehan les gens de sa navie et lor demanda sil
voloient illuec sejorner et ilz li dirent qil n'avoient aucun qui le
contredist et que s'il lui plesoit si establir, ils le tenroient a signor
et avoez qui bien est digne lor—mes petit apries sequerelerent, si
que ces que l'ung voloit, l'autre se desdisait a tant que les naviors
s'en retornerent en Normendie.

   " L'an M IIIe. septente et nœuf, lamirax Jehan li Roenois aparila
à son coust une naut moult grande et biele que il apiela Nostre
Dame de Boun Voiage, parce que ele estoit richement imagiée en
bosc et painte marviliensement. Mais il la mist detri et sor caue,
solement en Septembre, quar il savoit, comme dit est, que les
pluies tempestoises qui efondoient sor ces costes foraines, tres
mois paravant, estoient moult perilouses et q'il estoit mort de cele
pestilence et malage grant plaute domes naus lor masons come sor

l'aigue et l'air en cel saison est molt punais et brulant par un tonoire continuet. Adoncques morurent illuec (Dieu ait lors aames), Legier frere mounsire l'Amirax, Gervois, Sebille, Haibiers, Torcol, Tiebau, Doumare, Odon Cambers, los vaillant nots de Normandie, sans qu'ilz porent trover ung sol mire dans tot le pais. Mes li boun sire Jehan l'Amirax revint apriès Paske, en sa nes avœucques li remainant sa navie et grant plante d'or que li homs noirs li avoient douné.

"Li Roi, ki alors estoit à Diepe, envoia à messire Jehan et as compaignons li cuens de Pontiex et i ot message assés pour lor dire qu'il voloit les voir incontinent. Adoncques messire Jehan et cel sa navie s'en vinrent estament avecq le quens, et furent moult bien receus du Rois, de ses barons et damoiseles, quar de lor besoing estoient moult engries et cuidoient qu'ils estoient morts illuec.

" 'Biaux sires Prunauts, Dieu vous maintiegne tos !' fist le Rois et moult débonairement les festoya deus jors et firent boine ciere et ce fu joie, tant que nus poroit dire. Et li Rois requist messire Jehan l'Amirax qu'il li raconta les novieles et miervelles de la pais d'illuec ils s'en venoient. Quant li Rois ot oi ses grant proeces, les dons li fist et li donna une belle tierre. Par deseur le fit amirax sa navie, dont moult s'esjoit mounsire Jehan, qui pour tant jugia honour as Rois comme à signour. Si vos dirai que delà avint li non Prunaut à messire Jehan et que il le warda parce qu'il estoit moult preu, fier et hardi en fait de navigaige et homme de haut sens. Ançois li Roi volsit que sa progenye et lignye furent apielés Preunauts, comme fius de vaillant, preu et gentil navior. E ces dons confirma li Rois de son saiel sor cartes escrites, si que d'oir en oir il le doient tenir.

"Adoncques après que messire Jehan et tos ses compaignons furent molt festoiés, ils trerent vers Roan et chevaucoit mounsire Jehan le Pru navior avecques son escu pendu as cos, sor un pale-froi, molt richement arnacé et atorné et li autre come ilz peurent. L'arcevesque de Ruan et tote sa clergie, en oiant qu'il s'en venoit avœc tos cel sa navie ala encontre et lor fist moult honours, quar il savoit jà que mounsire Jehan estoit retorné et q'il estoit moult

aimé de Dieix et ses sains, kar il avoit edifié illuec petite kapiele,
et ke il wardoit por pastour frai Piere li Normant moult bon cler
pour doctriner ces paiens et mescreans à aimer Dieu, bien parler,
praiecier et por destruire la loi paienne.   En cel kapiele qu'on dist
Nostre Dame furent enfonis moult honorablement li naviors qui
departirent aluel (?) comme dist est, de la pestilence.   Adoncq à
Roan aveucq l'arcevesque vint à l'encontre messire Jehan et compaignons grand cevaucie des signors et si ot gens et manans à pié
assés, siergeans et borgois de Diepe, Kaan, Chieresborg, et de
totes les cités de Normendie, que là estoient venu pour veoir li
gentil amirax et ses prus compaignons.   Natent, la fême cest
amirax, dame belle et saige, e ele ert voirement la plus belle riens
qui fust oncques et estoit de grant lignaige en Saxonie avec Légier
son fiu et Erkenbous, frére á cestui, ambedui petits enfens, qui lor
boun pére acolêrent et beserent, et iceus signors, borgeois et
manans en grand lesse et tot ensemble li menerent jusques à son
ostel, car nul mot n'avoit dit de mounsire Legier et des autres qui
morts estoient en la tiere foraine.

   "L'an ensuivant messire Jehan Prunaut resta empres l'ost li
Roi, mes envoiia oultre mer sa nes nostre Dame come pieça, cil de
Diepe et Roan lor naus Saint Nicolas et l'Espérance.   Nostre Dame
ancra as liu qu'ils apeliérent la Mine par la grant plante d'or qui
saportoit de par entor.   Illœuc est icele Kapiele de la Benoiste
mere Dieix que mounsire Jehan fonda, come dit est, aveucq un
petit castiaus fort et fortelesce et une mason quarrée que i fit fere
sor un borg qui s'apiéle la terre des Pru-naus par remembranco
d'iceus et de lor amirax come aussi petit Diepe, petit Roan,
petit Germentruville et petit Paris, porce qu'il venus estoient de
Diepe, Roan et Paris.   Là aussi firent forz castiaux as lui q'on
dit Cormentin et Akra.

   " L'an miiij et dis se departit grant plante des Mariniers de Normandie et les merchants perdirent lors ricesses qui estoient
maugiées par les gueres qui lors estoient et en onze ans deus naus
a tot solement alerent à la costiere d'or et un por le grand Siest et
petit apriés les guerres estant moult estormes sur eaues come sur
tierre les besoignes des merchandises furent destourbées et
destroites."

The translation of this manuscript is as follows :—

"In the month of September, 1864 of the Incarnation of our Lord, those of Dieppe and Rouen equipped two ships, and had for admiral (or captain) Monsieur Jehan le Rouenois, a man of great renown in the land of Normandy, and sailed a long time on the sea, till Christmas, to a place called Ovideg, where those of Normandy had never as yet been, and anchored [par de là] to advance their affairs at a very hot place, which is called now-a-days Cape Bugiador, which belongs to the kingdom of Guinea. The Gilofs (so the people of these parts are called, who are quite black in their faces and skins, and quite naked except where covering is necessary) had never seen any white men, so that those who saw the ship were frightened, and all turned back together to the sea with a great number of their companions, to see these Normans, but did not dare to get into their ships till they were assured that the Normans did not wish to hurt or grieve them. The good sailors, who were all generous, gave them a profusion of little toys and presents, and made them drink good red wine, so that they rejoiced and emboldened them much. Then the blacks in their turn gave them ivory, skins of wild beasts, and other things of their country very strange to see. When their ship was full of precious commodities, and other fair things marvellous to behold, Messire Jehan, his brother Legier, and his shipmates, full of joy, made the blacks understand that they would return again in the following year, and that they would supply themselves with such merchandises as they promised them. Then they hoisted their sails, and spread them to the wind, and returned towards Normandy, and sailed till the month . . . .
(Several lines effaced by damp and illegible.)
the said sailors and their chief, Mesire Jehan le Rouenois, made their preparations for seeking commodities (?) . . . .
(Illegible lines.)
had four ships, and returned by this way, and dealt with the black men. But they could not long remain there, because the ships were much damaged by the rains, and great and horrible squalls, and darkness with an east wind, which came upon them. Then

Messire Jehan asked permission of the people of those parts to
take land and build several houses wherein to put his merchandises
and his men [?] in safety. Which the chiefs willingly granted
him, and helped them to build these houses, and then those of the
ships drew up their vessels on the coast. The chiefs of those parts
much desired the alliance of Messire Jehan, and from this time
began the commerce between the sailors of Normandy and these
black men. Then Messire Jehan assembled together the people of
his ships and asked them if they would sojourn there, and they
replied that they had nothing to say against it, and that if it
pleased him to establish himself there, they would have him for
their lord, and acknowledge him worthy of them; but a little
while after they quarrelled, so that what one wished the other
contradicted, and so the sailors returned to Normandy.

   " In the year 1879, Captain Jehan le Rouenois equipped at his
own cost a very large and beautiful ship, which he called *Notre
Dame de Bon Voyage*, because it was richly carved in wood and
marvellously painted . But he [la mist detri] and launched it only
in September, for he knew, as has been said, that the tempestuous
rains which poured down on these foreign coasts, three months
before, were very perilous, and that there had died of the pestilence
and illness a great number of men in their houses, as the water
and the air at this season have a bad smell and burn with continual thunder.   There died there (may God rest their souls)
Legier, brother to the captain, Gervois, Sebille, Haibiers, Torcol,
Tiebau, Doumare, Odon Cambers, all valiant sailors of Normandy,
without finding a single physician in all the country. But the
good sire, Jehan the Captain, returned after Easter in his ship,
with the fleet that remained to him, and a great quantity of gold
which the black men had given him.

   " The King, who was then at Dieppe, sent the Count of Pontieux
to Messire Jehan and his companions, and charged him to tell them
that he wished to see them immediately.   Then Messire Jehan and
those of his ships went instantly with the Count, and were very
well received by the King, his barons and gentlemen, for they had
been very grieved on their account, and thought that they were
dead in those parts.

" ' Fair sires Prunauts, God keep you all ! ' said the King, and kindly feasted them for two days, and they made good cheer, and there was joy such as cannot be described. And the King prayed Messire Jehan, the Captain, to relate to him the news and the marvels of the country whence they came. When the King had heard of his great prowess, he gave him gifts, and bestowed on him a fair estate in land. He also made him admiral of his navy, which greatly rejoiced Messire Jehan, who rendered honour to the King as his seigneur.

" So I will tell you that from this came the name of Prunaut to Messire Jehan, and that he kept it because he was very valiant, high-spirited, and bold in feats on the sea, and a man of great sense. Also the King desired that his progeny and lineage should be called Prunauts, as the sons of a valiant, gallant, and gentle sailor. And these gifts the King confirmed with his seal on charters, so that he might hold them from heir to heir.

" Then, after Messire Jehan and his companions were well feasted, they turned towards Rouen ; and Messire Jehan, the bold sailor, rode, with his shield at his side, on a palfrey richly harnessed and adorned, and the rest followed as they could. The Archbishop of Rouen and all his clergy hearing that he was coming with all his ship's company, went out to meet him and did him much honour ; for he knew already that Messire Jehan had returned, and that he was beloved by God and His saints, for he had built in those parts a little chapel, and appointed as its priest Friar Pierre, the Norman, a very worthy clerk, to teach the Pagans and unbelievers the love of God, to speak well, to preach, and to destroy the Pagan law. In this chapel, which was dedicated to our Lady, were buried very honourably the sailors who died * . . . as has been, said, of the pestilence. Then at Rouen, with the Archbishop, came to meet Messire Jehan and his companions, a grand cavalcade of lords and such high folks, and many peasants on foot, sergeants and burghers of Dieppe, Caen, Cherbourg, and all the cities of Normandy, who had come to see the gentle Captain and his bold companions. Natent,

* " Aluel," perhaps "à la Noël," at Christmas.

the wife of the Captain, a wise and beautiful dame, and she was certainly the most beautiful there ever was [?], and was of great lineage in Saxony, with Legier, her son, and Erkenbous, his brother—both little children—who embraced and kissed their good father and these lords, burghers, and peasants in great numbers, and all together brought him to his lodging: for no word had been said of Mounsire Legier and the others who were dead in the foreign land.

"The following year Messire Jehan Prunaut remained with the King's army, but sent beyond sea his ship, *Notre Dame*, as before ; those of Dieppe and Rouen their ships *Saint Nicolas* and *l'Espérance*. *Notre Dame* anchored in the place which they called La Mine, because of the quantity of gold which was found round about. Here is that chapel of the Blessed Mother of God which Messire Jehan founded, as we have said, with a little strong castle and fortalice, and a square house which he had made on a hill, which was called the land of the Prunaus, in remembrance of them and of their admiral, as also Petit Dieppe, Petit Rouen, Petit Germentru-ville, and Petit Paris, because they had come from Dieppe, Rouen, and Paris. There they built also forts [and] castles at the places called Cormentin and Acra.

"In the year 1410 many of the Norman sailors went away, and the merchants lost their wealth, which was devoured by the wars which then were, and in eleven years only two ships went to the Gold Coast, and one to the Great Siest ; and a little while after, as the wars raged at sea as well as on land, the mercantile affairs were disturbed and destroyed."

Some pages earlier in his book, M. Margry quotes an expression of my late honoured friend, the Vicomte de Santarem, that "*it is not by documents that may be discovered that positive history* ought to be put in doubt. Even if a document should happen to be discovered which was opposed to facts recognised as true, it would not be sufficient to upset the *unanimous testimony of cotemporaries*." Now while it would, clearly, be too much to say that light may not be, and often is, thrown upon known history by the production of newly-discovered documents, it is equally clear that any single document not only not corroborated, but contradicted and condemned by a flood of well-authenticated historical facts, must require uncommonly strong authentication to save it from the gravest suspicion. In the present case, however, I think I can produce sufficient reason to lead the reader to coincide *entirely* with the dictum of the Vicomte de Santarem.

For this purpose I will here briefly state the principal items of evidence which I had adduced to disprove the claims of the French to priority of discovery on the coast of Guinea, claims which as yet have never been authenticated by any document.

1. It is asserted that the absence hitherto of documents proving the conjoined explorations of the Rouenese and Dieppese to the coast of Guinea in the fourteenth century is explained by the destruction of the Dieppese archives in the English bombardment in 1694. To this one replies with the inquiry: How comes it that Rouen, which was not so bombarded, supplies no testimony on the subject? M. Margry now suggests that such documents *may* have been used in lighting pipes or covering jam-pots. Well, be it granted. Almost inconceivable though it be, we will suppose that jam-pots and a bombardment have ruthlessly denuded these two important cities of every shred of testimony, whether cotemporary or retrospective, to their having earned for themselves a distinction of which any nation might be proud. Then let us look for further evidence.

2. A Swiss doctor in 1617 adduces the statement of

Guinea negroes, *a hundred and thirty* years old, that the
French were there in the fourteenth century; and a Dutchman, Olivier Dapper (a man whose testimony is shown to
be worthless by his own mis-statement of perfectly well-
known facts in the history of Portuguese discovery, as well
as by the acknowledgment of a learned Frenchman, M.
Eyries, that his assertions mislead those who do not examine
for themselves) stated in 1668 that in the Castle of La Mina
was a ruined battery named Batterie Française, in which
were the first two figures of the date 13—, but the following
numbers could not be deciphered.

In answer to the latter statements, I have shown that the
French had really been on the Guinea coast in the beginning
of the last half of the sixteenth century, quite long enough
to allow of the existence of a ruined French battery, and
also to render possible the obliteration of a date (if date it
was) exposed to the annual corrosive action of three months
of rain.   But there is nothing to prove that the figures were
part of a date.   In opposition to the assertions of the superannuated negroes I adduce the evidence of the *Norman*
narrators of the expedition of Jean de Bethencourt (whose
estate lay only twenty-five miles from Dieppe) to the Canaries
at the period when the asserted Dieppese intercourse with the
coast of Guinea was in its zenith, in which it is declared
that " it is M. Bethencourt's intention, with the help of
God and that of Christian princes and people, to *open* the
way to the River of Gold.   It cannot be doubted that much
remains to be done which *might* have succeeded in times
past *if it had been undertaken.* . . . . and he *will spare no
pains to decide whether success is possible or impossible,*" &c.

3. I further adduce an indignant remonstrance of a
Dieppese Captain in 1539 against the arrogant extrusion of

the French from Guinea by the Portuguese on the score of the acknowledged priority of discovery of that coast by the latter,—a remonstrance *so* indignant, that if a claim *could* have been set up by the writer to priority of discovery by his own people, it assuredly would not have been wanting, whereas such a claim was not attempted by the French till a century and a half later.

4. I further show that one of the warmest advocates of these claims acknowledges that no specimens of Dieppese ivory carvings, which he *asserts* were made as early as the close of the fourteenth century, could be found of an older date than the close of the sixteenth century, a period at which I have shown that the Dieppese did really traffic with the Guinea Coast.

5. I further show that whereas the most beautiful and elaborate maps we possess of the beginning of the sixteenth century are *Dieppese,* not one of them exhibits the names of " Petit Dieppe " and " Petit Paris," asserted to have been given by their people to places on the Guinea Coast in the fourteenth century; whereas in 1631, five years after the Rouenese and Dieppese *did really combine* (in 1626) to traffic with that coast, we do, *for the first time,* find those names laid down on the Dieppese maps of Jean Guérand.

In the presence of a mass of historic evidence such as this, to which might be added an octavo volume full of proof, both positive and negative, laboriously brought together by the Vicomte de Santarem, assuredly a *solitary* document, uncorroborated and *unauthenticated,* will scarcely pass muster. It is unfortunate, doubtless, but unavoidable that the recent *exposé* of the spurious Newton and Pascal correspondence should cause the unexpected production of reputed early documents on important subjects to be treated with the most rigid investigation that even suspicion can suggest, and the reader will certainly demand such investigation at our hands.

As will be presently seen I have spared no exertion to trace the existence of this newly adduced document.

In the " search for old French texts " by M. de Rosny, observed by Mr. William Carter (p. xxii.), there was apparently a clue to this manuscript through the medium of the registers of the Reading Room of the British Museum. I there found that there were at that time two readers, and only two, of that name, but neither of them residing in Oxford Street. One, named simply William Carter, resided then as now in Philpot Lane, and never had any knowledge of M. Lucien de Rosny or of the MS. in question. The other, Mr. William George Carter, lived then in the Temple, and died in 1861 in Raymond's Buildings, Gray's Inn. But after an elaborate and interesting search, in which I traced and corresponded with all the surviving connections of this Mr. Carter, with results at first somewhat hopeful, but finally, as will be presently seen, almost conclusive against his being the person referred to, I applied to M. Margry himself, and received an answer, of which the following is a translation :—

> " 11, *Rue du Mont Thabor,*
> " *Paris, November* 20, 1867.

" Sir,—I have received the letter in which you ask me or M. de Rosny Foucqueville to be so good as to point out to you the means of finding Mr. William Carter, and the manuscript which was in his possession, and which I have published in my book on the French navigations from the fourteenth to the sixteenth century.

" I regret excessively my inability to give you this information.

" When M. Lucien de Rosny copied the document from the volume confided to him by Mr. Carter, he unhappily attached to it no other importance than that which a philologist studying the old French language would find in it. Consequently, not suspecting that he had in his possession a paper touching the honour of a nation, he did not feel himself called upon to take, nor expect ever to have to give, any pledge of his good faith against the remonstrances of that nation or the criticisms of the learned. When, in compliance with your request, I again questioned M. de Rosny on the origin of his document, he told me now, as before, that Mr. W. Carter, in 1853, when he met him, was living in Oxford Street, but as he did not go to his house, in which, as I believed and told him, he was in fault; as they met only in the British Museum in the Ethnological Room; and, finally, as M. de Rosny is a man full of reserve, he never learned from Mr. Carter either his number, or whether he lived in London or was only passing through it.

"This reply, Sir, is doubtless not calculated to satisfy you, but I can do no better. For you to have the same confidence as myself, you should see and hear M. de Rosny himself relate how he entered into conversation with Mr. Carter on the subject of botany; how the latter, learning by chance that M. de Rosny was specially occupied in the study of the old French language, begged him to acquaint him with the contents of an old manuscript in that language which he was unable to decipher, and which he brought with him the next day; how, finally, M. de Rosny, authorised by Mr. Carter to extract from it what he pleased, took from the volume, which was a collection of from about sixty to seventy leaves, bound in a sort of parchment of dark apple-green, the document in question, and an old carol of the fifteenth century, of which he has to-day brought me the copy.

"All this is said so simply, so honestly; M. de Rosny has always taken to himself so little merit for this discovery, which he did not appreciate till I had made him aware of its importance; he is so well known even to several persons in the Museum, among others to Mr. Franks, for his philological studies, in which he follows the steps of his maternal grandfather, M. Hécart, of Valenciennes, that really I feel almost ashamed of the sort of interrogatory to which your question obliges me to subject him.

"I am aware that in the country whence has proceeded the

scientific mystification of the voyage of Bartolomé Fuente, people
are not contented without seeing and touching, examining the
water-mark of the paper and the character of the writing.

" On this point I have nothing to say, I have put forward
honestly what I believed, and still believe, to have been communicated to me with equal honesty.

" Now whether Mr. Carter, who was a man of from fifty-five to
sixty years of age, with grizzled hair, and in feeble health, suffering in his legs, is dead, or compelled by his infirmities to remain
at home, or has left London, I do not know more than M. de
Rosny. This is no reason why his document may not one day be
found; and if the difficulty which M. de Rosny encountered in
reading the collection whence he extracted the document which
interests us, should again have the effect of making this collection
a dead letter in the hands of the heirs of Mr. W. Carter, as it
seems for a long time to have been in his own, allow me to say
that I shall congratulate myself on having, at the risk of being
attacked, taken advantage of a happy chance which has given me
occasion to publish a document which would only have appeared
to disappear again.

" In any case, Sir, I do not think I have given in my book the
last word which may be said on the subject which interests you.
I have heard that there is now in England a gentleman, a connois-
seur in documents on French discoveries in Africa anterior to those
which I have quoted. Where has he found them ? In the papers
brought from France by the English at the time of their expulsion ?
I know not. I only know the name of the gentleman, but this I am
not at liberty to publish, because he may perhaps himself intend to
win honour by these documents. Thus history now, as ever, is
remodelled piece by piece. Each one brings his portion to this
great piece of marqueterie. There are some which do not agree
with accepted history, but they are none the less true. In fact
Montaigne has observed that if he had in his possession the events
which are unknown, he could easily supplant the known ones in
every kind of example. I have already several times brought
examples to the support of this thought. But if, in the matter in

question, I have not the happiness of seeing you accept my conclusions, believe me, Sir, that I consider myself fortunate, in one
respect, at least, in the circumstance which places me in communication with a distinguished savant.

" I will conclude this letter with the assurance that you are at
liberty to publish it in extenso, if you think it necessary, either
for the purpose of finding Mr. Carter, or any other reason unknown
to me, but which, judging by your procedure towards myself, can
only lead to a courteous discussion, aiming at the discovery of Truth,
the supreme end of History. This permission will doubtless be,
at least in your eyes, a pledge of the good faith of those who have
advanced the fact which you wish to dispute, and I think that you
will also see in it a mark of the sentiments of consideration with
which

        " I have the honour to be, Sir,

      " Your very humble and very obedient servant,

                " PIERRE MARGRY.

" R. MAJOR, ESQ."
                ———

                  11, *Rue du Mont Thabor*,
                        Paris, ce 20, *Novembre*, 1867.

MONSIEUR,

J'ai reçu la lettre par laquelle vous me demandez à moi ou à M. de Rosny
Foucqueville de vouloir bien indiquer les moyens de rencontrer M. William
Carter et le manuscrit qui était en sa possession, manuscrit que j'ai publié dans
mon livre sur les Navigations Françaises du quatorzième au seizième siècle.

Je regrette vivement de ne pouvoir vous donner ce moyen.

Lorsque M. Lucien de Rosny a copié ce document dans le volume que lui a
confié M. Carter, il n'y a malheureusement attaché d'autre importance que celle
qu'y peut trouver un philologue, étudiant le vieux langage français. Il en est
résulté que ne se doutant pas qu'il avait entre les mains un papier touchant à
l'honneur d'une nation, il n'a pas cru devoir prendre ni avoir à donner un jour
aucune garantie de sa bonne foi contre les réclamations de cette nation ou contre
les critiques des érudits. Lorsqu'à votre demande, j'ai de nouveau interrogé
M. de Rosny sur l'origine de son document, il m'a dit aujourd'hui comme autre-
fois que M. W. Carter en 1853, lorsqu'il le voyait, demeurait à Oxford Street,
mais que comme il n'est pas allé chez lui, aussi que je le croyais et l'ai dit à tort,
comme ils se rencontraient seulement au British Museum dans l'Ethnological
Room; comme enfin M. de Rosny est un homme plein de réserve, il n'a jamais
su de M. Carter ni son numero, ni s'il était de Londres, ou s'il y était seulement
en passant.

Cette réponse, monsieur, n'est sans doute pas de nature à vous contenter, mais
je ne puis rien de mieux. Pour vous donner la confiance que j'ai il faudrait

voir et entendre M. de Rosny lui-même racontant comment il est entré en relations avec M. Carter à propos de Botanique ; comment celui-ci apprenant par
hasard que M. de Rosny s'occupait surtout de l'étude du vieux langage français
lui demanda de lui renseigner sur ce que contenait un vieux manuscrit en cette
langue indéchiffrable pour lui et qu'il apporta le lendemain ; comment enfin M.
de Rosny autorisé par M. Carter à en extraire ce qu'il voudrait, a pris dans ce
volume composé d'un recueil de 60 à 70 feuillets environ, recouvert d'un espèce
de parchemin vert pomme foncé le document dont il s'agit, plus un vieux noel
du quinzième siècle dont il m'a apporté aujourd'hui la copie.

Tout cela est dit si simplement, si honnêtement ; M. de Rosny s'est fait toujours si peu un mérite de cette découverte qu'il n'appreciait pas avant que je
lui en cusse fait voir l'importance ; il est si bien connu, même de plusieurs
personnes du Museum, M. Franck entre autres, pour ses études philologiques, où
il suit les traces de son grand père maternel, M. Hécart, de Valenciennes, qu'en
vérité je me sens presque honteux de l'espèce d'interrogatoire que votre demande
m'oblige à lui faire subir.

Je comprends que dans le pays d'où est partie la mystification scientifique du
voyage de Barthélémy Fonte on veuille voir et toucher, reconnaître la marque
du papier et le caractère des écritures.

Là dessus je n'ai rien à dire, j'ai livré loyalement ce que j'ai cru et ce que je
crois encore m'avoir été communiqué avec une égale loyauté.

Maintenant si M. Carter qui était un homme d'entre 55 et 60 ans aux
cheveux grisonnans et d'une santé faible, souffrant des jambes, est mort, ou que
les infirmités l'obligent à demeurer chez lui, ou qu'il ait quitté Londres, ce que
je ne sais pas plus que M. de Rosny, ce n'est pas une raison pour que son document ne se retrouve pas un jour et si la difficulté que M. de Rosny a rencontrée
à lire le recueil dont il a extrait le document qui nous intéresse devoit avoir une
fois encore pour effet de faire de ce recueil une lettre morte dans les mains des
héritiers de M. W. Carter comme il paraît l'avoir été longtemps dans les siennes,
laissez moi vous dire que je me féliciterai d'avoir profité, au risque d'être
attaqué, d'un heureux hasard qui m'a donné lieu de publier une pièce la quelle
n'aurait guères apparu que pour disparaître.

Quoiqu'il arrive, Monsieur, je ne crois pas avoir donné dans mon livre le
dernier mot à dire sur le sujet qui vous occupe ici. J'ai entendu dire qu'il y
avait en ce moment en Angleterre un gentleman connaissant des documents sur
des découvertes en Afrique faites par les français antérieurement à celles que
j'ai citées. Où les a-t-il trouvés ? est ce dans les papiers emportés de France
par les Anglais lors de leur expulsion ? je l'ignore, tout ce que je connais c'est le
nom du monsieur, mais je ne suis pas autorisé à le publier parceque ce gentleman a peut-être lui-même l'intention de se faire honneur de ces documens.
Ainsi l'histoire ici comme ailleurs se recompose pièce à pièce, chacun apporte un
morceau à cette grande marqueterie. Il y en a qui ne s'arrangent pas avec
l'histoire convenue, mais ce ne sont pas toujours les moins vrais. En effet
Montaigne a pu dire que " s'il avait en sa possession des événemens incogneus,
il pourroit très facilement supplanter les cogneus en toute espèce d'exemples."
J'ai déjà plusieurs fois apporté des exemples à l'appui de cette pensée. Mais si
dans le cas dont il s'agit aujourd'hui je n'ai pas le bonheur de vous voir accueillir

mes conclusions, croyez, monsieur, que je regarde comme heureuse au moins
par un coté une occasion qui me met en relation avec un savant distingué.

Je terminerai cette lettre en vous disant que vous pouvez la publier in extenso,
si vous le croyez nécessaire, soit pour retrouver M. Carter, soit pour d'autres
vues qui je ne connais pas, mais qui d'après votre demarche auprès de moi ne
sauraient être que celle d'une discussion courtoise, ayant pour objet la découverte
de la vérité, cette suprème fin de l'histoire.   Cette autorisation sera sans doute
au moins à vos yeux un témoignage de la bonne foi de ceux qui ont avancé
le fait que vous voulez contester et je pense que vous y verrez aussi la marque
des sentimens de considération avec lesquels,

J'ai l'honneur d'être, monsieur,

Votre très humble et très obéissant serviteur,

PIERRE MARGRY.

A. M. R. Major.

This letter placed the transaction, if possible, in a still
more unsatisfactory position, and on the 26th of November
I addressed the following lines to M. Margry:—

"*British Museum, November* 26, 1867.

"Sir,—I beg to offer you my best thanks for your obliging
letter of the 20th inst., but regret that it brings me no more
satisfactory account of the interesting manuscript lent to M. de
Rosny by Mr. William Carter.   It might greatly assist me in my
endeavours to find it if M. de Rosny would kindly tell me how
and where he restored to Mr. Carter the volume which he had
borrowed of him.

"Trusting that both yourself and M. de Rosny will pardon the
trouble I am giving in consideration of my earnest desire to do full
justice to a very important subject,

"I have the honour to be, sir,

"Your very obedient humble servant,

"R. H. MAJOR."

On the 4th of December I received an undated letter from
M. Margry, enclosing another from M. Lucien de Rosny, of
which the following is a translation:—

"*Levallois Perret, Banlieue de Paris, November* 80, 1867.

"Sir,—By his letter, which has just reached me, M. Margry
has acquainted me with your desire to become acquainted with the

MS. which Mr. W. Carter communicated to me when I resided
in London.   During the long period since I have ceased to live
in England I have lost sight of that gentleman, whom I only knew
from meeting him sometimes at the British Museum, not in the
Reading Room, but in the Ethnological Room, and in the different
collections of the Museum.   When I became acquainted with this
MS., from which I have copied some passages (less than I had
wished, for the cursive writing of the 16th and 17th century is
very difficult to read), I was obliged to confine myself to some
extracts, among which was that in which you are concerned, and
to which I attached no other interest beyond what I felt generally
for all ancient documents in the old French language.   I was
ignorant at the time that this passage would make so much impression on the minds of the readers of M. Margry, for, otherwise,
I would, as I have already stated, have taken every possible
precaution to guarantee the authenticity of my copy.

    " I have lost sight of Mr. Carter.   I only knew him as an
obliging and confiding man, for there was not established between
us any relationship of a durable friendship.   Accident brought us
together, and we separated in the same way.   M. Margry tells me
that in the researches made in the register of persons authorised
to frequent the British Museum, the name of Mr. Carter is not
found inscribed.   This does not astonish me, for he did not
frequent the Reading Room, and there is no occasion for any
authority to visit the collections of this establishment open to
the public.

    " It was not in the Reading Room that Mr. Carter communi-
cated his MS. to me, but *when leaving the British Museum* [under-
lined by M. de Rosny].   He had it in his pocket, and I followed
some time in conversation with him.   We separated in Oxford
Street, where he lived.   It was then between four and five o'clock.

    " I returned him his MS. in the British Museum, in the Gallery
of Antiquities, and he put it again in his pocket, if I remember
well, for it is now fourteen years ago.   This, sir, is all that I
know.   From this MS. I copied not only the document of which
we speak, but an old carol, and some lines on a *pronostic* accom-

panying a sneeze, for this *pronostic* interested me only because it entered into a monograph on the subject of sneezing which I was writing at the time.

" If it is in my power, sir, to give you satisfaction on other questions upon which you may be interested to make inquiries, I am at your disposal. I avail myself of this opportunity to ask you in my turn to do me the kindness to give the.inclosed letter to Mr. Franks, and to forward the other to the Superior of Trinity College, with a word of recommendation.

" Pray accept, sir,

" The assurance of my devotedness,

(Signed)   —————   " L. DE ROSNY."

Levallois Perret, Banlieue de Paris,

MONSIEUR,   30, *Novembre,* 1867.

Par sa lettre qui m'arrive, M. Margry m'entretient du désir que vous auriez de connaître le Msc. que M. W. Carter m'a communiqué lorsque je residais à Londres. Depuis bien long temps·que je n'habite plus l'Angleterre, j ai perdu de vue ce monsieur que je n'ai connu que pour l'avoir rencontre quelquefois au British Museum, non point au Reading Room, mais dans la Salle d'Ethnographie et dans les diverses collections de ce musée. Quand j'ai eu communication de ce manuscrit dans lequel j'ai copié quelques passages, moins que je ne l'eusse voulu (car l'écriture cursive du seizième et du dix·septième siècle est fort pénible à lire), j'ai dû me borner à quelques extraits parmi lesquels existe celui qui vous intéresse et auquel je ne donnais d'autre intérêt que celui que m'inspirent généralement tous les anciens documents du vieux langage français. J'ignorais alors que ce passage dût faire tant d'impressions sur l'esprit des lecteurs de M. Margry, car j'aurais pris toutes les précautions possibles pour garantir l'authenticité de ma copie comme je viens de vous le dire, monsieur.

J'ai perdu de vue M. Carter, je n'ai vu en lui qu'un homme obligeant et confiant, mais il ne s'est point établi entre nous des rapports durables et d'amitié. L'occasion nous a fait trouver ensemble ; nous nous sommes séparés de la même manière. M. Margry me dit que dans les recherches faites sur le registre des personnes autorisées à fréquenter le British Museum, le nom de M. Carter ne s'y trouve pas inscrit ; cela ne m'étonne nullement, car il ne fréquentait pas la Salle de Lecture et il ne faut pas d'autorisation pour aller visiter les collections de cet établissement ouvert (sic) au public. Ce n'est pas non plus dans la Salle de Lecture que M. Carter m'a communiqué son manuscrit, mais *en quittant le British Museum.* Il l'avait dans sa poche et je le suivis quelque temps en causant avec lui. Nous nous séparâmes à Oxford Street où ce monsieur devait habiter. Il était alors de 4 à 5 heures.

Je lui remis son manuscrit au British Museum dans la Salle des Antiques, et il le remit dans sa poche si je me le rappelle bien, car il y a bien 14 ans de cela.

Voilà, monsieur, le peu que je sais. J'ai copié dans ce manuscrit outre le document dont il s'agit, un vieu noël, et quelques lignes sur un pronostic accompagné d'un éternuement, car ce pronostic m'intéressait par cela seul qu'il rentra dans une monographie sur le culte de l'éternuement que j'écris en ce moment.

Si je puis, monsieur, vous être agréable sur d'autres questions que vous pourriez avoir intérêt à me faire je me tiens à votre disposition. Je profite de cette occasion pour vous demander à mon tour un service, ce serait de remettre la lettre ci inclue à M. Franck, et de faire parvenir l'autre au Supérieur de Trinity Collège avec un mot de recommandation.

Veuillez agréer,

Monsieur,

l'assurance de mon dévouement,

L. DE ROSNY.

Now it will be observed that M. Lucien de Rosny volunteers the observation that it was *not in the Reading Room* of the British Museum, where the addresses of students are kept, that he met Mr. William Carter, and further, he particularly states that that gentleman had not a reading ticket. This circumstance is to the last degree unfortunate, and when combined with the remarkable fact that the volume is lent to a perfect stranger by a perfect stranger, whose address is neither given nor asked, and in a part of the Museum where the chances of meeting again, or even of finding each other at an indefinite period after the extracts should have been made, were rendered the more uncertain by the attendants taking no cognizance of visitors, reduces the possibility of tracing the document to a minimum.

But it is further remarkable that M. Margry states in his book that the occurrence took place in consequence of Mr. Carter's "seeing M. de Rosny searching for old French texts." Now this search *could* only have taken place either in the Reading Room or in the Manuscript Department, and if in the latter, there would have been a twofold register of the students' names, one in the Reading Room, where the address would be also kept, and the other in the Manuscript

Room itself. Neither of the names occurs in the Manuscript
Room register, whereas in the Reading Room I find that
M. de Rosny received his reader's ticket on the 13th of
December, 1852, and that instead of there being, as M. de
Rosny states as the result of M. Margry's inquiry, no reader
at the time of the name of William Carter, there were the
two whom I have already mentioned. Mr. William George
Carter was a man so unusually reserved that as his executors,
his clerk, his housekeeper, and the legatee of his books, as
well as the purchaser of his books, Mr. Jones, the librarian of
the London Library, have of their own accord informed me,
was the last man in the world to accost a stranger, or, in
fact, to fall into conversation with any one. He never in
all his life took any interest in botany. Mr. Jones assures
me that there was no MS. among his books, and even if
there had been, I am informed by a letter from his clerk, Mr.
Tubb, written from Bishop's Sutton on the 2nd of November
last, only just in time, for he died the day after, that he
" should not think it likely he would lend his book to any
Frenchman, as I don't think he was an admirer of the
French."* It is true that the negative poles of a magnet
will attract each other, but it would be strange indeed if two
men, both so "pleins de reserve" as M. de Rosny and Mr.
Carter, were to gravitate to each other, and respectively
make and accept a loan of a valuable and curious volume
without the slightest regard to the most ordinary rules of
precaution.

* I cannot refrain from mentioning here a most extraordinary occurrence.
After a romantic and almost hopeless search of many weeks, I succeeded in
tracking the address of Mr. Tubb, Mr. Carter's clerk. In reply to my
inquiries, he wrote me a most obliging letter, containing the above sentence,
and informing me who was the purchaser of Mr. Carter's books, a point
which I had in vain endeavoured to ascertain elsewhere. Mr. Tubb's letter
was written on Saturday the 2nd of November, and on Sunday the 3rd he
died, even before his letter reached my hands.

I mention all these facts, which have cost me much labour in tracing, solely because in the Reading Room or Manuscript Room only *could* Mr. Carter have " seen M. de Rosny searching for old French texts," and to show that I have spared no pains to do justice to M. Margry's statements; because if I had failed in doing justice to him, I should thereby have also failed in doing justice to my readers and to myself. But, as we have seen, the Reading Room is now out of the question. In a matter so important, this position of the case is greatly to be regretted, for, on the one hand, M. Margry's original statement of Mr. Carter's seeing M. de Rosny searching for French texts would have involved a meeting in the Reading Room, where there would have been some sort of guarantee to the two strangers, not only of respectability, but of the chance of meeting again; while now that M. Margry shifts his ground, the transaction is based entirely on a quicksand, and one might as well hope to recover a sunken ship from the Goodwin as to regain this *ignis fatuus* of a manuscript.

Why should this cruel fate pursue all Dieppese documents ? Why should this solitary seventeenth century copy of a solitary document testifying to the fourteenth century glories of the Dieppese have lain *perdu* for two centuries, only to flicker for a few days before the eyes of a French *savant* and again to hide itself in its beloved obscurity ? Why also should the vampire which has brooded over the fame of Dieppe so remorselessly have clouded the intelligence of a French *savant*, who has shown his interest in early voyages by publishing in French the first letter of Columbus, as to blind him entirely to the importance of a document which cost him great pains to decipher, when that document most intimately affected the maritime glory of his own country ? Yet it was not till some seven years later that the document was shown to M. Margry, by whom at length the film was removed from the eyes of M. de Rosny. Of a truth, the whole story is a curious and notable one.

Thus strangely are we left at sea respecting this Mr. William Carter himself, while the document in his possession, reproduced by M. Margry, is but a copy by M. Lucien de Rosny of a copy, supposed by M. Margry to have been made in the seventeenth century. This is most unfortunate, for it disarms all criticism on the construction of the language, as being supposed of the fourteenth century. But it is far more unfortunate on another account—*there is nothing to prove the authenticity of the seventeenth century copy*, and, certes, such proof is eminently necessary to countervail the accumulated arguments which have been brought together in refutation of claims *which were not set up till just before the time* when this reputed copy was supposed to have been made.

But let us examine the document internally. Of two things one, 1. Either it was written to describe a genuine voyage in which we should look for consistency with geographical facts. 2. Or it has been concocted at some period or another with the view of establishing French precedence of the Portuguese in the discovery of the coast of Guinea, and it will be interesting to see whether the document betrays in any way such an intention. In both these respects the language of the document is damaging to its integrity. The text says, " In the month of September, 1364, those of Dieppe and Rouen equipped two ships, and had for admiral [or commander] Monsieur Jean le Roanois : . . . . . . and sailed a long time on the sea till Christmas, to a place called Ovideg [the Senegal], where

those of Normandy had never as yet been, and anchored
par de là to advance their affairs at a very hot place which
is now called Cape Bugiador, which belongs to the kingdom
of Guinea. The Giloffs (so the people of these parts are
called, who are black in face and skin, &c.) had never seen
white men, &c." Now this mention of Cape Boyador is
remarkably suspicious. To the sailors of the Peninsula
and the Mediterranean this cape had been the *ne plus ultra*
of exploration until the time of Prince Henry, for it was
difficult for small craft, hugging the coast, to round it.
For twelve years the Prince's sailors strove in vain to
accomplish this feat; but when once, by putting well out
to sea, they had conquered the difficulty, it for ever disappeared. Nevertheless the difficulty which *had* existed
has rendered the name of Cape Boyador conspicuous as
the bone of contention in the endeavours of the French to
wrest the honour of priority from the Portuguese. But
this difficulty is not even professed to have been encountered by the ships of Jean le Rouennais. It is therefore,
I repeat, remarkably suspicious that the name of this cape
should occur in a document adduced for the substantiation
of the French priority, inasmuch as the above-mentioned
difficulty being eliminated, there remained nothing in the
locality itself to tempt a navigator to have anything to
do with it, or his historian to mention it, *unless with conscious reference to the discussion.* But the text tells us that
" they *anchored there* for advancing their affairs." For the
sake of brevity, I will refer the reader to page 131 of the
present volume, that he may judge whether Cape Boyador
was a place to anchor at, unless indeed at the very tip
of the long spit of sand of which the cape consists, where
there is an exceptional little bit of anchorage, the exist-

ence of which is so valuable in the eyes of my friend M. d'Avezac;* but whether anchoring there would much "advance their affairs" is, I imagine, rather questionable. But why should I take all this pains when the precious document now adduced transports Cape Boyador itself to beyond the Senegal, a distance of some 700 miles from its true position, which gave Prince Henry's mariners so much trouble? We trace then in this passage a great geographical error, throwing suspicion on the genuineness of the voyage, and a very suspicious reference to a place, conspicuous in a discussion of far more recent times, but to the last degree unlikely, even if we overlook the geographical displacement, to have been visited in the manner described, and therefore equally unlikely to have been mentioned except with a view to the more recent discussion. It may legitimately be retorted that if they reached the Senegal, the true Cape Boyador was *ipso facto* passed. True, but the name embodied in itself a temptation to triumph, and every item of assertion must be canvassed in a document which is adduced at this late period in contravention of all surrounding history.

To these points of evidence I will add that, in the document just produced by M. Margry, occurs a repetition of an old French assertion that the Fort de la Mina was first built by the French in the fourteenth century. In reply, I adduce the honest avowal of a learned *Dieppese*, M. Bruzen de la Martinière, who, in his " Grand Dictionnaire Géographique," Paris, 1768, fol., under the heading of " Saint George de la Mine," uses the following words :—

" However, all the details related by the Portuguese,

---

* M. d'Avezac wrote a paper on the subject, for a notice of which see page 131.

circumstantially given in the 'Decades' of Barros, tend
to show that the Portuguese found no traces of a previous
establishment. The difficulties which were thrown in their
way when they wished to dig the ground for the foundations of their fortress are facts which do not correspond
with the story of those who place there a fortress built
previously. No mention is made either of a fortress or
church except what they themselves built. This is not
easy to reconcile. No doubt is entertained of the correctness of Barros, who has worked on excellent memoirs.
I could wish that Father Labat had at least pointed out
his authority for what he has stated, for Desmarchais,
whom he quotes, is not more to be trusted than he is on
such ancient facts, and they both need guarantees before
they can be believed on a matter of antiquity of several
centuries."

In all candour I contend, in corroboration of this most
just remark, that it is impossible to read the *naïves* and
simple descriptions by the Portuguese of their first discoveries of these coasts and of the construction in 1482
of their Fort de la Mina, the stones of which were carried
out ready cut from Lisbon, and to believe that, had they
found traces of any predecessors on those virgin coasts, no
word of such a phenomenon should have escaped them;
whereas, on the contrary, all is fresh and new, and corroborated, as I have shown in the text, by *French* testimony
generations before a French claim had ever been brought
before the world.

It is further stated that the French, in 1380, built the
strong forts of Cormentin and Accra. Now it is perfectly
true that, long before the loyal Sir Nicholas Crispe
(whose heart lies inurned beneath the bust of his royal

master in Hammersmith Church) had at his own cost erected the fair Castle of Cormentin, in consequence of the concession to him and others of the exclusive trade to Guinea for thirty years by letters patent from King Charles in 1629, that place had been the chief emporium of the trade on that coast, but had lapsed into disuse. Both there and at Accra there had doubtless been forts, which were requisite for the security of the commerce first carried on there by the Portuguese, and afterwards by the French; but I have written in vain the latter portion of the chapter which in this volume bears the title of the "Sea of Darkness," if the question as to whether of these two nations took precedence of the other is not definitively established.

With respect then to the documents now produced by M. Margry, the sum of the investigation yields a result which, unless further explanations can be given, is unavoidable, that, as all the surrounding evidence is not only not corroborative, but contradictory and condemnatory, an *unauthenticated* document, with internal indications of not being genuine, and represented by *a copy of a copy* which is itself not forthcoming, nor its possessor traceable, is worth absolutely nothing.

I close this Preface with the pleasant duty of tendering my warm thanks to those from whom I have received the kindest and most valuable assistance. To his Excellency the Count de Lavradio for most generous help in books beyond my reach, for a photograph of the statue of Prince Henry at Belem, for a copy of the Prince's signature, as well as for a variety of valuable information from his own

well-stored mind, I beg to offer the respectful expression
of my sincere gratitude. For similar precious assistance
I owe my thanks to my valued friend the Count de Rilvas,
Chancellor of the Portuguese Legation in London, who
has spared no pains to help me with his researches in
Lisbon. To the spontaneous generosity of another most
kind friend in Portugal, the Marquis de Souza Holstein,
Chamberlain to His Most Faithful Majesty, I am indebted
for the busts of King João and Queen Philippa from their
effigies on their tomb in Batalha, and from which are
drawn the portraits here given, as well as for a photograph
of the tomb of Prince Henry. To His Excellency the
Marquis de Sá da Bandeira, I have to acknowledge my
great indebtedness for the plan of Sagres and of the
monument which, to his lasting honour, was at his behest
erected therein to the memory of the illustrious Prince
whose residence had immortalised that desolate spot.
Others, from among whom must not be omitted my friend the
Chevalier dos Santos, will be sure that I am not unmindful
of their ever-ready kindness and assistance to me in the
course of this work.

PRINCE HENRY THE NAVIGATOR.

———◆———

CHAPTER I.

THE PURPOSE.

THE mystery which since creation had hung over the Atlantic, and hidden from man's knowledge one-half of the surface of the globe, had reserved a field of noble enterprise for Prince Henry the Navigator. Until his day the pathways of the human race had been the mountain, the river, and the plain, the strait, the lake, and inland sea; but he it was who first conceived the thought of opening a road through the unexplored ocean, a road replete with danger but abundant in promise. Although the son of a king, he relinquished the pleasures of the court, and took up his abode on the inhospitable promontory of Sagres at the extreme south-western angle of Europe. It was a small peninsula, the rocky surface of which showed no sign of vegetation, except a few stunted juniper-trees, to relieve the sadness of a waste of shifting sand. Another spot so cold, so barren, or so dreary, it were difficult to find on the warm and genial soil of sunny Portugal. Landwards the north-west winds were almost unceasing, while three-quarters of the horizon were occupied by the mighty and mysterious waters of the as yet unmeasured Atlantic.

In days long past there had stood upon the sister headland of St. Vincent, at about a league's distance, a circular
Druidical temple, where, as Strabo tells us, the old Iberians
believed that the gods assembled at night, and from the
ancient name of Sacrum Promontorium, hence given to the
entire promontory by the Romans, Cape Sagres received its
modern appellation. As may be imagined, the motive for
the Prince's choice could not have been an ordinary one.
If, from the pinnacle of our present knowledge, we mark on
the world of waters those bright tracks which, during four
centuries and a half, have led to the discovery of mighty
continents, we shall find them all lead us back to that same
inhospitable point of Sagres, and to the motive which gave
to it a royal inhabitant. To find the sea-path to the
" Thesauris Arabum et divitis Indiæ," till then known
only through faint echoes of almost forgotten tradition,
was the object to which Prince Henry devoted his life. The
goal which he thus set before himself was at an unknown
distance, and had to be attained through dangers supposed to
be unsurmountable and by means so inadequate as to demand
a proportionate excess of courage, study, and perseverance.

To be duly appreciated, this comprehensive thought must
be viewed in relation to the period in which it was conceived.
The fifteenth century has been rightly named the "last of
the dark ages," but the light which displaced its obscurity
had not yet begun to dawn when Prince Henry, with
prophetic instinct, traced mentally a pathway to India by an
anticipated Cape of Good Hope. No printing-press as yet
gave forth to the world the accumulated wisdom and experience of the past. The compass, though known and in
use, had not yet emboldened men to leave the shore and put
out with confidence into the open sea ; no sea-chart existed
to guide the mariner along those perilous African coasts ;
no lighthouse reared its friendly head to warn or welcome
him on his homeward track. The scientific and practical
appliances which were to render possible the discovery of
half a world had yet to be developed. But, with such objects

in view, the Prince collected the information supplied by ancient geographers, unweariedly devoted himself to the study of mathematics, navigation, and cartography, and freely invited, with princely liberality of reward, the co-operation of the boldest and most skilful navigators of every country.

We look back with astonishment and admiration at the stupendous achievement effected a whole life-time later by the immortal Columbus, an achievement which formed the connecting link between the old world and the new; yet the explorations instituted by Prince Henry of Portugal, were in truth the anvil upon which that link was forged; and yet how many are there in England, the land of sailors, who even know the name of the illustrious man who was the very initiator of continuous Atlantic exploration? If the final success of a bold and comprehensive idea outstep the life of its author, the world, which always prefers success to merit, will forget the originator of the very result which it applauds. This injustice is specially manifest in the case of Prince Henry, for the vastness of his conception on the one hand, and the imperfection of his appliances on the other, made the probabilities of success during his own life-time infinitely the more remote. It is in such cases that Fame needs to be awakened to her task. Thus slept for centuries the fame of Christopher Columbus; thus sleeps the fame of Richard Hakluyt, the pioneer of the prosperity of his country.

If it be the glory of England that by means of her maritime explorations the sun never sets on her dominions, she may recall with satisfaction that he who opened the way to that glory was the son of a royal English lady and of the greatest king that ever sat on the throne of Portugal. The importance of these personages is such as to demand a separate chapter.

CHAPTER II.

THE Infant Dom Henrique, better known in England as
Prince Henry the Navigator, was the fifth child and fourth
son of King João I., "of good memory" (also surnamed the
"Great," and "Father of his country"), and of Queen
Philippa, daughter of "old John of Gaunt, time-honoured
Lancaster." He was thus the nephew of Henry IV. of
England, and great-grandson of Edward III. He was also
a descendant of the last kings of the line of Capet, and
allied to the family of Valois.

Although in reality one of the oldest nations in Europe,
Portugal did not begin to assume a prominent position till
the accession of Prince Henry's father to the throne. It
had been the fate of that little country to struggle for six
centuries to throw off the yoke of its powerful and implacable enemies, the Moors. Reduced in numbers, subdued
and despised, the Portuguese yet found, in their desperate
patriotism, the materials for the final expulsion of their
oppressors. It was the realization of an impossibility.
But no sooner were the Moors ejected from the peninsula
than repeated efforts were made by Spain to effect the subjugation of Portugal, with whom she had been previously
united against the common enemy. To King João, the
father of Prince Henry, it was reserved to vindicate, under
frightful disadvantages, the honour of Portugal against
Spain—to establish the throne upon a solid basis, yet

more, to be the first to carry into the country of the Moors the sword of the avenger, and to prepare the way for those more expansive movements which were to issue from the genius of his son. With his accession to the throne commenced the glorious dynasty known as that of Aviz, which lasted two hundred years and embodied the period of Portugal's greatest dignity, prosperity, and renown. It is remarkable that King João was the youngest, and an illegitimate son of a sovereign who had three other sons legitimate, or accepted as such, who attained maturity. His father, Dom Pedro I., surnamed the Severe, by his marriage with Constance, daughter of João Manoel, Duke of Peñafiel, had two sons and a daughter. Of the sons, Luiz, the elder, died in infancy; the younger, Fernando, succeeded his father in 1367. By the beautiful but unfortunate Iñez de Castro, who, as Calderon says, was not a queen till after her death, Dom Pedro had three sons and a daughter. One of the sons, Affonso, died in infancy; the two others were João and Diniz, of whom we shall hear more presently. Besides these, he had by Theresa Lourenzo, a lady of noble birth, a natural son named João, Prince Henry's father, who, at the age of seven, received from his father the Grand Mastership of the Order of Aviz. Two years after the death of Dom Pedro, which took place on the 18th January, 1367, his eldest son and successor, Fernando, became, as great-grandson of Sancho IV., the rightful heir to the crown of Castile, on the death of Don Pedro the Cruel without legitimate offspring. That crown, however, was in the hands of Enrique of Trastamare, the illegitimate brother of the late king, a man by no means inclined to give up the kingdom he had usurped, unless under compulsion. Dom Fernando therefore formed an alliance with Don Pedro of Aragon, whose daughter Leonora he engaged to marry. Enrique the Bastard forthwith invaded Portugal, and a contest ensued which was only brought to a close through the intervention of Pope Gregory XI. by a treaty of peace signed at Evora, at the close of 1371, one of the conditions being that Fer-

nando should marry Enrique's daughter Leonora. Fernando
was thus betrothed to two Leonoras, the one of Aragon,
the other of Castile, and he now became passionately enamoured of a third Leonora, surnamed Telles de Meneses, the
wife of João Lourenço da Cunha, Lord of Pombeiro. The
five months within which, according to the treaty, Leonora of
Castile was to pass into Portugal had nearly expired, when
the king annulled the marriage of Leonora Telles, sent her
husband into Spain, and publicly took her to wife. This insult
to the King of Castile was followed by another, if possible,
still more flagrant; for, in defiance of the terms of the treaty,
King Fernando entered into an alliance with John of Gaunt,
Duke of Lancaster, who, having in 1370 married the eldest
daughter of Pedro the Cruel, laid claim to the crown of
Castile. The war that ensued was one of the most cruel
and deplorable that Portugal ever had to sustain, King
Enrique having sworn that he would not return to Castile
till he had reduced Lisbon to ashes. Happily, however,
Gregory XI. again became the mediator between the two
sovereigns, and a treaty of peace was signed in 1373, which
remained in force till after the death of Enrique in 1379.

Leonora Telles, who was as remarkable for her heartless-
ness and subtlety as for her marvellous beauty, had a sister,
Maria Telles, beautiful like herself, but, unlike her, endowed
with a pure, noble, and affectionate nature. To this lady the
king's half-brother, João, eldest son of Iñez de Castro, was
secretly married. Leonora, who hated them both, and feared
that they might one day succeed to the throne of Portugal,
took occasion first to intimate to the prince a wish for his
marriage with her daughter, Brites, and, secondly, to insinuate
charges against the chastity of his wife. The prince, incapable
of suspecting such infamy on the part of the queen, believed
the falsehood, and hastening to Coimbra, where the princess
was, killed her with his own hand. No sooner was the crime
accomplished than Leonora derided the assassin, who fled
for safety to Castile. The other son of Iñez de Castro, Dom
Diniz, was driven into exile for refusing, at a formal au-

dience, to kiss the hand of the adulterous queen, presented
to him by the king. Another object of the queen's mur-
derous designs was the king's illegitimate brother, the Grand
Master of Aviz, whose life she twice attempted by forging
the king's signature for his execution, and afterwards by
poison, but happily he escaped her malice. She now added
to the number of her crimes infidelity to the king himself.
Her paramour was Don Fernando Andeiro, a Castilian
subject, but a special favourite of the king, who had employed him to negotiate a secret alliance with the Duke of
Lancaster for the subjugation of Castile. On his return
from this mission he was, for some time, concealed in the
Castle of Estremos, the residence of the king and queen,
with the latter of whom he thus had frequent opportunities
for private interviews.

King Juan of Castile, Enrique's successor, hearing that Fernando was forming large armaments and expecting assistance
from England, lost no time in preparing for an encounter
with his perfidious ally, but after a few indecisive engagements a treaty of peace was concluded, one condition of which
was that the second son of the King of Castile should marry
Brites, the daughter of Fernando and Leonora de Telles.

In the interval King Juan's wife died, an event which
suggested to Fernando a yet more advantageous marriage
for his daughter, who, after having been affianced to many
princes, became the wife of the king of Castile himself.

The marriage treaty provided that if Fernando died without legitimate male issue, Brites should wear the crown
until the birth of her first legitimate child, on whom it
should then devolve, and that until it should attain its
majority at the age of fourteen, Leonora should be regent.
If Brites were childless, and died before her husband, her
father having also died without heirs, the crown of Portugal
should then devolve upon King Juan of Castile and his heirs.
Corresponding stipulations were adopted with regard to the
crown of Castile. "No treaty," says Nuñez de Leão, "was
ever more solemnly sworn to, or surrounded with greater

precautions, and none was ever worse kept." King Fernando's failing health prevented him from being present at the brilliant marriage of his daughter. He had at length become aware of the guilt of the infamous queen, but not having the courage to remove her paramour from the court, he called to his aid his illegitimate brother João, the Grand Master of Aviz, with whom he resolved upon Andeiro's death, but before this could be effected, the king fell dangerously ill, and was conveyed to Lisbon, where he died on the 22nd of October, 1383. As his daughter Brites was childless, the throne of right belonged to João, Duke of Viseu, the eldest surviving son of Iñez de Castro, but King Juan lost no time in seizing that unfortunate prince, and placing him in safe custody at Toledo. Leonora forthwith assumed the position of regent, but, on the demand of the King of Castile, was compelled to proclaim her daughter Brites as queen.

The Portuguese chafed at the thought that the Castilian yoke should be imposed upon them through the marriage of their princess with a king of Castile. Leonora and her paramour were universally detested; and not only the nobility, but the whole kingdom, were prepared to hail as their deliverer any one who should take the life of the latter. The two sons of Iñez de Castro being kept in safe custody by the King of Castile, the Grand Master of Aviz, who was the only son of King Pedro I. now in Portugal, at once saw in this favourable conjunction of circumstances a chance of obtaining possession of the crown.

Leonora was not blind to the same possibility, and by way of removing him, made him governor of the Alemtejo for the defence of the frontier. This was a crisis in his life. Andeiro's death had been secretly resolved upon by the leading nobles of the kingdom, and the hand of the Grand Master was by all regarded as the one to strike the blow. Accordingly, at the close of an interview with the queen in her palace, he led Andeiro into an antechamber, as if to speak with him, and there slew him. He then gave orders that the gates of the palace should be closed; and in pursuance of a preconcerted plan, his page, Gomez Freire, rode through the streets of Lisbon, crying out that his master was shut up in the palace, and in imminent danger of his life. The people, by whom he was much beloved, rushed in crowds towards the palace gates, threatening to force an entrance unless they were convinced with their own eyes of the Grand Master's safety.

When at length he made his appearance, and rode through the streets, the shouts of joy with which he was received told plainly how near he was to the realization of his most sanguine hopes. The people were enthusiastic in his favour, but many of the nobles who had sided with him while it was a question of getting rid of Andeiro, returned to Leonora, now that that favourite was removed. The queen had called to her aid her son-in-law, the King of Castile, and when the people of Lisbon reflected on the dangers to which they would be exposed if their city were to be at the mercy of Leonora and of the Castilians, the instinct of self-preservation drove them the more anxiously to look for protection and safety in the talents and energy of the Grand Master. They therefore declared their wish to recognise him as their protector and sovereign, and to place at his command the city and its revenues.

The approach of the King of Castile to the frontiers of Portugal left no alternative; and even the nobles were at length, though against their inclination, induced to give in their adhesion, and accordingly, an act, which constituted the Grand Master defender and regent of the kingdom, with powers little less than royal, was formally signed on the 16th of December, 1383.

In this new and difficult position, the Grand Master displayed talents equal to his responsibilities. To invest that position with befitting dignity, he styled himself in all official letters and ordinances, "João, by the grace of God, son of the most noble King Pedro, Master of the Order of Chivalry of Aviz, Regent and Defender of the Kingdoms of Portugal and the Algarves." He placed the royal arms upon the

cross of his order, so that only the extremities of the latter
were visible, thus skilfully blending the insignia of the Grand
Master of the order with those of the Regent of the kingdom.

He was prudent in the selection of his ministers of state,
among whom the most remarkable were his High Chancellor
João das Regras, and Nuño Alvarez Pereira. To the legal
acumen of the former he subsequently owed his crown,
while the latter, who was his well-loved friend from boyhood, stands pre-eminent in Portuguese history for his
valour, his piety, and devotedness to the king's service.
The Regent, however, was not blind to the fact that his half-
brother, Prince João, who was still a prisoner in Castile,
had a claim to the throne which took precedence of any that
he himself could advance beyond such as might emanate from
the expressed will of the people. Accordingly, he declared
that he held his authority on behalf of his half-brother, and
caused banners to be painted representing the Prince in a
dungeon, loaded with irons. By this means he secured the
good-will of the prince's partisans, and at the same time
intensified the people's hatred of the King of Castile, and
their attachment to his own family. The queen, who for greater
security had now withdrawn from Alemquer to Santarem,
perhaps the strongest fortress in the kingdom, issued letters
to the commanders of various strongholds, calling on them
to proclaim her daughter Brites queen, and urged on the
King of Castile the necessity of forthwith enforcing her
rights by the sword, thereby only the more exasperating
the popular fury.

The people's devotion to the Regent made him strong
within the frontiers of Portugal, but an enemy was approaching who would have to be encountered and repulsed by force
of arms. The Regent addressed himself with energy to the
needful preparations, and appealed successfully to the different towns of Portugal for aid. He also sent an embassy
to the King of England, requesting assistance and promising
future reciprocation, and suggested to the Duke of Lancaster, who was then at the court in London, that if he

wished to obtain possession of the crown of Castile, it was now the fitting opportunity, when Portugal was ready to assist him. The English were delighted with the proposal. Money and men were forthcoming on the moment. Troops were dispatched forthwith, and King Richard's reply was in the highest degree encouraging.

The Regent's next anxiety was to provide for the security of Lisbon, in the event of its having to sustain a siege. This charge was assigned to Nuño Alvarez Pereira, who with unfailing activity collected stores, and in spite of all opposition conveyed them into the city. The King and Queen of Castile had already entered Portugal, and had received from Leonora a formal renunciation of the crown in their favour. This measure, which emanated from Leonora's hatred of the Grand Master, brought over many of the nobility to the side of the King of Castile, who thus found himself in possession of numerous strongholds of the kingdom. Before long, however, a disagreement arose between Leonora and the king, as to the appointment of the chief Rabbi of Portugal, and the Queen became so irritated, that she attempted the assassination of her son-in-law. Her designs being discovered, she was placed in a convent at Tordesillas, near Valladolid, where she ended her days.

If by this removal of Leonora the king secured a positive gain, he incurred at the same time a more than corresponding loss in the withdrawal of the support of his adherents in Portugal. Lisbon was now the focus both of his hope and his anxiety, and with the view of effectually reducing it, he blockaded it from the sea, while his forces ravaged the Alemtejo and endeavoured to hem it in by land. It was absolutely necessary to check at once the advance of the land force, and the Grand Master entrusted this important charge to the gallant but youthful Nuño Alvarez Pereira, who, in spite of his great inferiority in numbers, resolved to give them battle. The undertaking was a critical one, but the religious enthusiasm of Pereira gained for him the day. After addressing his soldiers in fervent language,

he dismounted and knelt in prayer. His men followed his
example, and when they arose from their knees and attacked
the enemy's cavalry, which constituted the main strength of
their army, the onslaught was so tremendous, that the Castilians fled in the utmost disorder. The effects of this victory,
known by the name of Atoleiros, from the field where it was
won, were immense. Many who had hesitated to attach
themselves to the cause of the Regent now readily gave in
their adhesion, nor did the indefatigable Pereira cease his
exertions till he had rendered futile all the efforts of the
Castilians to subjugate the Alemtejo.

King Juan now devoted all his thoughts to the siege of
Lisbon. He had received large reinforcements from Castile,
but delayed the attack till the arrival of his fleet from Seville.
The Grand Master meanwhile lost no time in refitting the
vessels which were lying in the harbour of Lisbon. The
hearts of all were in the cause. Lorenzo, Archbishop of
Braga, lance in hand, and with his episcopal costume over
his armour, rode from point to point, encouraging and
urging all to assist in the work. If a priest excused himself on account of his orders, he answered, that he also
was a priest, and an archbishop to boot. Lisbon was soon
invested both by land and by sea, but through the foresight
of the Grand Master, it was well supplied with provisions,
its walls repaired, and its seventy-three towers well stocked
with arms and projectiles. The people had full confidence
in their chief. All took their part in the work of defence,
and the utmost order prevailed, though the city was crowded
with refugees. For five long months the king was foiled in
all his efforts to take it. The only hope now left was to
reduce it by famine, and it seemed most likely that this
dreadful scourge would effect the king's object. Pallid faces
and the groans of those who were perishing of starvation told
a piteous story of the condition of those within the walls,
yet none thought of surrender. But amid the ranks of the
besiegers stalked a yet more deadly enemy, the plague, which
carried off almost two hundred Castilians daily. In this

direful position of affairs, each party obstinately waited
to see which would be the conqueror, the famine or the
plague; till at length, when symptoms of the malady began
to show themselves on Queen Brites, the king struck his
camp; and on the 5th of September took his departure for
Torres Vedras, uttering bitter execrations on the city which
had thus successfully resisted him. On the 14th October he
crossed the frontier, not in triumph, but as it were with a
funeral procession; for in the van of his army were carried
on biers the bodies of many noble victims of the plague,
whose remains had been preserved that they might be buried
in the tombs of their ancestors. The gloom inspired by
the black trappings of death was unrelieved either by the
gladness of success or by the consciousness of glory won.
Sadness and silence were the companions of that homeward
march. Meanwhile at Lisbon the joy was that of men
restored from death to life. The people were bent on
solemn acts of fervent thanksgiving to the Almighty, and
the bishop and clergy in their sacerdotal vestments, the
Regent, the nobility, and the populace testified their united
and humble thankfulness by walking in reverent procession
with bare feet to the convent of the Holy Trinity to offer
to God the incense of their praise and gratitude. To none
were the glad tidings of this happy event more welcome
than to that truest of friends and patriots, Nuño Alvarez
Pereira. With his usual fearlessness, he sailed down the
Tagus from Palmella in a light skiff through the enemy's
fleet to offer his congratulations. At his instigation a re-
newal of the act of homage to the Grand Master by all the
nobles, knights, prelates, and municipal authorities, took
place on the 6th of October, in the royal palace, where the
Grand Master resided. Mortified at his failure, the King of
Castile now attempted the life of the Regent by the hand of
an assassin, but the plot was discovered.

Soon after this the Cortes were assembled at Coimbra.
The safety of the kingdom rendered necessary the appointment of a responsible chief, and it was evidently the wish of

the people to proclaim the Grand Master King. Some of
the nobles, however, thought the only legitimate course was
that the Grand Master should be Regent for his half-brother
Dom João, or in case of his death for the Infant Dom Diniz,
who had been declared legitimate children of King Pedro.

At this juncture the Grand Master had the good fortune
to possess in the Chancellor João das Regras an advocate
who served him as well with his tongue as his faithful friend
Nuño Alvarez Pereira had already done with his sword.· The
Chancellor's main purpose was to show that the throne was
without an heir, and that by the laws of Lamego * the
choice lay with the people. He first asserted that Brites,
Fernando's daughter, was illegitimate, and further, that she
and her husband had, by making a violent entry into
Portugal, broken the treaty by which the terms of the succession had been settled. He then dwelt on the doubtful
legitimacy of the children of Iñez de Castro, and further
declared that they had forfeited all right to the throne by
joining their country's enemies. In conclusion, he argued
that the Portuguese possessed the power of choosing their
own king, and that there was no one who by his birth,
abilities, and devotion to his country, so well deserved to be
raised to the throne as the Grand Master of Aviz. The
discussions which ensued were set at rest by the Chancellor
producing the written refusal of Pope Innocent VI. to
recognise the legitimacy of the children of Iñez de Castro.
His success was complete, and on the 6th of April, 1385,
the Grand Master was proclaimed King to the joy of the
whole nation.

Among the individuals to whom the king held himself
most deeply indebted, Nuño Alvarez Pereira stood pre-
eminent; and on him therefore, though but twenty-seven,
two years younger than himself, he conferred the highest
military rank in the realm, that of Constable. The remark-

* It was at Lamego, in the province of Beira, that the first Cortes of the
kingdom were convoked in 1143, by the King Affonso I., and the fundamental
laws of the constitution drawn up.

able combination in him of courage and religious enthusiasm gained for him in after-years the title of the Holy Constable. His invaluable qualities were soon to be brought into active operation. Intelligence arrived of a fresh invasion by the King of Castile. Pereira at once set out with all the troops at his command for Santiago, and collecting men as he proceeded, made himself master of various places which held allegiance to the King of Castile. When the King at length joined his forces to those of the Constable, in the province of Entre Douro e Minho, he obtained possession of the most important places in that province, and the Castilian party found itself daily more and more straitened.

The struggle now began to assume more alarming proportions, and it became evident that the decisive hour was approaching. The Castilians had crossed the frontier by Almeida, and were advancing upon Viseu. The Portuguese marched to meet them with three hundred lances, a small band of regular infantry, and a number of peasants. They were drawn up at half a league's distance from Trancoso, by which place the Castilians would have to pass. The latter had been pillaging for several days, and the large quantity of booty made them anxious to avoid the enemy, but the Portuguese intercepted them. A deadly engagement ensued, which lasted from morning till afternoon. The Castilians had the superiority in numbers, and the humiliation of defeat was not to be endured. The Portuguese were on their own ground, and had the thought of hearth and home to stimulate their antagonism to their ancient foes. It was not till the four hundred chosen lances of Castile were laid low in death that the obstinate engagement was brought to a close.

The actual loss amongst the flower of the Castilian nobility was great, but the blow to the *morale* of those who remained was perhaps even more important. On the other hand, this well-won victory of Trancoso encouraged the Portuguese for those heroic efforts which were still to be required of them. The King of Castile now determined to bring the whole of his forces into Portugal, and to engage King João in one decisive battle, a plan which was opposed by his more prudent advisers for reasons among which the king's health was by no means the lightest. King João and the Constable lost no time in collecting such forces as could be mustered, and happily at this time three large ships arrived at Lisbon from England with about five hundred men-at-arms and archers. The greater part of them were mere adventurers. There were no knights amongst them, but they were led by three squires named Northberry, Morberry, and Huguelin de Hartsel, whereas two thousand French knights had joined the Castilian army. On the 14th of August the Portuguese army took up an advantageous position in a plain at a league's distance from Porto de Mos. When the first ranks of the Castilians came in sight, they did not offer battle to the Portuguese, but marched in the direction of Aljubarrota, where they halted. The older and more prudent officers of the Castilian army advised that they should remain quietly where they were, as the soldiers were fasting and fatigued with the march. This prudent counsel was overruled by the impatience of the young soldiers, who clamoured for an instant encounter.

Historians differ as to the respective strength of the two armies in this important battle, but there can be no doubt that the Castilians were very superior in numbers, in experience, and in equipments. They had also the advantage of possessing ten pieces of cannon called "trons," the first ever seen in Spain. The movement of the Spaniards towards Aljubarrota had necessitated a change in the position of the Portuguese army. The ground occupied by King João was a level plain covered with heather, and as his force was small, it was divided into only two lines. In the vanguard was the Constable with only six hundred lances. In the right wing was a goodly band of gentlemen who, as a point of honour, had resolved to defend to the death the spot on which they might be placed. This division bore the name of the " Enamorados," or " Volunteers," and was distinguished by a green banner. The left wing consisted of Portuguese and foreigners, among whom were some English bowmen and men-at-arms. Behind the men-at-arms in both wings were bowmen and infantry, so placed as to give ready help to the cavalry. The King, with seven hundred lances and the royal standard with the guard appointed for its defence, were in the rear-guard, behind which was a strong barricade formed with the baggage which was begirt by foot soldiers and bowmen. It was evening, and the men had suffered much from the necessity of remaining under arms all day beneath the full blaze of an August sun, especially as, from reverence for the vigil of the Assumption, few of them ate or drank. But the example of the King and the Constable quite sustained the courage of all. On both sides the trumpets sounded for the charge; the war-cries of " Castile and Santiago," and " St. George for Portugal," rang through the air, and the two armies met with a heavy shock.* The Portuguese van-guard at first suffered terribly from the arrows of the Castilian bowmen. The Castilian light horse endeavoured, though in vain, to penetrate the baggage waggons, but the force of the battle was soon concentrated round the banner of the Constable, the Castilians directing their principal efforts against the division of " Enamorados," who suffered the most. When the King perceived that the foremost ranks were penetrated, and that the Constable was hard beset, he pressed forward with the rear-guard and the royal banner. The contest became fiercer and more deadly every moment, King João himself kindling the courage and valour of his troops by surprising proofs of his own strength

* In accordance with the Portuguese historians Manoel de Faria and Duarte Nuñez de Leão, the armies are here made to meet in the open plain. Froissart, on the contrary, says that, in pursuance of the advice of the English, the King of Portugal made a stronghold of the church of Aljubarrota which was on a small eminence beside the road, and surrounded by large trees, hedges, and bushes. Trees were cut down and so laid that the cavalry could not pass them, leaving one entry not too wide, on the wings of which they posted all the archers and cross-bows. The men-at-arms were on foot, drawn up beside the church, where the king was.

and intrepidity. In the height of the combat the royal standard of Castile was thrown down, and at the disappearance of the banner which had served them as a rallying point, some of the Castilians began to give way. When the King of Castile saw his standard overthrown, and his soldiers seizing any horses they could find to flee upon, he resolved to look to his own safety before the battle was entirely lost. His keeper of the household, Pedro Gonsalvez de Mendoza, who had foreseen the result of a contest, entered upon against his own advice and that of the most experienced knights of the council, had steadily remained by his master's side to help him in the moment of necessity. That moment had now arrived. Setting the king upon a strong horse in exchange for the mule which, after leaving his litter, he had ridden on account of his illness, he led him from the field, and then, in spite of the king's remonstrances, returned to the fight. "The women of Guadalaxara," said he, "shall not reproach me with having led their husbands and sons to death, while I return home safe and sound." Accordingly he fought his way into the thick of the battle, where he fell like a true-hearted soldier as he was, whilst his master rode for his life, tearing his beard and cursing the day that he had entered Portugal. Meanwhile the Portuguese bowmen and the infantry who protected the baggage, having been taken in flank by the Castilian light horse, the King ordered the Constable to hasten to their assistance. The Portuguese were already successfully defending themselves, and on the appearance of the Constable the Castilian cavalry ceased from the attack. The wings were now able to bring all their strength upon the Castilian van-guard, and complete its overthrow. The Castilians, finding that their king had fled, lost all hope, and favoured by the darkness, took to flight. To this day there is shown in Aljubarrota a baker's shop, which tradition records to have been at that time a bake-house, in which Brites d'Almeida, the baker's wife, slew with her oven-peel no less than seven Castilian soldiers.

This famous battle of Aljubarrota was fought on the 14th

of August, 1385. It was a day the proud memory of which is deathless in the annals of Portugal; for, apart from its incalculable importance to the permanent well-being of that country, the battle then fought was as remarkable for the display of chivalrous courage as any that has been recorded in the history of modern Europe. In accordance with the custom of the period, King João remained three days and three nights upon the field, until the fetid exhalations from the bodies of the slain obliged him to withdraw. The booty taken from the Castilians was immense. The king's tent, with all its furniture, the silver triptych belonging to the portable altar of the Castilian army, which is still to be seen in the sacristy at Guimaraens,* were taken as well as the king's sceptre, which was long preserved in the now extinct Convento do Carmo at Lisbon, built by the Constable Nuño Alvarez de Pereira. It was near the site of this famous battle that the king afterwards erected the beautiful convent of Batalha, as a mausoleum for himself and his posterity, and here are still preserved the helmet and sword worn by him on that eventful day.

Meanwhile the King of Castile had fled, accompanied only by a few servants. At midnight he arrived sick and exhausted at Santarem, about twelve leagues from Alju-barrota, where, still alarmed for his safety, he took a boat the same night, and descending the river, reached the port of Lisbon on the 15th of August. Thence he sailed in safety to Seville, where he took the precaution of landing during the night of the 22nd of August. In his despondency at the great calamity which had befallen him, he is said to have worn mourning for seven years.

Taking advantage of the depressed condition of Castile, the Constable now resolved to carry his arms into the enemy's country, and thus afford the King an easier oppor-tunity of reducing to subjection the north of Portugal, many

* In the same sacristy is shown the pelote worn on this occasion by the King of Portugal, and a large Bible then taken was given to the Abbey of Alcobaça, and is now in the Bibliotheca Nacional at Lisbon.

towns in which were still in the hands of the Castilians. In the month of September, 1385, he levied troops at Evora to the number of 1,000 lances and 2,000 infantry,* intending to cross the frontier and attack Valverde. The Castilians, in order to be in advance of the Portuguese, immediately assembled a large force from the towns of Andalusia, part of which they sent across the Guadiana, while part remained behind in reserve.

By dint of hard fighting the Constable forced the passage of the river, but only to find a second force of 10,000 awaiting him on the other side. His position was now in the highest degree perilous, but his exhaustless energy and marvellous presence of mind again worked wonders. But not to his own efforts only did he trust the result of so important an engagement. While the battle was at its height, and all apparently depended on his presence, he for a while disappeared from the field. Two messengers dispatched successively in search of him found him on his knees in prayer. He paid no attention to their representations, but at length, when his prayers were concluded, arose, his countenance bright with confidence, and returned to the fight. Seeing the banners of the enemy on the summit of a neighbouring eminence, surmounted by the standard of the Grand Master of Santiago, he ordered his own standard-bearer to plant his colours by the side of the other, he himself cleaving his way through the masses which well-nigh smothered his little band, till he encountered a worthy adversary in the Grand Master himself. The combat was short ; the Grand Master fell mortally wounded, and his fall and the overthrow of his standard gave the signal for the flight of the Castilians. The Constable pursued them till nightfall, and on the

* At Garcie, a trumpeter presented himself with a challenge from the Castilian nobles, accompanied by a certain number of scourges on the part of each of them. The Constable received them with his habitual composure, and sent a graceful message of thanks to the Castilian grandees for the challenge which they had sent him, and more especially for the whips, with which he promised himself the pleasure of chastising them all. To the herald he gave a hundred golden dobras (about £360).

morrow retraced his steps towards Portugal. The disaster which Castile had experienced at Aljubarrota was thus speedily followed by a scarcely less crushing blow at Valverde. Most of the Portuguese towns occupied by the Castilians soon surrendered to the King, and, in order to reduce the rest to submission, he was making preparations for levying a considerable army when news arrived that the Duke of Lancaster was on the point of proceeding to Spain to prefer his claim to the crown of Castile, in right of his marriage with the Princess Constance.

From early times an alliance, cemented by numerous political and commercial treaties, had existed between England and Portugal, and the elevation of the Grand Master of Aviz to the throne and his victory over the King of Castile had supplied his ambassadors with reasons for suggesting to the Duke of Lancaster that the opportunity was favourable for carrying out his own designs upon Castile.

Accordingly, on the 20th of July, 1386, the Duke arrived at Corunna with 2,000 lances, 3,000 archers, and a fleet of 180 galleys, accompanied by the Duchess Constance, their daughter Catherine, and Philippa, the duke's daughter by a former marriage. Without delay an interview was arranged between him and the king, at which the latter undertook to assist him in the conquest of Castile, and bound himself to supply and maintain 2,000 lances, 1,000 cross-bowmen, and 2,000 foot soldiers, for eight months, while the Duke, on his part, pledged himself, in the event of success, to cede to the King of Portugal several considerable places on the frontier, and to repay the expenses of the campaign.

By way of sealing this new compact, it was agreed that the King should receive one of the duke's daughters in marriage. It was the wish of the Portuguese that the King should choose Catherine, with a view to his thus becoming the heir presumptive to the crown of Castile, but the King himself, both from policy and from real preference, chose the Princess Philippa. Having first obtained from the Pope the necessary dispensation from the vow of chastity which

he had taken as Grand Master of the Order of Aviz, he was
married to Philippa with great pomp, and to the extreme
delight of the people, on the 2nd of February, 1387. The
young king had endeared himself to his subjects by his
well-proved heroism and wisdom, while Philippa, who was
one year his junior, was as remarkable for the modest
dignity of her bearing as for her beauty, both qualities
well befitting the grand-daughter of Philippa of Hainault.

On the 25th of March, the King, having levied a larger
contingent than he had engaged to furnish, brought them
as a reinforcement to the Duke of Lancaster, whose force
sickness and frequent skirmishes had already reduced to 600
lances and a small number of archers. He could not
conceal from himself, however, that the campaign offered
little prospect either of glory or of ultimate success. The
Castilians showed no disposition to recognise his father-in-
law as their sovereign, and his remaining force was dwin-
dling away from sickness and want of provisions, while the
resources of the King of Castile were comparatively great.
He therefore represented to the Duke of Lancaster that two
alternatives only remained, to levy more soldiers in England,
or to come to a compromise with the King of Castile. On
many grounds the duke preferred the latter course, to which,
moreover, other circumstances were leading him. When, on
first landing, he had summoned the King of Castile to
acknowledge his right to the crown, the king had proposed
a marriage between Catherine, the duke's daughter, and her
cousin the Prince Royal of Castile. This proposal, though
far from displeasing to him, the duke had set aside at the
time on account of his alliance with the King of Portugal,
but it still remained open.

Meanwhile it had become necessary to effect a retreat.
On the 15th of May the Allies re-entered Portugal by way
of Almeida, but on his way to Coimbra, to visit his daughter,
the Duke was met at Trancoso by a deputation from the
King of Castile, offering terms of peace, and again proposing the marriage between the Princess Catherine and the

Prince Royal. The King engaged to grant to the princess certain towns in Castile as a dowry, and to pay the duke 600,000 florins as indemnity for the expenses of the war, as well as an annuity for life of 40,000 florins, if the Duke and Duchess of Lancaster would renounce all claim to the crown of Castile. The duke was invited to discuss these proposals with the plenipotentiaries of the King of Castile at Bayonne, which at that time belonged to England, and there, towards the end of September, the compact was formally agreed upon, the Princess Catherine being fourteen years of age, and her affianced husband ten. On this occasion, the king gave the Infant the title of Prince of Asturias, a title which has ever since been borne by the heir presumptive to the crown of Spain.

King João, whose chief wish was to secure the independence and promote the internal well-being of Portugal, far from desiring to set up any claims of his own to the throne of Castile, saw in the proposed marriage that for which he was so anxious, a prospect of peace, and his hopes in this respect were soon realized. So rapid was his success in 1389, in recovering some of the Portuguese towns which had given in their adhesion to Castile, that the King, alarmed at his progress, proposed a truce of six years, which was agreed to. On the 9th of October, 1390, the King of Castile died from the effects of a fall from his horse at a tournament. During the following reign, Enrique, King Juan's successor, persuaded Queen Brites to cede her right to the crown of Portugal to the Infant Diniz, who had taken refuge in Castile, and who now, with two thousand lances and a number of Portuguese malcontents, entered the province of Beira, calling himself King, and promising large rewards to those who should render him obedience. No one declaring for him, he soon found it prudent to withdraw again to Castile. The King of Portugal continued to obtain important successes, and negotiations for peace were opened, but failed, solely through the exorbitant demands of the Castilians. At length, mainly through the influence of his

wife, Queen Philippa's sister Catherine, King Enrique was
induced to agree to a cessation of hostilities for ten years.
At his death in December, 1407, Queen Catherine became
Regent during the minority of her son, and a definite treaty
of peace was concluded on the 30th October, 1411.   Meantime the friendship between Portugal and England had
become most closely cemented.   The riband of the newly-
established Order of the Garter had been conferred upon
King João, who was the first foreign sovereign to receive it,[*]
and the sovereigns of Portugal and England agreed, that in
any treaty between either of them and Castile, the other
should be included.   Whilst Portugal was thus gaining
importance, King João and Queen Philippa became the
parents of a noble family of children, whose names and
order of birth are as follows :—

1st. Branca, who was born in Lisbon on the 13th July,
1388, but lived little more than eight months.

2nd. Affonso, who was born in Santarem, on the 30th of
July, 1390, and who, according to Fernam Lopez, lived two
years, though Cunha, in his history of Braga, where he was
buried, gives the 22nd of December, 1400, as the date of his
death.

3rd. Duarte, who was born in Viseu, on the 31st of October,
1391, and received his name of Duarte (or Edward) in
memory of his great-grandfather, King Edward III. of
England, and who succeeded his father on the throne of
Portugal.

4th. Pedro, who was born in Lisbon, on the 9th of
December, 1392, and of whom much will be said in the
following pages.

5th. Henrique, the " Prince Henry," of the present
biography, born in Oporto on Ash Wednesday, the 4th of
March, 1394.

6th. Isabel, born on the 21st of February, 1397, afterwards married to Philip le Bon, Duke of Burgundy, who

---

[*] He succeeded Sir William Arundel, who died August, 1400.

established the order of the Toison d'Or, in honour of the occasion.*

7th. João, born in Santarem, the 13th of January, 1400, afterwards Grand Master of the Order of Santiago.

8th. Fernando, born in Santarem on the 29th of September, 1402, whose patient endurance of suffering in Morocco won for him the designation of " The Constant Prince."

King João had also two illegitimate children, Affonso Count of Barcellos, who married the daughter of the Constable Nuño Alvares Pereira, from which union sprung the royal house of Braganza, and Brites, who married on the 26th of November, 1405, Thomas Fitzallan, Earl of Arundel.

---

* It is almost certain that Jan van Eyck, the perfecter, if not the inventor, of painting in oil, was attached to the embassy sent to Portugal to solicit the hand of Isabel for the Duke of Burgundy. In 1836, when King Ferdinand, then Prince of Saxe Coburg, was on his way to Portugal to marry Dona Maria da Gloria, he made a short stay at Brussels, and, at a court fête given on the occasion, the Queen of the Belgians appeared in a costume faithfully copied from a portrait in which his Excellency the Count de Lavradio, who negotiated the marriage, thought that he recognised the portrait of the Princess Isabel by Van Eyck.

CHAPTER III.

CEUTA.

A.D. 1415.

Now that Portugal was at peace with Castile, it began to attain a high degree of prosperity, and King João, though dreaded by his neighbours, was beloved by his people. The glory identified with his name served as a stimulus to the ambition of his sons, the three eldest of whom, Duarte, Pedro, and Henry, were now of age, and had been admirably trained by their father in every chivalrous accomplishment. The princes were anxious to receive the honour of knighthood; but, as this was a distinction only to be gained at the point of the sword, the King proposed to hold a succession of tournaments during an entire year, to which knights of all nations, and of the highest renown in feats of arms, should be invited. His minister of finance, João Affonso de Alemquer, represented to him the useless expenditure inseparable from such a plan, and suggested that an invasion of the Moorish city of Ceuta would offer a far more honourable and fitting opportunity for conferring the rank of knighthood upon the princes, while it would be carrying the sword of the avenger into the country of their former conquerors, and opening a door to the advance of Christianity. The King yielded to the representations of his minister and the wishes of his sons, to whom the idea of winning their spurs at a tournament was most distasteful.

Desiring to obtain as much information as possible respecting the strength and position of Ceuta, he had recourse

to the following stratagem. He sent Affonso Furtado de
Mendoza and Alvaro Gonsalves de Camelo, prior of the
hospital of St. John of Jerusalem, as envoys to Sicily, to
ask the hand of the Queen in marriage for Dom Pedro, and
as the vessels must necessarily pass near Ceuta, where ships
of different nations were in the habit of stopping, he desired
the envoys to make the most of the opportunity to examine
the place. Accordingly, under pretence of taking in provisions, which, in their character of ambassadors, they were
permitted by the Governor to do, they remained four days
in the city, carefully noting everything about which the
King needed information.

They then proceeded to Sicily, and delivered their message
to the Queen, but with no successful result. On their return,
when they had reported the issue of their mission to Sicily,
the King desired Mendoza to state what he had ascertained
respecting Ceuta. His only answer was an assurance that
the King would be successful in the proposed enterprise,
and, when pressed for his reasons, instead of reporting his
observations, he told a story of a prophecy uttered to him
when a boy by an old Moor, and already partly verified, that
a king named João, a natural son of the late king, should
be the first of his country to gain dominion in Africa.

The King then applied for information to the prior of St.
John, but it seemed that he was fated to be answered only
in enigmas, for this envoy declared his inability to afford
any details unless he were supplied with two loads of sand
and two bushels of beans. When after some demur these
singular materials were produced, the prior formed the sand
into a representation of the seven hills from which Ceuta or
Septa takes its name, described the double wall on the
landward side, with its towers and curtains, and represented
with the beans the apparent number and position of the
houses, and what was all important, indicated the most
convenient spot for the safe and expeditious landing of the
troops.

The King warmly commended his zeal and sagacity, and

after consultation with the Queen and the Constable, at once
commenced his preparations for the expedition. The Kings
of Aragon and Granada immediately took the alarm. To
the former, who sent messengers to him requesting a frank
avowal of his intentions, King João replied that he had no
idea of attacking Aragon, but that on the contrary, in case
of necessity, he was ready to protect it. The Moorish King
of Granada at first sought to allay his fears by asking the
intervention of the King of Aragon, but receiving a contemptuous answer, he sent envoys direct to King João
himself, begging an assurance of peace under the King's
seal, so that commercial intercourse between the two
countries might not be interrupted. The King replied that
he would take time for consideration. The envoys now had
recourse to Queen Philippa, and besought her, in the name
of Riccaforna, Queen of Granada, so to use her influence as
to induce the King to remain at peace, promising, in requital,
to send her choice and costly gifts for the nuptials of her
daughter. To this Queen Philippa, who, as the old chronicler
says, being English by birth, held both Jews and Moors in
detestation, gave the following dignified reply:—" I know
nothing of the methods which your queens may resort to in
dealings with their husbands, but with us it would be
regarded as an indecent thing for a wife to interfere in her
husband's affairs, especially in such as have to be debated in
council. As regards the present which your queen has so
liberally offered me, I thank her and accept her good wishes,
but beg her to dispose of her gifts elsewhere as she may
please, for, when the time comes for my daughter to be
married, she will have no lack of costly ornaments."
The King at length informed the envoys that he had no
intention of invading Granada, but as he would not give
the assurance in writing, they took it for granted that
their worst fears were to be realized, and hastened back to
report their apprehensions, whereupon the King of Granada
garrisoned and provisioned all the towns on the sea-coast.
Rumours of these preparations reached King João, who very

reasonably supposing that they might also reach Morocco, gave out that he was about to declare war against the Count of Holland, to whom he sent an envoy with instructions secretly to inform the Count of the truth, but openly to threaten war. Important as it was to allay the suspicions of the neighbouring powers, it now became necessary to announce the truth to his own subjects. He first summoned the peers to Torres Vedras, and declared to them the various reasons for which he had determined to attack Ceuta. The announcement being received with applause, the King issued a proclamation representing that the fleet was prepared for his sons, and that those who might desire to volunteer to go out with them should declare in writing how many armed men they would supply, and that stores would be taken on board at Lisbon and Oporto. When the fleet was completed, and while the soldiers were busily engaged in lading the vessels, a pestilence broke out in both cities. The King by no means relaxed his efforts on this account, but personally superintended at Lisbon the preparations for the expedition. At Oporto Prince Henry, armed with full authority from his father, equipped seven triremes, six biremes, twenty-six ships of burden, and a great number of pinnaces, with which he set sail for Lisbon, where he joined Dom Pedro, who was awaiting him in the roads with eight galleys. When on the point of departing they met with an unexpected obstacle yet greater than any they had hitherto encountered. The Queen had been attacked with the pestilence, and a letter from Dom Duarte summoned his brothers to Sacavem, where she lay. It was evident, on their arrival, that her end was fast approaching, but the sight of her sons revived her.

The old Italian chronicler, Matteo de Pisano, relates minutely the scenes which followed. The Queen had had three swords made, richly set with precious stones, for the purpose of presenting them to her three sons when they were knighted. On the day after their arrival the Queen solemnly addressed them in the King's presence, giving each a portion of the true cross with her blessing. In presenting the sword to

Dom Duarte, she impressed upon him his duties as a king, especially that of ruling justly. To Dom Pedro she gave, as his knightly duty, the charge of protecting the honour of helpless maidens and widows, and to Prince Henry she commended the care of the soldiery. On the thirteenth day of her illness she suddenly inquired, " What wind blows so strongly against the side of the chamber?" and when told by her sons that it was the north wind, she said, " It is the wind most favourable for your departure, which will doubt-less take place on the feast of St. James." This proved a true prophecy, though it seemed at the time scarcely possible, for the feast of St. James would fall only six days after. The Queen died on the 19th of July, 1415, to the sincere grief of the people ; for while sharing for twenty-eight years the throne of the most highly gifted of the kings of Portugal, she had exhibited qualities which would have placed her amongst the most noble of her sex in any country or in any age. To do good was with her a necessity of existence, and her choicest pleasure was in stilling conten-tions and reconciling disputants. The virtue of abstinence she carried to an excess, for, from a deeply-seated sense of devotion, she fasted so severely as to seriously undermine her health. In the details of domestic economy she took as much interest as the humblest among her subjects, and encouraged similar habits in the ladies who were about her person. Such an example was calculated to produce, as in fact we find that it did produce, a notable effect on the bearing, manners, and tone of the nobles of the court. But of all the occupations of the Queen, that in which she took the greatest delight was the training and instruction of her children, in which she communicated to them much of the lofty tone of her own exalted character. She also possessed the faculty of developing their understandings in a manner which was remarkable for the period, and their history shows how eminently qualified she was to be the mother of princes and heroes.

The nature of the Queen's disease, together with the heat of

QUEEN PHILIPPA.

FROM THE RECUMBENT STATUE OVER HER TOMB
AT BATALHA.

the weather, rendered it necessary to hasten the interment of her remains, and on the following day the funeral was celebrated with great pomp in the monastery of Odivellas, but her body was ultimately removed, on the 14th August, 1434, to the chapel erected by King João at Batalha, for the joint sepulture of himself and his beloved queen.

After the funeral Prince Henry joined the King at Restello, whither the nobles had induced him to retire for safety from the pestilence. There was much discussion as to the time for the departure of the expedition to Ceuta, but the King overruled the opinion of some who thought that there ought first to be a period of public mourning, by saying that an immediate departure would best carry out the Queen's expressed wishes. Accordingly, the expedition started with a favouring wind on the 25th of July, that feast of St. James which had been indicated by the dying Queen. Many distinguished adventurers from England, France, and Germany took part in the enterprise. A baron of the last-named country took with him forty knights, and a wealthy Englishman, whose name is difficult to recognise under the transmitted form of Menendus, Mondo, or Mongo, brought four vessels laden with provisions. The armament was an unusually large one for the period. Of the number of vessels and of fighting men, Azurara, the contemporary chronicler, says nothing; but Zurita, in his "Annals of Aragon," informs us that the fleet consisted of 33 galleys, 27 triremes, 32 biremes, and 120 smaller vessels, with 50,000 men, of whom 20,000 would seem to have been soldiers, and the remainder oarsmen and mariners.

The armada anchored in the Bay of Lagos at nightfall of Saturday the 27th. On Sunday morning the King disembarked, with all the chiefs of the expedition, and heard mass in the cathedral, after which Father João de Xira, the Preacher-Royal, read the Bull of the Crusade granted by the Pope in favour of those who should be present at the conquest of Ceuta. On the 30th the King departed for Faro, where he was detained by a calm until the 7th of August, and where Prince Henry had an opportunity of displaying

great presence of mind; for the lantern of his vessel having
caught fire during the night, and there being imminent
danger that the flames would spread to the ship, he, though
suddenly aroused from sleep, with much risk to himself,
seized the burning lantern and threw it into the sea. On
the afternoon of the 10th the armada anchored at Algeziras,
a place belonging to the King of Granada. Ceuta was to
have been attacked on the 12th, and the fleet was already in
full sail when a strong wind arose, which, combined with
the action of the current in the strait, carried the large
vessels nearly to Malaga, so that only the galleys and smaller
craft reached Ceuta, where many of them anchored.

This city, in old times called Septa, had been partly constructed and fortified by the Emperor Justinian. It was the
principal port of Morocco, being the centre of commerce
between Damascus, Alexandria, and other eastern places,
and the nations of Western Europe. Its position was one
of great importance, for in all the invasions of Spain and
Portugal, it had been the point of muster for the Moorish
armies and the rendezvous of the corsairs. It occupied
the western portion of a peninsula nearly three miles in
length, jutting out almost due east from the mainland. It
was divided into two unequal parts by a wall, the smaller
and westernmost part terminating in the citadel, which
covered the neck of land by which the peninsula was joined
to the continent. The portion of the peninsula eastward of
this wall was called Almina, and contained the outer and
larger division of the city, as well as the seven hills from
which Ceuta derived its name, by far the highest of which
was at the easternmost extremity, and was surmounted by a
fortress called El Hacho. On the north side of the peninsula, from the citadel to the foot of this last-mentioned hill,
the city was protected by another lofty wall. Eastward of
this hill was a small bay named Barbazote,* in which

---

* I do not find the harbour of Barbazote laid down on any map, but from the
descriptions I conjecture it to be Desnarigado Bay.

CEUTA

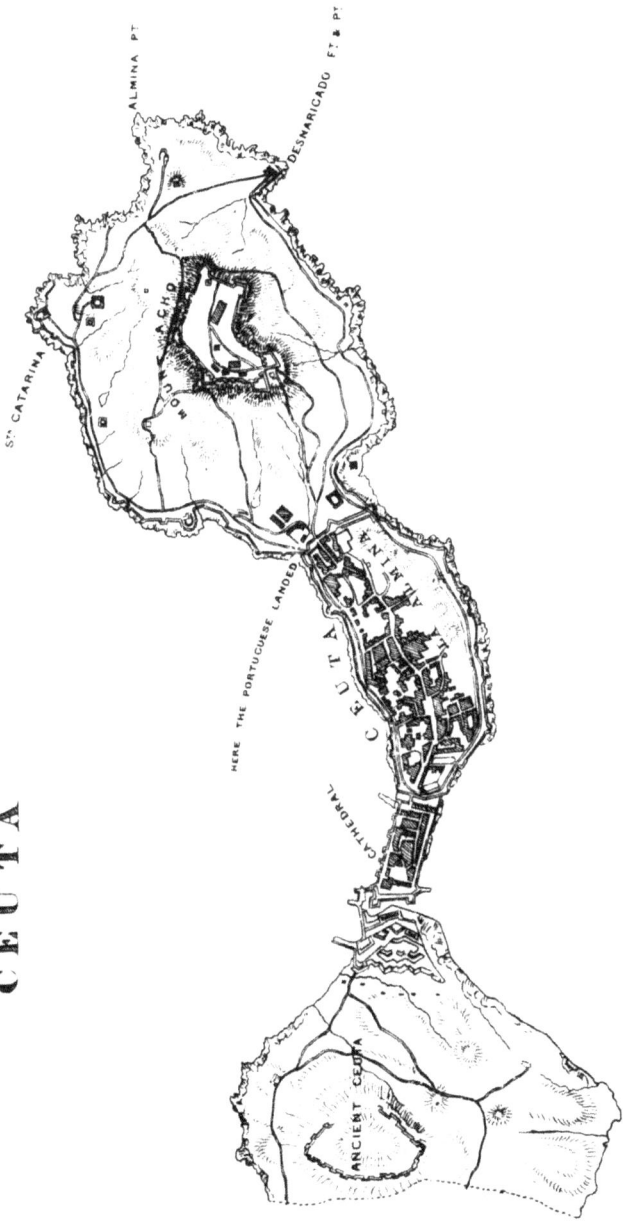

ALMINA PT

DESNARICADO FT. & PT

MOUNT ACHO

Sta CATARINA

HERE THE PORTUGUESE LANDED

CEUTA

TRINITY

CATHEDRAL

ANCIENT CEUTA

tolerably large vessels could lie at anchor sheltered from the
west winds and but little exposed to missiles from the
northern wall.   Here the King determined to await the
arrival of the vessels which had been driven out, intending to
effect a landing immediately on their return.   After much
delay, Prince Henry succeeded in bringing them up, but a
violent tempest frustrated the King's plan by compelling
him to seek another anchorage, for while the large vessels
turned with difficulty the point of Almina, the current
caught the smaller craft which moved more slowly and
carried them towards Malaga.   This apparent disaster, which
in the minds of the superstitious awakened doubts as to the
success of the enterprise, actually contributed in no small
degree to that success.

The first appearance of the strangers had caused great
anxiety to the Moors, who lost no time in preparing for
defence, and obtained help from the sovereign of Fez and
from other neighbouring chiefs to the extent, it has been
said, of one hundred thousand men.   When, however, the
Moors saw the fleet a second time dispersed, they imagined
that it would be impossible again to bring it together, and the
Governor of Ceuta, Zalá ben Zalá, accordingly dismissed the
auxiliaries, and contented himself with the ordinary garrison.

The Portuguese themselves were discouraged, and, but for
the determination of the King, the Princes, and the Constable,
would have abandoned the expedition.   Prince Henry having
again collected the fleet, preparations were resumed for the
attack, which was at length ordered in the following manner.
Prince Henry was to anchor off Almina with all the vessels
he had brought from Oporto, and to be ready at daybreak
on a signal from the King to land his men with all expedition.
The King himself with the main body of the fleet was to
anchor opposite the castle.   The Moors would naturally
flock to the point where the greater part of the fleet lay, and
Prince Henry would thus be able to land with comparatively
little hindrance, while, if the Moors should turn to oppose
him, he would be supported by the King's division.

These movements of the fleet greatly alarmed the Moors. Zalá ben Zalá was so convinced that the issue of the struggle would be disastrous, that, but for the counsel of a few of his confidential advisers, he would have fled. In order to produce an impression that Ceuta was a very populous city, he now gave instructions that the wall on the side where the fleet lay should be crowded with men, and that lighted candles should be placed in all the windows of the houses. The effect was brilliant, but, as might have been anticipated, in no way alarmed the Portuguese. At daybreak the King, in spite of a severe injury which he sustained in descending from his galley into a boat, visited the fleet, and gave his instructions to each commander, encouraging all with the certain hope of victory. In accordance with a request made by Prince Henry at Lisbon, he forbade any one to set foot on shore until the Prince himself should have landed. Meanwhile Zalá ben Zalá was overcome with terror, from which the younger Moors sought in vain to arouse him. They therefore took matters into their own hands, and, while Prince Henry and his men awaited the appointed signal, issued from the city shouting their war-cry and defying the enemy. Seeing this, João Fogaza, comptroller of the household of the Count de Barcellos, could brook no further delay, and disregarding the King's injunction, put off with several armed men for the shore. The first who touched the soil was Ruy Gonsalvez, a man renowned for his daring, who attacked the Moors so desperately that they recoiled enough to allow of others landing. This hastened the movements of the Portuguese, and, after some opposition from the Moors, Prince Henry and Dom Duarte effected a landing with about three hundred men. Two only accompanied Prince Henry in his boat, namely, Estevão Soarez de Mello, and Mem Rodriguez de Refoyos. The Moors poured out in great numbers from the town, and a long and fierce contest ensued, in which the latter were driven back to the Almina Gate which opened on the landing-place, and through which they entered and the Portuguese with them. The first who passed through was Vasco Eannes de Cortereal, closely followed by Dom Duarte, and thus they continued charging the Moors till they reached the gates of the city. Here Prince Henry offered to resign the command to Dom Duarte, but the latter would not accept it. Prince Henry therefore, having put his men in military order, proposed to await the arrival of their father, as he had commanded, but Dom Duarte overruled this, suggesting the advantage of their continuing to avail themselves of the evident panic of the Moors. After retreating before the first onset, however, the Moors made a stand, being protected by the walls and encouraged by their champion, a gigantic negro who fought naked and used no weapons but stones, which he hurled with terrible force, and with one of which, while the combat was at its height, he struck Vasco Martinez de Albergaria, a nobleman of Prince Henry's household, full on the helmet. The Portuguese staggered under the violence of the blow and stood for a moment half stunned, but recovering himself, he broke his way through the ranks of the enemy and thrust his spear into the side of the giant. When the Moors saw their champion fall, they fled in confusion into the city, the Portuguese entering with them. Prince Henry's most anxious care now was to secure possession of the city gates, not only for the sake of facilitating the entry of his countrymen, but also in order to prevent those who had already entered being hopelessly shut in by the enemy. The two princes, with the Count de Barcellos, their illegitimate brother, and about five hundred men occupied a mound within the city, and there fixed Prince Henry's standard, the spot being favourable for defence, should the Moors renew the engagement. In consequence of the smallness of their force they were not free from anxiety, lest, before fresh troops arrived, the soldiers might be tempted to begin plundering, which would give the Moors an opportunity to collect in sufficient strength to shut the gates, and so render their position in the highest degree perilous. But reinforcements came in with great rapidity from that part of the fleet which

Prince Henry had commanded, and some of the Moors in
their alarm announced to Zalá ben Zalá in the citadel, that
the city was taken. Some took to flight with their wives
and children. Zalá ben Zalá, overwhelmed with dismay,
came from the citadel in the hope of checking, if possible,
the advance of the enemy through the narrow streets until
the citizens could pass the wall to the western, or inland,
side of the city, where, if anywhere, they might receive help
from their neighbours.

Among the new comers was Vasco Fernandez de Ataide,
who, despising the easy entrance through the open gate by
which the Prince had entered, called together his men and
attacked another gate which was carefully kept by the Moors;
but his attempt was fruitless, for while he was striving to
force the gate with axes, the Moors attacked him with stones
and darts, and he was compelled to return, himself mortally
wounded and eight of his men slain. When Prince Henry
perceived that the greater number of his men had arrived, he
thought it better to waste no more time on the spot where
he had waited with Dom Duarte, and gave orders to the
captains to occupy various parts of the city, so that no
opportunity might be afforded for the panic of the Moors
to subside, or for them to reorganize their forces. Dom
Duarte took possession of Cesto, the highest of the hills
overlooking the city, and Prince Henry of the main street.

Meanwhile the King, who had now inspected the fleet, gave
orders for a general landing, and receiving news of the rapid
victory of his sons, offered up thanks to God for their success.
He then advanced with his retinue towards the town, and,
supposing from the quantity of plunder which was being
carried on board the ships that nothing more remained to
be done, seated himself near the gate.

In the interim the Moors seeing the Portuguese intent on
plunder and approaching in utter disorder very near to the
citadel, had attacked them with such fury that they fled in
confusion. The Moors thought this the moment for avenging
their injuries, and endeavoured to drive the enemy completely

out of the city and close the gates. Prince Henry allowed the flying Portuguese to pass him, knowing that if he checked the foremost, those in the rear would be exposed to great danger. He himself was left with but a handful of men, but seeing that the position of affairs was critical, he opposed the Moors with such vigour that he put them to flight with great slaughter. Pursuing them however too eagerly, he found himself alone with the enemy, and would certainly have been cut off had not the narrowness of the road in a great degree protected him. For a short time he had to sustain the conflict quite alone till, his soldiers coming to his assistance, the Moors were again put to the rout. While his men pursued them, Prince Henry rested in a house which the Portuguese had converted into a store for the goods which they had brought on shore, but the fugitives having received reinforcements, the Portuguese were again driven back as far as the house where the Prince was. In vain he endeavoured to rally them, they were worn out with the heat and thirst, and out of the many whom he addressed not more than seventeen remained with him. With these few he boldly met the on-coming enemy, and forced them to retreat through the gate which led into the inner part of the city, and which could be secured on both sides. After a long and violent struggle Prince Henry succeeded in clearing this gate, thereby securing his return to his troops.

But evening was now coming on and the Portuguese began to seek their respective leaders, from whom they had been separated in the turmoil of'the day; and many were the anxious enquiries for Prince Henry, whose gallantry had won all hearts; and it was rumoured that at the head of a handful of men he had made his way to the above-mentioned gate, and fighting to the last had there met his death. The King hearing of this, said with a calm and unmoved countenance, " Such is the end which soldiers must expect."

In another part of the town, Dom Duarte was deliberating with Dom Pedro and some other nobles as to the means of storming the citadel, and sent a message to Prince Henry desiring his presence. This Prince Henry at first refused, for he waited to see if the Moors would return to the conflict, but when a second messenger urged on him that it was now evening, and that if the citadel were stormed no further work would remain to be done, he joined the council.

Meantime the Moors, who feared that they would be unable to defend the fortress, after consultation with Zalá ben Zalá, determined on flight. Each man loaded himself with as much as he could carry, and having constructed a testudo at the western gate, which opened landward, they silently retired with their wives and children to the neighbouring towns and villages.

It was now sunset, and the Portuguese, having resolved to attack the citadel at daybreak, sent out a reconnoitering party, who, finding no sentinels on guard, suspected that the Moors had deserted. On hearing this, the King, who now had entered the city, sent a knight, named João Vaz de Almada, to attempt an entrance into the inner part of the city, and if he found the citadel abandoned, to place the standard of St. Vincent, the patron saint of Lisbon, upon the highest tower of the fortress.

When Almada reached the gate in the wall which divided the city into two parts, he found it shut, and ordered his men to hew it down. While they were so doing, two Moors, who had waited to see the end, told them in Spanish to spare themselves the trouble, for that they would open the gate. Almada then entered the citadel and placed the standard on the highest tower.

Meanwhile the King, hearing that Prince Henry was alive and present at the council, sent to summon him to his presence. The King's face grew bright with joy as his son approached, and he welcomed him with the proposal that as he had borne himself so gloriously that day in the midst of so many well-tried veterans he should receive the honour of Knighthood in precedence of his brothers. Prince Henry, however, besought the King, that as his brothers Dom Duarte and Dom Pedro took precedence of him in age, they might also do so in honour. The King commended the wisdom of his son's reply, and gave orders that at daybreak all the bishops and priests who were present with the army should assemble in the great Mosque, and consecrate it as the site of the Cathedral of the city. On the following day this was done, and the three Princes presented themselves before the King, in full armour, each bearing unsheathed the sword which the Queen had given him, and with all due solemnity they were invested with knighthood, each in the order of his birth.

The night had been passed in the greatest watchfulness. When in the morning the Portuguese entered the city, it lay before them in unbroken stillness. They encountered nothing but the dead bodies of the slain, and some few old men, women, and children, who still lingered near the homes which they loved, even at the risk of becoming slaves to the victorious Christians. The spoil was most abundant in gold and silver, and jewels of great price, with stuffs and drugs in great quantity, but the destruction and waste were immense. The morning was stormy, with rain and hail, and such was the recklessness of the troops that, mingled with the streams of water which flowed down the streets, were oil, honey, spices, preserves, and butter, with fragments of the great jars which had hitherto contained them. This waste was afterwards the subject of much vain regret, when it was found that the provisions in the city were enough to have maintained for a considerable time the garrison necessary to hold it in subjection. The spoil which fell to the share of the nobles was very rich. Dom Affonzo, Count de Barcellos, with princely taste took for his plunder more than six hundred columns of alabaster and marble from the gates and windows of the palace of Zalá ben Zalá and the other chief buildings of the city. From one square was taken an entire vaulted roof of elaborate gilt work, which together with the columns was afterwards used in the construction of the count's palace at Barcellos.

The Moors were now seen ascending the mountains carrying their wives and children, whom after awhile they left in charge of the old men that were unable to bear arms, whilst they themselves returned to the walls of the city, challenging its present occupants to fight, rather with a passionate desire of vengeance than with any hope of recovering what they had lost. Dom Duarte rode forth with a large company and speedily repulsed them, and as, when they again returned, they found the gates shut, they withdrew, uttering such wild sad wailings of anguish and despair as moved even the hearts of their enemies to pity.

On the Sunday after this important victory, the principal mosque having been purified, the King with his sons and the Grandees proceeded thither to the sound of martial music to hear the first mass. They were met at the entrance by a large number of priests in rich vestments, and the sound of the instruments was answered by two bells which were found in the highest tower of the mosque. How came they there? was the natural question of those who knew that the Mahometans were not in the habit of using bells. The answer was not without its interest. Some years before, the Moors had attacked and pillaged the city of Lagos, and carried off these bells, which had been carefully but vainly concealed, and which now again spread far and wide the summons to attend a Christian service. Many Moors of both sexes were witnesses from a distance of this sudden and for them heart-rending transformation of a structure which but two days before had been the scene of that exclusive devotion which regarded the approach of a Christian foot as a desecration meriting death. When the hated sound of those bells reached their ears they stood aghast, as if under the influence of a hideous and unnatural dream.

While the most solemn services of the church were being celebrated in the newly consecrated cathedral, messengers from the King were hastening in different directions with the news of the victory, the fame of which rapidly spread throughout Europe, for it was felt to be one that promised important consequences.

The conquerors were naturally desirous not to prolong

their stay upon the coast of Barbary, and the King, though by no means inclined to resign into the hands of the Moors so important a conquest, was anxious to resume the government of his country. The majority of the Portuguese doubted the possibility of holding the place, and one Grandee, Martin Affonso de Mello, whom the King had selected as commander, declined the honour, though it was a greater than had been offered by the King to any subject in any of his enterprizes. While it was uncertain who was to accept the charge of the place, Dom Pedro de Meneses, Count de Viana, of the noble house of Villareal, happening to have in his hand a stick of Zamboa wood,* uttered the exclamation, " By my faith, with this stick alone, I feel myself man enough to defend these walls against every Morisco of them all." What then appeared an empty boast became afterwards a valuable reality. The King took him at his word, and this stout-hearted knight remained the first commander of Ceuta, and had the honour of being told by the King that he should require of him no other pledge than that which was afforded by his high character and noble birth. Faria y Sousa, who wrote two centuries later, records that this staff was still, in his time, preserved at Ceuta, and placed in the hands of every governor on the occasion of his taking the command of the place. The valiant Dom Pedro held the governorship for nearly two and twenty years. He was engaged in frequent contests with the Moors, but proved himself well able to maintain with honour that dignified but responsible position. From that day to this, the Moors have never recovered possession of the city.

The King left Ceuta with the fleet on the second of September, 1415, and a few days afterwards, anchored amidst the exultant welcomes of his people in the port of Tavira, on the coast of Algarve. At Tavira the King summoned his sons to him, and declared his wish to reward them for the great service which they had rendered him. To Dom Duarte, who

* A variety of the orange tree. Faria uses the words—Azebuche, Azebo, and Azambugeyro.

was to succeed him on the throne, he had nothing greater
to offer, but upon Dom Pedro he conferred the titles of
Duke of Coimbra and Lord of Montemoro Velho, Aveiro,
and other territories which thence, as forming the appanage
of his rank, took the name of the Infantado, a designation
which still remains. The title of Duke had not previously
existed in Portugal. Prince Henry was made Duke of
Viseu and Lord of Covilham.

In Tavira the King discharged with many thanks and
ample presents those who had volunteered their assistance,
and dismissed with liberal payment the foreign vessels
which had been employed in his service. Among these
were twenty-seven English ships, which, touching at the port
of Lisbon on their way to the Holy Land, had at the King's
request joined him in his expedition against the Moors.

This enterprize, which had in the first instance been
undertaken mainly with the view of affording a worthy
opportunity for the young Princes to earn knighthood,
proved in many ways of great importance. It was a severe
reprisal upon the Moors, who had for so many centuries
inflicted their hated dominion on the Peninsula, and it
transmuted Ceuta, from being the chief emporium and key
of the Mahometan states, into the very bulwark of
Christendom against them. But further, and this is especially note-worthy, as a successful naval enterprize it was the
parent of those grand achievements which made the close of
the fifteenth century memorable in the history of the world.

For three years the Count de Viana was able to hold the
Moors in check with the forces which had been left in his
command, but in 1418 it appeared necessary for him to
seek aid from the mother country. Ceuta was surrounded
inland by a large army of Moors, and was attacked from the
sea by the forces of the King of Granada, who had sent a fleet
of seventy-four sail, and numerous troops, under the command
of his nephew Muley Saïd, to attack the city from the sea.
Fortunately the munitions in the arsenals of Lisbon were
abundant, so that the King was able to despatch a strong

force under the command of Prince Henry, who took with him his brother Dom João. At the same time, Dom Duarte and Dom Pedro proceeded to Algarve, in order that they might be ready to reinforce Prince Henry in case of necessity. As the latter was entering the mouth of the strait, he was met by a pinnace, bringing him written information from the Count de Viana that Muley Saïd had already taken possession of the eastern part of the Almina, in combination with the army already *in situ,* while the galleys blockaded the port. The glory of destroying the navy of Granada did not however fall to the lot of Prince Henry, for, before he could reach Ceuta, the Count de Viana had sallied forth at the head of his small but stout-hearted garrison against the position which Muley Saïd had taken up on Mount Hacho. The brave Moor met the Count with a desperate resistance, which though it was sufficient to secure his honour, could not win for him success. At the commencement of the engagement, his galleys, which had sailed out of the bay, came in sight of the Portuguese fleet, of the approach of which the Moors had given repeated signals from both sides of the strait. The whole of the fleet of Granada took the alarm and fled, only one galley remaining to aid the escape ʹof Muley Saïd with a small handful of men. By the time the Princes landed the action was at an end.

The Princes remained two months in Ceuta hoping that the Moors would make an effort to regain their lost city, but they waited in vain. During this time Prince Henry, who chafed at the thought of returning to the presence of his father without having achieved a single act of distinction, conceived the bold idea of taking Gibraltar by storm. Although he was opposed by the almost unanimous opinion of the council, yet he determined to make the attempt, and set sail accordingly. Fortune however did not favour the undertaking. A storm arose which drove the fleet towards Cape de Gat, where it remained fifteen days, and on their return to Ceuta the Princes received a letter from their father with positive orders for their return to Lisbon.

CHAPTER IV.

"Talent de bien faire" was Prince Henry's adopted motto, and human wit perhaps could scarcely suggest a better. In his time the word "talent" conveyed not, as now, the idea of "power" or "faculty," but of "desire," and the appropriateness of the motto to Prince Henry himself has in it something remarkable. Its principal characteristic is aspiration, and when the exertions of the Prince's life have been depicted they will be found to have been great indeed in effort, but great only in ultimate, not in immediate, result, the most indisputable evidence of a life devoted to the "Talent de bien faire."

Azurara states (page 40) that the renown of the Prince became so high in Europe that he was invited severally by the Pope, the Emperor of Germany, and the Kings of Castile and of England to undertake the command of their respective armies. These offers will most probably have been made after the taking of Ceuta, where the Prince had so greatly distinguished himself as a soldier. In all probability it was in 1420 or 1421 that he received this invitation from the Pope (Martin V.), after the embassy sent to him by the Greek Emperor Manuel Palœologus asking for his assistance against the Turks. The Emperor of Germany spoken of was Sigismund, whose close relations with the court of Lisbon and with the ambassadors of Portugal in the Council of Constance would enable him to form a correct opinion of the eminent qualities of the Prince. The Kings of Castile and of England of whom Azurara speaks are John II. and Henry V. (See Santarem's note to Azurara, p. 40.) When Prince Henry, after the capture of Ceuta, set his mind upon the conquest of Guinea he sent every year two or three vessels to examine the coasts beyond Cape Non, the limit of Spanish exploration, yet none of his ships for many years had the hardihood to round Cape Bojador. It is recorded by Barros, the great historian of the Portuguese, when describing the effect of a storm which assailed one of the earliest of these expeditions, that " the Portuguese mariners of that time were not accustomed thus to venture on the open sea, all their nautical knowledge being limited to coasting in sight of land." Hercules was yet in his cradle. The little nation had but just succeeded in strangling the snakes of Moorish and of Spanish oppression. So far it had done bravely. It had thrown off the yoke and was able to draw breath. What wonder if having achieved such victories it felt its pulse beat strong for greater and yet nobler efforts. True, the ocean was a new and formidable antagonist. Other nations mightier than they had tempted the same danger but had withdrawn disheartened from the contest, and their unavailing efforts, so far from diminishing, enhance the glory due to that persistent bravery which yielded neither to difficulty nor danger. But the inspiration and encouragement to this perseverance emanated from Prince Henry himself.

It was not however to the exploration of the West Coast of Africa only that the thoughts of the Prince were directed. The hope of reaching India by the south point of Africa was a yet higher object of ambition. The political decay of the Roman Empire had not been accompanied by any decrease in that love of luxury and profusion which necessitated commerce and navigation. The civilization and trade of the world had simply fallen into the hands of new masters. The vast dominion acquired by the followers of Mahomet gave them the control of a gigantic commerce.

Not indeed that maritime communication possessed for them any charms : the contrary was the case, and the timidity of their navigation was peculiarly remarkable, but their overland caravans were the means of carrying on a traffic which extended from the Mediterranean to India, and from the heart of Africa to Astrakhan and the countries of the north. One of the most important of these caravan routes was that which traversed the great African desert, and introduced into the Mediterranean the slaves and gold dust, the ivory, and malaguette pepper that were procured from the negroes.

In the middle ages a variety of causes conspired to direct the attention of European nations to the East. The Crusades, mischievous as they were in their primary effects on the nations from which they emanated, not only made them acquainted with distant countries but also with that oriental luxury which supplied a stimulus to the cultivation of mercantile relations with those countries. Another event which had great influence in inciting the western states of Europe to maritime discovery was the war between the Moors and the inhabitants of the Peninsula. The vast mercantile operations of the Arabs had filled Spain with the rich productions of the East, and the luxurious habits of the Moorish courts of Seville and Granada were imitated by the Catholic princes of Aragon and Castile. But as hostilities between the conquerors and the conquered daily became more obstinate and implacable, the lack of these objects of luxury began to be felt by the latter, to whom, at least amongst the wealthy, they had become necessities. So that it may be fairly inferred that the expulsion of the Moors from the Peninsula was one of the great stimulants to the search for a passage to India by the sea. In this expulsion the Portuguese took the lead, and were consequently the first to feel the effect of the incentive. The conquest of Ceuta was the first step towards the desired object, and Prince Henry with his love of study, his chivalrous courage, and zealous nature, was exactly the man to pursue that object with the perseverance of a fixed determination.

The geographical position of Portugal was eminently suggestive and encouraging. The large revenues of the Order of Christ, of which the Prince was the Grand Master, provided him with resources for which he could imagine no more worthy employment than the conquest and conversion of the heathen, and the general extension of the knowledge of the human race, with its concomitant commercial advantages. During his stay in Africa he gathered important information from the Moors respecting the populous nations of the interior and of the coast of Guinea. We have evidence of the nature of the enquiries instituted by Prince Henry in the cotemporary accounts of Azurara and Diogo Gomez. From the latter we learn that the Prince gained information of the passage of large caravans from Tunis to Timbuctoo and to Cantor, on the Gambia, which instigated him to seek those countries by the way of the sea. From Azurara we learn, as will be seen hereafter, that he gathered from Azanegue prisoners information of the position of certain palms growing at the northern mouth of the Senegal, or so-called Nile, by which he was enabled to give instructions to his navigators for the finding of that river.

But while Prince Henry was thus anxious to inform himself respecting the geography of Africa, he was no less anxious as Grand Master of the Order of Christ to further the cause of Christianity in that country. After the destruction of Carthage the chair of the Primate of Africa remained vacant for centuries, although individual bishoprics contrived to secure for themselves a continuance of existence even under the dominion of Mohammedanism. When St. Francis first established his order in 1208, with a view to the revival of the Christian faith, it was with him a subject of ardent desire to send missionaries to Marocco, and it was not long before a bishopric was there established, the bishops of which belonged to his order. Agnellus was consecrated the first Bishop of Marocco and Fez in 1233.

From that time Christianity was propagated with inde-fatigable energy throughout the north of Africa. Churches were erected and the right of free celebration of Christian worship was a frequent item in commercial compacts, but until the taking of Ceuta, Marocco remained the only bishopric of the Catholic Church in Africa.

As regards the West Coast of Africa, very little indeed had up to this time become known to explorers. Ibn Khaldoun in the preceding century had placed the limit at Cape Non, but Ibn Saïd had related the chance arrival of some Arabs at the Glittering Cape (Cape Branco) two centuries before, and it is certain that Cape Bojador was known as early as 1375, for it is laid down under the form of Bugeder in the Catalan map of that date. But here was in very truth the limit of known coast. We have not sufficient evidence to show the exact extent of the information which the Prince was enabled to gather respecting the interior of the country, but we are not entirely deprived of the means of forming what may probably be an approximately correct judgment on that point. The seaports on the North Coast of Africa had long been the medium of conveying to Europe the valuable commodities brought from Nigritia, but as these were brought over by land, and not by sea, it is manifest that much had to be learned by enquiry respecting the nations and the countries from which they were supplied. To become acquainted not only with the Moors and their immediate neighbours to the south, but also with the lands both on the Eastern and the Western Coasts beyond the Great Desert, was the object of the Prince's desire. And it must be acknowledged that the chances of gaining approximately accurate local knowledge from the Arabs was greater than could be looked for from Europeans, for while the former took diligent notice of individual narratives of travel, and industriously availed themselves of the geographical information which they acquired, the latter made a secret of many of their commercial connections, and even treated with mistrust the details of explorations which were openly made known, whether by Arabs or by Christians. The one great source and even limit of the knowledge of African geography was commerce, and the kingdoms in the interior with which this commerce took place were Melli, Gana, Tekrúr, Takedda, Burnu, and Kanem. The most important of these was Melli, comprising the cities of Kabra, Timbuctoo, and Kuku on the Joliba. Of Timbuctoo some knowledge was already possessed in the Spanish peninsula, inasmuch as there appear to have been frequent communications between it and the kingdom of Granada. Leo Africanus, himself a native of Granada, who was born at the close of the century which witnessed Prince Henry's explorations, speaks of the Stone Mosque and Royal Palace of Timbuctoo, the only two remarkable buildings in the city, as having been the work of an experienced architect of Granada; and Ibn Batuta, writing in the century in which Prince Henry was born, mentions as one of the curiosities of Timbuctoo the tomb of Abu-Ishac-es-Sahili, a famous poet of Granada, who died at Timbuctoo in 1346. The old accounts leave us in much doubt in respect of the geography of the several kingdoms we have referred to, though many points have been settled in more recent times. The kingdom of Melli extended eastward as far as the city of Muli in Lemlem, and westward to the oasis of Waleta. An important map which at the close of last century came into the possession of Cardinal Borgia, and which is supposed to have been of Arab workmanship of about the year 1410, consequently just before Prince Henry was making his researches, contains the indications of three stations of a caravan track in the interior, viz., Teget (Teigent), Tagost (Audagost), and Tagaza, as well as the towns of Gana, Tocoror (Tekrúr), and Melli. It will be seen in a subsequent chapter that indications of the formation of the Western Coast, but with no local names proving actual intercourse with places thereon, had been derived from merchants reaching the ports of the Mediterranean from the interior as early as the first half of the fourteenth century. The most southern point to which a name was

given along the Western Coast was Ulil, where was a
natural deposit of salt which was carried thence inland to
Gana and other cities of the Blacks. Mr. Cooley, who has
devoted so much learned labour to enquiries respecting the
geography of these countries, has no hesitation in assigning
this position to the Bay of Arguin, where, he says, " the
natural deposits of salt, the little island or peninsula, and
the abundance of large tortoises," which form the description of Ulil, " are all found together." The learned Dr.
Friedrich Kunstmann, however, who has also carefully examined the subject,* carries the position of Ulil down as far
south as the island of Bissao, at the mouth of the Rio
Grande, in which he finds the requisite characteristics, with
the additional fact that in the neighbouring island of
Bulama is found abundance of amber, which is a concomitant
item in the description given by Ibn Saïd. There is a piece
of collateral evidence which seems very strongly to corroborate this conclusion of Dr. Kunstmann's, for whereas in the
eleventh century Ulil is described as the common boundary of
commerce and of *creed*, we learn from Valentin Fernandez,
within half a century after Prince Henry's time, that " the
negroes of the Rio Grande towards Cape Verde are for the
most part Mahometans, with some idolaters amongst them,
but on the other side of the river all are idolaters."

With respect to the immediate inducements which actuated
the Prince in exploring the land of Guinea we have five reasons
supplied to us by Azurara. The first was the desire to know
the country beyond Cape Bojador, of which till then neither
from writing nor tradition had any certain knowledge been
gained. The second was that, if any Christian nations and safe
harbours should be found in those lands, mercantile relations
advantageous both to Portugal and to the natives might be
established, as no other European people were as yet known
to have commercial intercourse with them. The third reason
was that he had been led to suppose that the Moors were in
greater strength in that part of Africa than had been generally

* In his work entitled " Afrika vor den Entdeckungen der Portugiesen," p. 11.

believed, and that there were no Christians there. The Prince
was therefore naturally anxious to learn the extent of the
power of his enemies. The fourth reason was that in all his
contests with the Moors he never found a Christian King or
potentate come.forward from that country to help him. He
was therefore anxious to learn whether there were in those
parts any Christian princes, who for the love of Christ would
help him against the enemies of the Faith. The fifth reason
was the great desire which he had for the extension of that
Faith and the salvation of the souls of such vast numbers
then lying in a state of perdition. To these five Azurara
adds a sixth from which he believed that the other five proceeded. This reason was an astrological one, " for as," he says,
" his ascendant was Aries, which is the house of Mars and is
the exaltation of the Sun, and his lord is in the eleventh house
accompanied by the Sun, and inasmuch as the said Mars is
in Aquarius, which is the house of Saturn and in the house
of hope, it signified that he should be engaged in mighty
conquests, and especially in the search for things hidden from
other men in conformity with the craftiness of Saturn in
whose house he is. And his being accompanied by the Sun
and the Sun being in the house of Jupiter, showed that all
his acts and conquests should be loyally done and to the
satisfaction of the King his sovereign."

The more entirely to enable himself to carry out his objects
without embarrassment, he took up his abode, with the King's
permission, on the Promontory of Sagres in Algarve, of which
kingdom he was made Governor in perpetuity after his return
from the succour of Ceuta in 1419. From a passage in the
cotemporaneous MS. of Azurara it has been inferred that
he did not betake himself to that secluded and barren promontory until after his return from Tangier in the year
1437,* but it is to be remarked that in that passage " his

* Despois da viinda de Tanjer, o Iffante comunalmente sempre estava no regno
do Algarve, por rezom de sua villa, que *entom* mandava fazer. " After the return
from Tangier the Infant commonly remained in the kingdom of Algarve, on
account of his town which he then was having built."—*Azurara*, p. 105.

town" is spoken of, whereas from another cotemporaneous manuscript, now lost, but the genuineness of which I hope hereafter to establish, we are informed that he originally named it Terça Nabal; quasi, Tercena Nabal or Naval Arsenal ; and it was only subsequently, when, as at the time of Azurara's writing (see Azurara, page 34), it came to have strong walls, and houses were being continually added to it, that it received the name of " Villa do Infante," or the " Town of the Infant." From these facts, combined with the importance of the proximity of the port of Lagos for the dispatch of the Prince's vessels, it would seem reasonable to conclude that the belief of the majority of Portuguese historians that the Prince established himself at Sagres after his return from Ceuta in December, 1418, is correct. In fact, if the genuineness of the missing manuscript just alluded to be, as it is hoped, successfully established, the fact is certain, inasmuch as it is there distinctly stated by a living witness. This remarkable position had not been without its occupants in yet older times. M. Ferdinand Denis informs me that at the period of the terrible earthquake of 1755, which covered both Sagres and Lagos with ruins, there were buildings on the promontory of Sagres as old as the eleventh century. When its occupation by Prince Henry brought it into notice the Genoese offered a large sum for the site for the establishment of a colony, which the Portuguese government prudently refused.

By the great kindness of His Excellency the Marquis de Sá da Bandeira, late Minister of Marine, I have been favoured with a drawn copy of an official survey of this interesting promontory, of which the accompanying plate is a reduction.

In this secluded spot, with the vast Atlantic stretching measureless and mysterious before him, Prince Henry devoted himself to the study of astronomy and mathematics, and to the dispatch of vessels on adventurous exploration.

I have heard it whispered that the greatest Portuguese historian of the day has expressed a doubt whether it can

PLAN OF THE FORT AND PROMONTORY OF SAGRES.

Taken by Captain Lourenço Germack Possollo on the occasion of the erection of a
Monument to Prince Henry in July 1840, under the auspices of his Excellency the
Viscount (now Marquis) de Sá da Bandeira, then Minister of Marine, by whom the
Copy from which the present reduction is made was kindly communicated to the Author.

Copied in the "Archivo Militar" by J. C. Bou de Souza in 1863.

REFERENCES.

a   *Tower now serving as a hay-loft, below
which is the entrance to the Forts and in
which above the doorway inside is placed
the Monument to the Prince.*

b   *Old walls brought into the Construction of
the new residences.*

c   *Remains of the original Mother Church.*

d   *Remains of the Barracks destroyed in 1793.*

e   *Stables.*

f   *Powder Magazine built on the ruins of a
circular edifice probably the Observatory built
by Prince Henry.*

g g   *Traces of walls entirely razed to the ground.*

h   *An excessively deep Cavern communicating
with the Sea.*

i i   *Batteries at the edge of the Promontory.*

k   *A Pedestal on which formerly stood a Cross
It is here that the Promontory commences.*

l   *Bay of Belixe.*

m   *Bay of Sagres.*

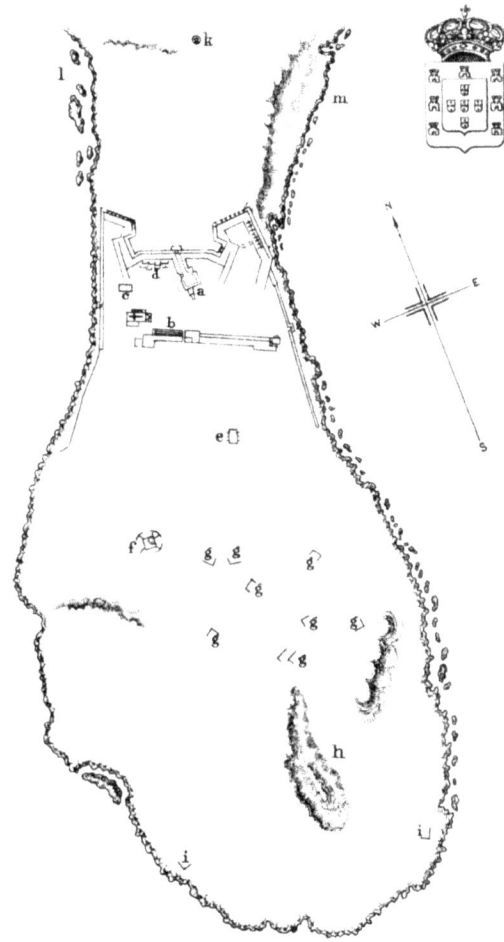

Scale of Half an English Mile.

1 Furlong   2   3   4

Edwd Weller, litho Red Lion Square

be proved that Prince Henry established at Sagres a school for the cultivation of cartography and the science of navigation. There can however be no doubt that Barros distinctly asserts that "in his anxiety to secure a prosperous result to his efforts at discovery on the West Coast of Africa the Prince devoted great industry and thought to that object, and at a large expense procured the services of one Mestre Jacome from Majorca, a man very skilful in the art of navigation and in the making of maps and instruments, whom he sent for to instruct the Portuguese officers in that science." This distinct assertion, combined with the fact that the Prince erected an observatory at Sagres, the first set up in Portugal, would seem to leave but little doubt upon the subject, even if the well-known love of mathematical study which the Prince shared with his brothers Dom Duarte and Dom Pedro were not enough to lead to the conviction that such a pursuit would naturally occupy the active attention of one who had located himself on that desert promontory for the very purpose of prosecuting the exploration of unknown coasts.

At the same time I am anxious in this chapter to vindicate the honour of Prince Henry from the aspersion of falsely attributed praise. The picture of a worthy life can only be marred in its beauty and fall short in its teaching if it be not exhibited in the light of truth. In a subsequent chapter I shall endeavour to show that the detraction from the glory of Prince Henry on the ground of prior discoveries along the West Coast of Africa are utterly untenable. In another chapter I shall show that the Prince's navigators had really been preceded in the mere fact of discovery, though not of colonization, of the Atlantic islands. I now proceed to show that the honour of originality in the invention or introduction of more than one important appliance in the art of navigation has been incorrectly assigned to him. It has been honourably and justly said by M. d'Avezac that "the historical glory of Portugal is based on enough of real merits to render it needless for her to dispute the legitimate share of other nations. Once again let me say: the

Portuguese were certainly not the first to undertake the enterprize of finding the great maritime route to India, but they were the first to persevere in it, and they were the first to attain that object. That is their share of honour. It is a fair one enough to render it needless for them to claim what belongs to others." This most true and graceful verdict is worthy of the distinguished *savant* who pronounced it, and no one can have traced his learned and ingenious arguments in confutation of what he believed to be false praise of Prince Henry without feeling convinced of the good faith in which every word was written. It has been a pleasure when our convictions have, as in the case of which we are about to speak, been in accordance with his own.

Pimentel first, and Montucla after him, declared that the invention of hydrographic plane charts was due to Prince Henry. A greater mistake could not have been committed. The very first charts that were constructed upon the base of a geonomic graduation were in a certain sense of this kind. As far as we know, Eratosthenes (two centuries B.C.) was the first who constructed such. Hipparchus, perhaps a century later, reconstructed the maps of Eratosthenes with meridians convergent at the poles, and Marinus of Tyre (second century A.D.) reverted to the plane chart. But not to dwell on maps of high antiquity, there can be no question that there existed sailing charts on the cylindrical projection, in which for convenience in navigation the meridians were made parallel to each other, before Prince Henry was born. Garçao-Stockler in his " Ensaio historico sobre a origem e progressos das Mathematicas en Portugal " recognizes the existence of a map of the kind bearing the date of 1413, which is mentioned by Don Joaquin Lorenzo Villanueva in his " Viage literario á las Iglesias de España," tomo 4, carta 28, p. 24, as existing in the year 1802 in the Carthusian Monastery of Val de Christo, near Segorbe, in Valencia. But while recognizing the existence of the map, he disputes the correctness of the date because the map contains the Atlantic islands

which, as he believed, were not discovered till afterwards by
Prince Henry's navigators. The reader will be prepared to
set aside this argument when he has read what I shall have to
say respecting the discovery of these islands in the chapter
of "Glimpses of Light." Had the distinguished Portuguese mathematician been aware of the facts there stated he
would doubtless have avoided resorting to so eccentric and
improbable an alternative. In short, to set the matter
at rest, the Catalan map of 1375, on which those discoveries
are repeated, is of the class of nautical maps of which we
are now speaking, and although the map discovered by the
P. Villanueva bears the inscription "Mecia de Vila Destes
me fecit in ano MCCCCXIII.," Señor de Navarrete asserts in
his "Historia de la Nautica" (Madrid, 1846, 8o.) that it is but
a repetition of the third sheet of the aforesaid Catalan map.
At the same time it should be remembered that to have
introduced on fresh tracks in the Atlantic those nautical
appliances which had already been employed in the Mediterranean, and by careful study and perseverance in recording
new observations to have led the way to subsequent improvement of those appliances, is a merit which needs no
superfluity of praise to commend it to the admiration of the
thoughtful. The celebrated Portuguese mathematician,
Pedro Nuñes, in a work in defence of the sailing chart,
which I have not seen, but which is quoted by Garção-
Stockler, makes the following interesting statement respecting the early navigations of his countrymen. He says,
"Now it is evident that these discoveries of coasts, islands,
and mainland were not made without nautical intelligence, but
our sailors went out very well taught and provided with
instruments and rules of astrology (sic) and geometry,
which, as Ptolemy says in the first book of his geography, are
things with which cosmographers ought to be acquainted."
There is no doubt that old travellers delineated geometri-
cally on maps the places which they had visited, and that,
as discoveries and observations increased, improvements
were made in map-making. Amongst these improvements

was the delineation of the sphericity of the earth, but this very improvement, while valuable for the student of geography, offered many embarrassments to sailors who had problems to solve and courses to calculate. Thus, the meridians had of course to be represented by straight lines or by curved lines meeting at the pole, and the course of a vessel not sailing directly under the equator nor under a given meridian would also be represented by a curved line. It became therefore necessary to devise a form of chart for nautical purposes on which, the meridians being parallel, the lines of the rhumbs or points of the compass could be drawn straight. The necessity produced the desired result, and the sea charts so made were known as *plane* charts, and, though they naturally involved a geometrical inaccuracy, the deviation from correctness was almost imperceptible in the short voyages of the period. It was not till the close of the sixteenth century, when extensive oceanic voyages had become frequent, that Gerard Mercator proposed to remedy the inconvenience by elongating the degrees of latitude towards the poles in the same proportion as the degrees of longitude decrease on the globes. He was however unable to determine the law of this prolongation, which was discovered about 1590 by Edward Wright, an Englishman, and publicly made known by him in 1599.

Very few details are left to us of the astronomical instruments used in the time of Prince Henry. The altitude of a star was taken by the astrolabe and the quadrant by means of an alidade, or ruled index, having two holes pierced in its extremities, through which the ray passed. The quadrant hung vertically from a ring which was held in the hand. We do not know how these instruments were graduated, but it is to be presumed very roughly. The astrolabe, the compass, timepieces, and charts were employed by sailors in the Mediterranean at the beginning of the fifteenth century. The learned Count Libri in his great work on the History of the Mathematical Sciences in Italy, Paris, 1838, tom. 2, page 220, quotes in corroboration of this statement the

Guerino Meschino, said to have been written at the beginning of the fourteenth century.*

The earliest allusion to the use of the compass in the
middle ages yet discovered occurs in a treatise De Utensilitus by Alexander Neckam, a native of St. Albans, who as
early as 1180, when he was but twenty-three years of age,
had become famous as a Professor in the University of Paris.
For the treatise in question we are indebted to the learned
researches of our distinguished and indefatigable antiquary,
Mr. Thomas Wright. It is given in a privately-printed
" Volume of Vocabularies," illustrating the manners of our
forefathers from the tenth century to the fifteenth, in the
rather obscure language given at foot.† The earliest account
of the mariner's compass previously known was contained in
some often-repeated lines of a satirical poem, entitled the
" Bible," by Guyot de Provins, in which he wishes the
Pope were as safe a point to look to as the North Star is to
mariners, who can steer towards it without seeing it by the
direction of a needle floating in a straw on a basin of water,
after being touched by the magnet. Nothing can more
clearly prove than these two passages that the compass was
in use in the West at the close of the twelfth century. But
to show how limited that use must have been, even more
than half a century later, it is only necessary to refer to a
passage in the description of a visit paid by Brunetto Latini,
the tutor of the immortal Dante, to Roger Bacon at Oxford,

* Pero li naviganti vanno con la calamita, securi per lo mare, e con la stella
e con lo partire della carta et de li bossoli de la calamita.

† " Qui ergo munitam vulthabere navem, albestum habeat, ne desit ei beneficium ignis. Habeat etiam acum jaculo suppositam, rotabitur enim et circum-
volvetur acus donec cuspis acus respiciat orientem, sicque comprehendunt quo
tendere debeant naute cum cinossura latet in aeris turbacione, quamvis ad
occasum nunquam tendat propter circuli brevitatem." But a fuller description
of the compass is given in another of Neckam's books, the treatise De Naturis
Rerum, lib. 2, cap. 18, (MS. Reg. 12. G. xi. fol. 53 verso). "Nautæ etiam mare
legentes cum beneficium claritatis solis in tempore nubilo non sentiunt, aut
etiam cum caligine nocturnarum tenebrarum mundus obvolvitur, et ignorant in
quem mundi cardinem prora tendat, acum super magnatem ponunt quæ cir-
culariter circumvolvitur usque dum, ejus motu cessante, cuspis ipsius septentri-
onalem plagam respiciat."

apparently in the year 1258.   When driven out of Florence
by the Ghibeline faction, Latini had sought an asylum with
the Earl of Provence, brother-in-law to King Henry the Third.
He came over to England with the King's brother, Richard,
Earl of Cornwall, then newly elected King of the Romans,
in the quality of preceptor to Henry d'Almaine, Richard's
eldest son. From England he addressed descriptions of what
he saw to the poet Guido Cavalcanti, who also had been one
of his pupils. These interesting letters, written in the French
patois of the Romansch language, were translated by a cor-
respondent of the Monthly Magazine in 1802, under the
title of " Extracts from the Portfolio of a Man of Letters."*
He says :—

    " The Parliament being summoned to assemble at Oxford
(probably the Mad Parliament in 1258), I did not fail to see
Friar Bacon as soon as I arrived, and [among other things]
he shewed me a black ugly stone, called a magnet, which
has the surprising property of drawing iron to it; and upon
which if a needle be rubbed, and afterwards fastened to a
straw, so that it shall swim upon water, the needle will instantly turn towards the Pole-star : therefore, be the night
ever so dark, so that neither moon nor star be visible, yet
shall the mariner be able, by the help of this needle, to steer
his vessel aright. [La magnete piere laide et noire. Ob ete
fer volenters se joint. Lon touchet ob une aguilet. Et en
festue lon fischie. Puis lon mette en laigue et se tient
desus. Et la point se torne contre lestoille. Quant la nuit
feit tenebrous et lon ne voie estoile ni lune, poet li mariner
tenir droite voie.]

    " This discovery, which appears useful in so great a degree
to all who travel by sea, must remain concealed until other
times; because *no master-mariner dares to use it* lest he
should fall under a supposition of his being a magician ; nor
would even the sailors venture themselves out to sea under
his command, if he took with him an instrument which

* The Monthly Magazine, or, British Register. Vol. xiii. Part 1, p. 449.
London, 1802.

carries so great an appearance of being constructed under
the influence of some infernal spirit. A time may come
when these prejudices, which are of such great hindrance to
researches into the secrets of nature, will probably be no
more; and it will be then that mankind shall reap the
benefit of the labours of such learned men as Friar Bacon,
and do justice to that industry and intelligence for which he
and they now meet with no other return than obloquy and
reproach."

Thus far we find the mariner possessed of a contrivance
which, without the moral hindrance to its use referred to by
Brunetto Latini, might possibly be used at sea, but certainly
only under favourable conditions. It is clear that as yet it
was known as an article of curiosity rather than one of practical utility. At what time it became effectively serviceable by
being fitted into a box and connected with the compass-card,
we have as yet no historical data to show, but we are told
by Antonio Beccadelli, surnamed Il Panormita from his
birth-place Palermo, and who was a cotemporary of Prince
Henry, that sailors were first indebted to Amalfi for the use
of the magnet. " Prima dedit nautis usum magnetis Amalphis"; and, " Inventrix præclara fuit magnetis Amalphis."*
The former line was better calculated than the latter to win
honour for the Amalfitan, Flavio Gioja, who is therein referred to. We have already seen that the *invention* of the
magnet was certainly not due to him, for by common consent the period at which he flourished was the beginning of
the fourteenth century, but if the honour described in the
former line of having *given sailors the use of the magnet* might
be taken in its severest meaning, we might gather that he
supplied what was hitherto wanting, viz., the box and fittings
which made the compass available. Be this as it may, we
have certain evidence of the practical use of the needle at sea

* The former of these lines is quoted from Il Panormita by Henricus Brenc-
mannus in his Dissertatio de Republica Amalfitana, and Klaproth has added
the latter. We must therefore presume that they are genuine, but I have sought
for them in vain through the verses of that elegant Latinist, but most impure
poet, to whom they are ascribed.—R.H.M.

before Prince Henry's time, not only from the above lines of Antonio Beccadelli, but from the words of Prince Henry himself, as will be hereafter seen when he is urging on one of his navigators to the rounding of Cape Bojador.

It was in the reign of Affonso the Fourth that the sciences of mathematics and astronomy first began to be studied in Portugal, the Moors and Jews being the most eager students, and they principally in judicial astrology. It is not however till the time of Prince Henry that we meet with the names of individual cultivators of those sciences. His brother the King Dom Duarte himself gave proof of the interest he took in meteorology by the following observations of the aspects of the moon made by him and preserved amongst his writings in the Carthusian Convent at Evora.* He says that " when the new moon is entirely red, it signifies much wind. If its topmost point be dark, it means rain. If it sparkle like water raised by oars, it shows that there will soon be a storm. If dark in the middle, it shows that there will be fine weather when the moon reaches the full."

It seems highly probable that the chair of mathematics in Lisbon was established by Prince Henry himself, since by a deed dated 12th October, 1431, he conferred on the University of that city, which had no house property, ·some houses which he purchased of João Annes, the king's armourer, for four hundred coroas velhas, while it is known that in 1435 that chair did exist, and that the subject was one in which he took especial interest.

A most valuable coadjutor of the Prince in the prosecution of these studies was his elder brother the Infant Dom Pedro. Excellently educated, as indeed were all the children of Queen Philippa, he was an accomplished student of the ancient languages and mathematics. In 1416 †, the year after the taking of Ceuta, this Prince was seized with the desire to

* See Sousa. Provas. Tom. 1, p. 540.

† The old chroniclers assign the date of 1424 to the Prince's departure on his travels, but his modern biographer, the Abbade de Castro, has found reason to place that event in 1416.

gain enlightenment by travel through the principal countries
of Europe and Western Asia. And accordingly on the first
Sunday after Easter, with the King's permission, he set forth
with that object attended by a small suite of only twelve
persons. He first visited his uncle the King of Castile at
Valladolid, who not only welcomed him with a present of five
thousand gold pieces, but escorted him in person a league forth
of the city. The King also gave him for a companion an
interpreter named Garcia Ramires, who had travelled in many
countries and was a notably able linguist. His first destination was to Palestine, whence, after visiting the Holy
Places, he proceeded to the Court of the Grand Turk and to
that of the Grand Sultan of Babylonia, where he met with a
magnificent reception. He thence passed to the Court of
Rome, where Pope Martin V. welcomed him with the highest
distinction and at his request conceded to the Kings of
Portugal the important prerogative, afterwards confirmed by
a bull dated June 16th, 1428, of receiving the rite of
coronation by unction in the same manner as observed in
the crowning of the Kings of England and France. This
grace was subsequently confirmed to King Duarte, King
João's successor, by Pope Eugenius in the year 1436.
The Prince also visited the Courts of the Kings of Hungary and Denmark, and Sousa states, on the authority of
the History of Bohemia by Æneas Sylvius, afterwards
Pope Pius II., that in company with Eric X., King of
Denmark, he served the Emperor Sigismund to such
good purpose in the war against the Turks, and also in
the war against the Venetians, that he granted him in
reward the Marca Trevisana.* After peace was established
between Sigismund and the Venetians the Prince went to
Venice, and there received from the Republic, in compliment
to him as a traveller and a learned Royal Prince, the priceless
gift of a copy of the travels of Marco Polo, which had been
preserved by the Venetians in their treasury as a work of

* I do not find the passage, but the deed of endowment was seen by Duarte
Nuñes in the archives of the Torre do Tombo.

great value,* together with a map which had been supposed
to have been either an original or the copy of one by the
hand of the same illustrious explorer.   The Prince then
proceeded to England, which he much desired to see on
account of its being the country of the Queen his mother.
His reception by Henry the Sixth was marked by every
demonstration of honour and regard that could be shown
by a powerful monarch to so near a relative.   On the 22nd
of April, 1427, the Prince was elected a Knight of the
Garter in place of Thomas Beaufort, Duke of Exeter, who
had died on the 27th December, 1426.

At the end of twelve years' travel Dom Pedro returned in
1428 to Portugal, where his safe arrival after so many wander-
ings caused the liveliest joy not only to the King his father
and his brothers, but to the whole population, by whom he
was henceforth spoken of as the Prince " that had travelled
over the seven parts of the world."   Unfortunately we possess
only a fabulous narrative of this most genuine peregrination
drawn up by one of Dom Pedro's own companions, named
Gomez de Santo Estevan.   This is the more to be regretted
as journeys of such length through distant countries were
seldom in those days made by Royal personages.   On his
return Dom Pedro devoted himself like his brother Prince
Henry to scientific studies, among which the art of cartography took a leading place, and there is little doubt that to
the genius and attainments of his elder brother Dom Pedro,
Prince Henry owed much of encouragement and enlightenment in his pursuit of geographical investigation.   The
Marco Polo MS. and the map brought from Venice would
doubtless act as a potent stimulus to these investigations.
We are unfortunately much in the dark as to the character
of this map, but according to Antonio Galvam it " had all
the parts of the world and earth described.   The Streight
of Magellan was called in it the Dragons taile : the Cape
of Bona Sperança, the forefront of Afrike (and so foorth of

---

* A Portuguese translation of this work was made and edited at Lisbon in
1502 by the same Valentin Fernandez of whom mention has been already made.

other places :) by which map Don Henry the King's third
sonne was much helped and furthered in his discoveries."
Galvam further states that he was told by Francisco de Sousa
Tavares that, in the year 1528, Dom Fernando, the son and
heir of King Manoel, showed him a map* which was found
in the Cartorio, or study, of the Royal Monastery of Alcobaça,
which had been made more than a hundred and twenty years
before, on which was laid down all the navigation of India, with
the Cape of Good Hope as it was now known. " If it be so.'
he proceeds to say, " there was as much or more discovered in
times past than now." This Francisco de Sousa Tavares was
the executor of Antonio Galvam, and the editor of his " History
of the Discoveries of the World," so that if any mistake had
been made by Galvam in first writing down this fact, the editor
would have been able to correct it. By not doing so he has
made the assertion his own. And such being the case the
closing remark of Galvam respecting the evidence of these
two maps* seems primâ facie not only reasonable but in the
highest degree damaging to that claim which it is the object
of this work to assert on behalf of the Portuguese, and par
excellence of Prince Henry, to the glory of having opened
the way to India by the Cape of Good Hope. This difficulty
has been ably met by a learned Portuguese writer, Antonio
Ribeiro dos Santos,† not by any endeavour to escape from, but
by enlarging the field of, the apparent danger. He shows
that similar indications occur upon maps yet earlier, as for

---

* If one may be guided by what is said in the first paragraph of Book 4 of
Cordeyro's Historia Insulana, p. 97, the map brought back by Dom Pedro
and the one which was formerly in the Cartorio of Alcobaça are identical; for
though, after speaking of the one brought by Dom Pedro, he says that in 1528
Dom Fernando showed Antonio Galvão *another* map found at Alcobaça, he two
lines later says that the latter must have been the one brought back by Dom
Pedro. By this it is clear that the word "another" simply implies "also." This
conclusion is confirmed by the fact that in this very paragraph, which consists of only one sentence, what the word "another" would make to mean two
maps, is thus spoken of as only one, "and of such map our discoverer Prince
Henry must have availed himself together with the information received from
the Venetians for giving instructions for the discovery of these new islands."

† See Memorias de Litteratura Portugueza. Tom. 8, pp. 275 et seq.

example on that of Marino Sanuto of about the date of 1320,
on a famous map still preserved in the Camaldolese Monastery
of S. Miguel de Murano near Venice, of about the date of
1380, supposed to be a copy of one brought from China by
Marco Polo. To these he adds two of a later period, though
anterior to those recognized discoveries of the Cape which
resulted from the expeditions of Prince Henry, viz., that of
the Venetian Andrea Bianco of 1436 and of the renowned
geographer Fra Mauro of the above-mentioned Camaldolese
Monastery of the date of 1459. But of these maps and how
far they were indicative of actual exploration I shall have
occasion to speak fully in a subsequent chapter.

Much doubt has been entertained as to the year in which the
Prince first dispatched a vessel on an exploratory expedition.
Some have even made it as early as 1412, but there appears
no sure foundation for such a supposition. From an expression which occurs in a bull of Pope Nicholas V. of the
date of 1455, it would be inferred that he commenced his
enterprize, when about coming of age (ab ejus ineunte
ætate), which would be in the year 1415. All seem to agree
in acknowledging the fact that when in Ceuta in that year
the Prince gathered important information from the Moors
of Fez and Marocco respecting the Arabs who lived on the
borders of the desert, as well as respecting the kingdom of
the Jaloffs near Guinea. He knew that the countries on
the North of Africa were enriched by commerce with that
country, and derived therefrom a considerable quantity of
gold. In this, as a step to yet greater purposes of advancement, he saw a source of prosperity for his own country,
which in itself was worthy of new efforts at exploration.
The earliest date assigned by any authority of the same
century to an expedition fitted out by him is that of this
selfsame year of 1415. It occurs in a narrative recounted
many years after the Prince's death to the celebrated German
knight, Martin von Behaim, by Diogo Gomez, almoxarife
or superintendent of the palace of Cintra, who had himself
been an explorer under the orders of Prince Henry, and had

been much about his person. He states, that, in 1415, a
certain noble Portuguese gentleman, named João de Trasto,
was captain of an expedition, fitted out by the Prince. He
was driven by stress of weather upon that part of the island
of Great Canary, which was named Telli, the fruitful. In
endeavouring to return, he encountered strong currents
between the islands, so that it was with great difficulty
that he made his way home. There is however so much
that is manifestly inaccurate in other statements of Diego
Gomez respecting the early voyages which he narrates from
hearsay, that we cannot be perfectly sure that the date here
applied to the earliest expedition is correct. Be this as it
may, it is certain that after his return from Ceuta, the
Prince made a practice of sending out an expedition every
year as far as was possible along the coast of Africa. Some
have attributed to his sailors the credit of first passing Cape
Non, which as its name imports had in old times been
regarded as the limit of safe or even possible navigation;*
but this is plainly wrong, for Cape Boyador, which really did
form that limit, is distinctly laid down on maps of the
fourteenth century, before Prince Henry was born. These
various expeditions which resulted in no immediate advantage called down upon the Prince much obloquy from
the nobles, who complained of an amount of useless
expenditure, from which meanwhile they were in no sense
the losers. But vituperations fell harmless upon one who
was consciously influenced by a noble purpose which could
only be effected by perseverance. At length an event took
place which silenced clamour for a while, and greatly encouraged the hopes of the Prince, but this must form the
subject of a separate chapter.

* The proverb ran "Quem passar o Cabo de Não, ou voltará ou Não:"
" Whoever passes Cape Non will return or *not*."

CHAPTER V.

1418—20.

THE discovery of the islands of Porto Santo and Madeira in 1418—20 was the first fruit of Prince Henry's explorations, and until the year 1827 the belief had prevailed for nearly three hundred years, that those islands were then discovered for the first time and then also received their respective names. True, a vague rumour obtained in some quarters, especially in the islands themselves, that the discovery had been made fortuitously by an Englishman named Machin at the close of the previous century, but great discredit was thrown upon this story by many, and none knew for a certainty what to believe. Happily the means have fallen within my power to establish the truth of this latter story, but in a subsequent chapter it will be shown that even earlier still, namely, in the beginning of the fourteenth century, the discovery was made in which the present names of the different islands of the group originated. It is to the great Portuguese historian De Barros that we owe the diffusion of the erroneous belief that the group first received those names and was for the first time discovered by the Portuguese in Prince Henry's time in 1418—20, and in making that statement he exceeded the authority of the ancient chronicler Azurara from whom he, by his own acknowledgment, derived his materials. He tells us that " two squires of Prince Henry's household named João Gonsalvez Zarco and Tristam Vaz, anxious for fame and desirous of serving their master, had set out on an exploring expedition to the coast of Guinea, but were taken by a storm off Cape St. Vincent and driven to the island of Porto Santo, which name was *then given by them* to the island on account of its saving them from the dangers of the tempest." The favourable report of the newly found island of Porto Santo induced Prince Henry forthwith to send out to colonize it, and after a while a dark spot was descried on the horizon by the colonists, which on examination proved to be what is now called the island of Madeira. In speaking of this De Barros, in like manner, says " *they gave it that name which means ' wood,'* on account of the thick forests with which it was covered," a statement equally incorrect with that already noticed. The real origin of these names will be described in another chapter. At present we have to speak of Machin's expedition and the process by which Zarco himself was led to his reputed accidental discovery. The story is one of the most romantic that has ever been dignified with the name of history, and has been told a hundred times in as many different shapes ; but the following is a digest of it as related by the possessor of the original manuscript account.

In the reign of Edward III., a young man of good family named Robert Machin had the misfortune to become enamoured of a young lady, the wealth and rank of whose parents were so far superior to his own that they treated his pretensions with disdain. To avoid his importunities they obtained from the King an order for his imprisonment, and in the interval united their daughter to a nobleman whose station was more suited to maintain the dignity of their family. As the lady whose name was Anne d'Arfet or Dorset reciprocated Machin's affection, he was no sooner released from prison than he determined on carrying her off. By the aid of a friend who contrived to gain admittance as groom into the lady's family, which was established at Bristol, this plan was finally effected, and from Bristol they set sail together in a vessel which Machin had already provided and manned for the purpose.

The intention was to sail for France, but a north-east wind carried them off that coast and, after thirteen days' driving before a tempest, they caught sight of an island on which they landed. They found it uninhabited, but well wooded and watered and eminently suited for habitation. For three days they enjoyed the peacefulness of security, and while some explored the interior, others in the ship examined the contour of the coast, but on the third night were overtaken by a storm and driven on the coast of Africa. The anxiety and suffering which the unhappy lady had undergone found their culmination in this disaster, and after three days of total mental prostration she expired. She was buried at the foot of the altar which had been erected in gratitude on their arrival, and, on the fifth day after her death, Machin also was found dead on the grave of his mistress. The survivors buried him, and then embarked in the ship's boat and, on reaching the coast of Africa, were carried before the King of Marocco, by whom they were thrown into captivity. In the same unfortunate circumstances they encountered their missing companions who had previously been carried away in the ship.

Among their fellow-captives was one Juan de Morales, a native of Seville, a good seaman and originally a pilot, to whom they gave a description of the land they had discovered. Now on the 5th March, 1416, died Don Sancho, the youngest son of King Ferdinand of Aragon, and by his will he left a large sum for the ransom of Christian captives from Marocco. Amongst those who were redeemed was this Juan de Morales, but the vessel which brought him over was captured by the Portuguese navigator João Gonsalvez Zarco. From pity however the latter liberated the unfortunate captives, reserving only Morales, whose experience in nautical matters he thought might be of service to his master, Prince Henry. This Zarco had, as we have already seen from Barros, gone out in company with Tristam Vaz Teixeyra, to explore the west coast of Africa, and had been driven by a storm on the island of Porto Santo. This appears to have occurred at the close of 1418 or at the beginning of 1419. From Morales he heard the account of Machin's discovery, and, with the permission of the Prince and under the guidance of Morales, he set sail and made the important discovery of the island of Madeira, to one half of which he gave the name of Funchal and to the other that of Machico.

This story was first given to the world in full detail by the graceful Portuguese writer, Francisco Manoel de Mello, in his "Epanaphoras de Varia historia Portugueza," published at Lisbon in 1660. He declares it to have been founded on an original narrative by Francisco Alcaforado, a squire of Prince Henry, who was with Zarco in this famous voyage, and which narrative De Mello states that he preserved as a precious jewel, and which had come into his possession by an extraordinary channel. As much suspicion has been thrown upon its truth I have been at great pains to investigate its history. Although the library of Manoel de Mello is preserved in the Bibliotheca Nacional at Lisbon, the manuscript of Alcaforado, which has been diligently searched for by my own request at the instance of a distinguished Portuguese nobleman, the Count de Rilvas, has never been found. The suspicion occasioned by this circumstance was increased by my finding that in De Mello's library was a copy of Antonio Galvão's "Treatise on the Discoveries of the World," written about the year 1555, and in which this story of Machin had been for the first time told in print, although in a far less detailed manner. This book had become so extremely scarce in the course of half a century that Hakluyt, who possessed an anonymous translation of it made by some "honest English merchant," strove for twelve years to find a copy of the original, sending to Lisbon for it, but in vain.* The suspicion excited by the absence of the Alcaforado manuscript from, and the presence of Galvão in, De Mello's

---

* What Hakluyt failed to do I had the good fortune to succeed in for the benefit of the Society which bears his name. Mr. John Carter Brown of Providence, Rhode Island, lent me a copy which was edited for the Hakluyt Society in 1862 by Admiral Drinkwater Bethune.

library, induced me to seek further, and at length I succeeded in obtaining from Munich an extract from an unpublished Portuguese manuscript, containing this story, which having been written in 1508 was earlier even than Galvão by half a century. It is the production of a German printer and compiler resident at Lisbon, the Portuguese form of whose name was Valentin Fernandez.* A comparison of the two narratives of De Mello and Fernandez presents the following differences. De Mello takes Machin to Madeira at once, while Fernandez takes him first to Porto Santo and then to Madeira. De Mello makes Machin die at Madeira; Fernandez makes him spend six months in cutting a canoe out of a large tree, in which he lands on the coast of Marocco, whence he is sent by the King of Fez to King Juan of Castile. The Spanish sovereign, however, was so closely engaged in a war with Portugal that the matter was neglected, and meanwhile Machin died.

That such differences should exist is intelligible, when we consider that De Mello's story is his own embellished compilation from Alcaforado, and that of Fernandez is also his own account drawn up from a source of which we are ignorant. The question is whether the story can, from them, be shown to be true in the main. This can be done in two ways, first by establishing De Mello's truthfulness as to the Alcaforado manuscript from internal evidence; and secondly, by showing that, even if that manuscript were a myth, the story nevertheless existed in a record earlier than any to which De Mello had access. First, there are certain facts which, when brought side by side, confirm the truth of De Mello's statement that he really possessed the now missing Alcaforado manuscript. De Mello's narrative based upon that manuscript gives not only the story of Machin, but a detailed account of Zarco's subsequent discovery, in which Alcaforado is said to have been present. Now Barros, writing a century before

---

* He also appears as Valentinus de Moravia in a Life of Christ which he published in 1496 in association with Nicolas de Saxonia.

De Mello's time, distinctly declared that in his day Zarco's descendants possessed a detailed account of his voyage, and De Mello himself informs us that by marriage he had become the representative of the Zarco family. Should this combination of facts presenting such strong presumptive evidence be held to fall short of positive proof, and if it be assumed that De Mello drew his information not from any manuscript by Alcaforado, but from Galvão, there yet remains the fact that the earlier manuscript of Valentin Fernandez was out of the reach both of Galvão and of Mello; and the truth of the story is thus distinctly established by its appearance in an earlier document derived from totally independent sources. Soon after the compilation of that document in 1508, it passed into the hands of the celebrated Conrad Peutinger (the fortunate possessor of the famous Tabula Peutingeriana), and remained in his possession till he died in 1547. During the whole of this period the noble but unfortunate Antonio Galvão, whose account, drawn up in 1555, was the earliest hitherto printed, was engaged in the East, either sword in hand or suffering in a prison, so that his account is shown to have been derived from independent sources, and the two separate documents point to the existence of another of a yet earlier date testifying to the truth of Machin's discovery. But further, Fernandez' account has remained in Germany ever since, so that when Francisco Manoel de Mello drew up his narrative in 1660, though he possessed a copy of Galvão's then rare book, he could have had no cognizance of the earlier statement of Fernandez, but, as he relates much more than either one or the other, it follows that he derives his additional matter from an ampler source, or that that source was a myth, and the additions a forgery. But if we bear in mind his own statement that he did possess such an original manuscript, which he says came to him by an extraordinary channel, an expression explained by his becoming the representative of the Zarco family through matrimonial alliances, and the distinct assertion by Barros about a century before, that that

family possessed a detailed account of Zarco's voyage which
is comprised in De Mello's story, suspicion of De Mello's
truthfulness, never otherwise impugned, becomes more indefensible than credulity.

So much for external evidence. The internal evidence is
no less conclusive. Although Azurara and Barros are silent
on the subject, the accounts of Fernandez, Galvão, and De
Mello, which I have shown to be independent of each other,
concur in deriving the local name of Machico from the name
of the Englishman Machin. Now none of Machin's crew
were left behind, and the importance attached to Zarco's
re-discovery in 1419-20 proves that the Portuguese had not
colonized the island when, some seventy or a hundred years
before, it was discovered, as I shall presently have to show,
by their own vessels under the command of Genoese captains. It follows therefore, although it has been nowhere
distinctly so stated, that the names of Machico and Funchal
must have been newly given by Zarco and Vaz at the time
of the partition of the island between them. The etymology
of the word Funchal is exclusively Portuguese. It signifies a place where fennel (in Portuguese, funcho) grows, and
the name is distinctly declared to have been given from that
plant having been found there in great quantities. The
entirely different Spanish form of the word "hinojo," and
the Italian form "finocchio," prove that the name could not
have survived from any previous Spanish or Italian discovery. And since no Englishman remained on the island
to preserve the name of Machin, the conclusion seems inevitable that, at the time of the partition, the Portuguese
showed their recognition of Machin's previous discovery,
communicated to them by the Spaniard Juan de Morales, by
themselves giving the name of Machico to the place where
they found the grave and cross, and other indications of Machin's tragic adventure. Further, it is past belief, that Manoel
de Mello, himself a Portuguese, should gratuitously detract
from the glory not only of his own country but of his own
family, by setting forth that his ancestor had been preceded

in a grand discovery by an Englishman, and, even more, had been guided to that discovery by a Spaniard, if it had not been true. I think, therefore, that henceforth the story of the accidental discovery of Madeira by Machin must be accepted as a reality, but the question arises as to the date. By the misreading of a passage in Galvão, the date of 1344 has been erroneously assigned to the event and repeated by many. That year is mentioned in connection with an entirely different occurrence which Galvão states was in the reign of King Peter IV., of Aragon [1336 to 1387], and then adds, "in the midst of this time also the island of Madeira was discovered by an Englishman named Macham, who was driven out of his course by a tempest, and anchored in the harbour now called Machico after his name." De Mello states that the adventure occurred during the reign of Edward III., ending 1377. It is clear, however, from their own statements, that neither of these writers was very precise in his chronology.

But to return to Zarco, who, although his discovery was not original, had accomplished a feat of very great importance and added honour to a name which he had already greatly distinguished. He had won his spurs at Ceuta, and had continued to serve bravely in the other African expeditions. He is also supposed to have been the first who introduced artillery on board the Portuguese vessels. In the June of 1420 he set sail for Porto Santo with two vessels, accompanied by João Lourenço, Ruy Paes, Alvaro Affonso, Gonzalo Ayres Ferreira,* and Francisco Alcaforado, the author of the narrative. On arriving he had his attention called to a dark line which was visible on the horizon towards

---

* We learn from Cordeiro's Historia Insulana, liv. 3, cap. 15, that in a charter of Prince Henry's dated 1430, this Ferreira is mentioned as a companion of Zarco. He was the first who had children born in Madeira. The eldest he called Adam and the second Eve. From him descended the noble family of Casta Grande of Madeira and the Ferreiras of San Miguel, who also derive from the Drummonds and the Royal Stuarts. It may here be observed that Prince Henry, as I am informed by the Count de Rilvas, was careful to institute family registers at that early period in the island of Madeira.

the south-west, an appearance which had astonished those whom he had left in the island. The pilot Juan de Morales conjectured that this would be the island they were in search of, and suggested that the thick fog was occasioned by the action of the sun on a soil covered with forests. After a stay of eight days, Zarco sailed in the direction of the fog, and as he approached it found that it diminished in extent and intensity towards the east; and, steering in this direction, he reached a point of low land to which he gave the name of Ponta de San Lourenço. Doubling this he coasted along the southern shore, and came to high land covered with thick wood from the shore to the top of the mountains, where the fog still rested.

The next day Ruy Paes was sent with a sloop to explore the coast. He found it answer to the description given of it from memory by Morales, and at length discovered the tomb with the epitaph and wooden cross which had been left by Machin's party, but no human being did he encounter. Zarco took formal possession of the island in the name of the King of Portugal, Prince Henry, and the order of Christ.

He then went on board his sloop, and accompanied by Alvaro Affonso in command of the other vessel, made an exploration of the coast. He soon fell in with four fine rivers of very pure water, to one of which he gave the name of Rio do Seyxo or river of the flint, which name still remains. From a valley further on, which was full of trees, he collected several samples of the different woods, and at the point of the river which flowed through it he set up a great wooden cross, which gave the name of Santa Cruz to the town afterwards built on the spot. Further on there arose from a point of land a great number of jackdaws, which caused him to name it "Ponta dos Gralhos" (Jackdaw Point). The name survives in the form of Cabo do Garajão. Two leagues further was another point, which with the first formed a spacious and commodious gulf, into which several valleys opened; the first was clothed with majestic cedars, and down the second flowed a broad river, which offered a convenient place for landing. Gonçalo Ayres was sent with some soldiers to explore the interior. He brought back word that from the top of the mountains he could see the outline of the whole island. The river has borne the name of that explorer ever since. On the west of the valley, the beach, which was broad and unsheltered, was one vast field of fennel, whence they called it "Funchal," the name which it has ever since retained. It is observable that the Portuguese instead of seeking grand names for their colonies contented themselves with preserving those which existed already, or adopting those which nature supplied. Some islets, opposite this "Funchal," offered an excellent roadstead where Zarco anchored to take in wood and water, and summoned the crews on board for the night.

Next day the sloops set sail with the view of doubling the westward point of the bay of Funchal. On that point they planted a cross and gave it its present name of Ponta da Cruz, or Point of the Cross. Beyond it extended a beautiful beach, to which they accordingly gave the name of "Praya Formosa." This ended in a considerable torrent, the beauty of which tempted the curiosity of two soldiers from Lagos; they went to reconnoitre it, and imprudently attempted to swim across it, but would certainly have been drowned, had they not received prompt assistance. This circumstance caused the torrent to be named, as at present, the "Ribeira dos Socorridos."

Continuing still to advance, Zarco came to a little creek sheltered by a rock, and entered it with the sloops; his arrival disturbed the repose of a troop of sea wolves or phocas, which fled into a cavern at the foot of the rock, which was their dwelling-place. This "Camara dos Lobos" (Chamber of the Wolves) was the terminus of Zarco's exploration of the coast. After taking in a good supply of water, wood, plants, and birds at Funchal, he returned to Portugal, where he arrived at the end of August.

The King received him with great distinction, bestowed

on him the title of Count of Camara dos Lobos, and gave him the hereditary command of his new discovery. He returned in the May following with his wife, his son, and all his family, and landed at the port of Machico, the name of which, given in remembrance of Machin, still survives. On the spot where the unfortunate Englishman was buried, he founded a chapel dedicated to the Saviour. He then went to Funchal, where the bay offered a better anchorage, and there founded a city, which rapidly increased in size, and in which his wife founded the Church of St. Catherine.

The entire island was divided between Zarco and Tristam Vaz, so as to form two Captaincies of about equal extent. The northern half, with Machico for its capital, was given to Tristam, and the southern, with Funchal for its capital, and the three Desertas, to Zarco.

Soon after Zarco had established himself at Funchal he erected a church, which from the great quantity of flint found on the coast he named Nossa Senhora do Calhão, or our Lady of the Flints, but as, inland from thence, the forests were so thick that they could not open a road, he had it set fire to, and it is stated by Gaspar Fructuoso that for the incredible period of seven years the fire was unextinguished. However this may have been, it seems clear from a formal act signed by Prince Henry on the 18th of September, 1460, a few months before his death, by which he endowed the order of Christ with the spiritualities of these islands, that it was not till he was thirty-five years of age that he began to colonize the island of Madeira and Porto Santo, which would be in the year 1425.

The province of Machico was richly wooded, and we learn from Azurara how, twenty years later, this wood was imported into Portugal by Prince Henry in such quantity that a great change took place in the architecture of the country, lofty houses being substituted for those which had previously been built in the Roman or Arabic style. The north of the island produced large quantities of corn and honey. The sugar cane was introduced from Sicily, and the first sugar grown in the whole island was in Machico. Prince Henry imported from Candia the Malvasia or Malvoisie* grape, and in Machico the best wine was produced. Hence under the corrupted form of the name we have our Malmsey Madeira. It will be seen in a subsequent chapter how this grape had thriven in the island in the course of thirty years.

On the return of Zarco and Vaz from their first discovery of Porto Santo, they suggested to the Prince the desirableness of colonizing the island. The Prince greatly approved of the idea, and provided them with the requisites for the colonization, and among those who offered to accompany them, was a gentleman of the household of Prince João, named Bartollomeu Perestrello. He had in a cage a pregnant rabbit, which had been given him by a friend. She littered during the passage, and with her young ones was taken to the island. Unfortunately the race increased so rapidly that they consumed everything that was planted by the colonists. On returning the following year after a short absence from the island, the colonists found the rabbits increased to such an extent that in spite of all their efforts to destroy them, they produced no sensible diminution of their numbers. Perestrello then returned greatly discouraged to Portugal, Zarco and Vaz having by this time discovered Madeira, and received from Prince Henry that island in partition between them. The Prince however subsequently caused Perestrello to return to Porto Santo, of which he gave him the governorship, and although the multitude of rabbits entirely prevented all vegetable cultivation, yet the island nourished a considerable number of goats, and the dragon-tree grew in abundance, so that they were able to export dragon's blood to Portugal and many other places. We shall meet with the family of Perestrello established in Porto Santo at the close of the century, when we come to speak of Christopher Columbus.

* Originally from Monemvasia or Napoli di Malvasia in the Morea.

CHAPTER VI.

1434—1436.

THE last years of the reign of King João, after the taking of
Ceuta, were employed in the peaceful pursuit of the internal
prosperity of his kingdom, and the dynasty of Aviz was
now firmly established. Even the warlike constable, Nuño
Alvarez Pereira, who had never known defeat, had retired
in 1423 to his magnificent Convent do Carmo, and, adopting
the habit of a monk, laid aside all his titles, and, by his own
desire, was addressed by the simple name of Nuño. Had he
followed his own inclinations, he would have existed on the
alms of the charitable and have made a pilgrimage as a
mendicant to Jerusalem.

For ten years more the kingdom enjoyed profound peace,
when in 1433 the King's health began to fail, and he went by
direction of his physicians to Alcochete, a village on the banks
of the Tagus, the air of which was considered more suitable for
him than that of Lisbon. But as his weakness increased
and he became convinced that his end was approaching, he
desired his sons to take him to Lisbon, for he did not think
it befitting that he should remain to die in an obscure place,
and in the house of a private individual, as he was so near
to the capital of his dominions. He was therefore removed
to the palace of Alcaçova, where he breathed his last on the
14th of August, 1433,—being the eve of the assumption of
the Blessed Virgin, and the anniversary of the battle of
Aljubarrota,—in the 77th year of his age and the 49th of his

reign. His subjects mourned for him as for a father. Nor is this difficult to understand. For him they had suffered much, and willingly sacrificed life and substance, while on his part the wisdom, skill, and courage which had made these sacrifices only the offerings of a willing loyalty, had procured for them a condition of prosperity and dignity which they had never before enjoyed.

The King had directed by his will that he should be buried in the convent of Batalha, in the noble tomb which had been already constructed for himself and Queen Philippa. *

King João was a man of a firm and resolute countenance, of large and well-proportioned frame, and of great strength, as shown by some pieces of his armour still existing, such as his helmet and battle-axe, which latter only a man of great power could have wielded. He was a man of remarkable self-control, and never allowed his features to betray emotion even in the extremes of joy or sorrow. His magnanimity was remarkably shown in the readiness with which he pardoned and restored to his favour those who offended him or who had conspired against his life. In his gifts he was always open-handed, and those who served him well either in peace or in war he rewarded almost always beyond their expectation. He was the founder of a great number of the buildings in Portugal, most remarkable for beauty and magnificence; as for example, the splendid palaces of Cintra, of Lisbon, of Santarem, and of Almeirim; the sumptuous church of our Lady of Batalha, not far from the site of the battle of Aljubarrota; the church of Peralonga of the order of St. Jerome, the first of that order founded in the kingdom, and the monastery of Carnota, of the order of St. Francis, near Alemquer.

He was a man of great piety, and was the first sovereign

* The portraits of King John and Queen Philippa given in this volume are drawn from casts from the statues on their tomb, expressly made for the author by order of his kind and valued friend His Excellency the Marquis de Souza Holstein.

who ordered the Hours of the Blessed Virgin to be translated into the Portuguese language, that all might make use of them in prayer. He also had the Gospels and the life of Christ and other spiritual books translated into the mother tongue. As Grand Master of the order of Aviz, he had the Royal escutcheon placed upon the green cross of the order, as a memorial of the care which as Grand Master he maintained over the kingdom. This is seen in the coins of his reign and those of his successors, until altered by King João II. Being also Knight of the Garter, to which order he was the first foreign sovereign admitted, from devotion to St. George, its patron saint, whose name was at all times his battle cry, he bore for his crest the dragon, the saint's well-known symbol. He was a man intellectually in advance of his age. One of the latest acts of his life, was a requirement that all public ordinances should be dated from the Christian era, instead of from the era of Cæsar, as had until that time been the practice; the alteration involved a difference of thirty-eight years, the era of 1460 corresponding with the year of our Lord 1422.

During the later years of his life the military ardour of his earlier days was allowed to give place to purposes of usefulness, and while he cultivated the chivalry that he loved, in the character and habits of the youthful nobility, he devoted himself to the internal improvement of his kingdom. With so many claims upon their reverence and their love, well might the Portuguese in after years speak of him as the "Father of his country" and "El Rey de boa memoria," "the King of happy memory."

The court of King João adopted for the most part English habits and usages, and the intercommunication between the two countries was much more extensive than it had previously been. The adoption of the French language as it was used at the English court and the devices and mottoes adopted by the King's sons attest this influence. The King himself was an exceedingly accomplished Latin scholar, and wrote in that language with remarkable skill and good taste. Many

KING JOÃO I.,

OF GOOD MEMORY.

FROM THE RECUMBENT STATUE OVER HIS TOMB
AT BATALHA.

passages of the " Leal Conselheiro " of his successor King Duarte show that the princes had conversations with the King their father and other well-instructed persons on various literary subjects, and discussed the rules and in- structions for making good translations of classical works. We find also that King João I., in his address to those who remained behind at Ceuta in 1415, quoted the "Regimento dos Principes " of Fr. Gil de Roma, and reminded them that he had often read it in his chamber. And so in that age of discoveries the reading of the "Wonders of the World" and the " Voyages of Marco Polo," brought from Venice by Dom Pedro, would doubtless give the greatest delight to the dis- tinguished men who were trained in the households of Prince Henry and his illustrious father and brothers. It has been generally believed that the King on his death-bed exhorted Prince Henry to persevere in his laudable purpose of pro- secuting the extension of the Christian faith amongst their hereditary enemies in the as yet unexplored regions of Africa. Such an injunction would fall with redoubled force upon a mind whose views, religious, patriotic and scientific, were already so strongly directed to that object. For a long series of years the Prince had with untiring perseverance continued to send out annually two or three caravels along the West Coast of Africa. Cape Non was passed, but the increasing violence of the waves that broke upon the dan- gerous northern bank of Cape Boyador had till now prevented his sailors from rounding its formidable point. As yet they feared to venture out of sight of land and risk their lives upon the unknown waters of the Sea of Darkness.

One of the earliest acts of King Duarte after ascending the throne was to testify his satisfaction with Prince Henry's efforts in the progress of discovery by making him a donation of the islands of Madeira Porto Santo, and the Desertas, by a charter given from Cintra on the 26th of September, 1433, and in the following year, by a charter dated from Santarem on the 26th of October, he granted the spirituality of these to the Order of Christ, of which the Prince was the Grand Master.

Each time that the Prince sent out a fresh expedition he stimulated his explorers with promises of increased reward, to aim at excelling their predecessors in throwing light on this dark subject. Accordingly, in 1433, the year of his father's death, undismayed by so many years of disappointment, he again sent out a squire of his, Gil Eannes, a native of Lagos, but with the usual bad success, for he reached no further than the Canary Islands, where he took some captives and returned home. In the following year the Prince strongly urged him to make another effort, at any rate to pass Cape Boyador, which if he could do, it would suffice for that voyage.

It is manifest that fanciful alarms suggested by sailors from other countries were superadded to the real dangers of the ocean to deter the Prince's mariners, for in his injunctions to Gil Eannes we find him thus remonstrating with him for giving heed to such fables:

" If," he says, " there were the slightest authority for these stories that they tell, I would not blame you, but you come to me with the statements of four seamen who have been accustomed to the voyage to Flanders, or some other well known route, and beyond that have no knowledge of the needle or the sailing chart. Go out then again and give no heed to their opinions, for, by the grace of God, you cannot fail to derive from your voyage both honour and profit."

The Prince was a man of commanding presence, and his injunctions had great weight with Gil Eannes, who now firmly resolved that he would not appear again before his master without bringing a good account of his errand. Accordingly, disregarding all danger, he put well out to sea, and succeeded in doubling the Cape. Although the exploit was in truth but a small one in the eyes of those who afterwards had gained greater experience, yet the hardihood of it was thought much of at the time, for if the first who reached that Cape had done as much, he would neither have been praised nor thanked, but the greater the sense of danger that others had attached to it, the greater was the honour that accrued to him who overcame it.

The Prince was as good as his word, and Gil Eannes on his return was handsomely rewarded. He informed the Prince that he had landed, but had found no human beings or signs of habitation, but as he thought he ought to bring back some evidence of his having been on shore, he presented to the Prince some plants that he had gathered, which were such as were called in Portugal St. Mary's Roses.

The Prince in consequence fitted out in the following year, 1435, a larger vessel than he had yet dispatched, called a varinel, or vessel with oars, in which he sent out Affonso Gonsalves Baldaya, his cup-bearer, together with Gil Eannes in his barque, and they passed fifty leagues beyond the Cape. They found no habitations, but only some traces of men and camels. Either in obedience to their orders or from necessity they returned with this report, but did nothing further. They named the place which they had reached Angra dos Ruivos, or Gurnard Bay, on account of the great number of those fish which they caught there.

These traces of men and camels satisfied the Prince either that there was a population at no great distance, or that there were travellers who came to the coast. Accordingly, he again sent out Baldaya in the same varinel, and recommended him to proceed as far as he could, and to do his best to capture one of the people, so as to gather some information respecting the natives. Baldaya passed seventy leagues beyond the point previously reached, making a hundred and twenty from the Cape, and here found what might be the mouth of a large river with many good anchorages, and the entrance of which extended eight leagues along the shore. This was what has ever since been known as the Rio d'Ouro, but it is only an estuary.

Here they cast anchor, and as Affonso Gonsalves had brought with him two horses, given him by the Prince for the purpose, he sent out two young men to reconnoitre and see whether they could discover any signs of villages or travellers. To make this task the easier they wore no armour, but simply took their lances and swords by way of defence, for in the event of their meeting any people in numbers, their best chance of safety would be in their horses' heels. The lads were but about seventeen years of age, but although they had no notion what sort of people or wild beasts they might encounter, they boldly set out and followed the course of the river for seven leagues.

They came at last upon a group of nineteen men, neither wearing armour nor carrying any weapons but azagays. When the lads saw them they rode up to them, but the men, although so many, had not the courage to meet them in the open field, but for safety collected near a heap of stones, and there withstood the onset of the youths. They fought till evening warned the latter to make their retreat and return to the vessel.

It is difficult to imagine what those men must have thought of this sudden appearance of two boys, of complexion and features so different from their own, mounted on horseback, and armed with weapons which they had never seen before, and withal so courageously attacking a great number of men.

The two Portuguese lads wounded several of their antagonists, and one of them was himself injured in the foot. "I afterwards knew one of these boys," says the old chronicler, " when he was a noble gentleman of good renown in arms. His name was Hector Homem, and you will find him in the chronicles of the kingdom well proved in great deeds. The other was named Diego Lopez Dalmeida, a nobleman of good presence, as I have heard from those who know him." They reached the ship towards morning, and took some rest.

At daybreak Affonso Gonsalves took some of his people with him in his boat, and ascended the river accompanied by the boys on horseback. They came to the place where the natives had been on the day before, hoping to fight with them and capture one of them, but after the boys had left them, they had decamped, leaving the greater portion of their poor property behind them. This Affonso Gonsalves took and put on board his boat, as an evidence of what had been done, and, judging that it would be of no use to continue the pursuit, returned to his ship. They named the bay Angra dos Cavallos, or Bay of the Horses. Near the mouth of the river they found an immense number of phocas, amounting, as some reckoned, to five thousand. They killed as many as they could, and loaded the ship with their skins.

Nevertheless Gonsalves was not contented, because he had not taken one of the natives. He therefore proceeded fifty leagues further to see if he could not capture some man or woman or child in order to gratify the Prince's wish. Accordingly he continued his voyage till he came to a headland where was a rock which looked like a galley, for which reason they called that port ever after the port of Gallee. Here they landed and found some nets which they took on board. These nets were a novelty, for they were made of the bark of a tree of such a texture that without any tanning or admixture of flax it could be woven excellently well, and made into nets or any other cordage.

Hence Affonso Gonsalves returned to Portugal, but without having been able to gain any certain knowledge whether those people were Moors or heathen, nor what was their manner of life. This took place in the year 1436.

The result may at first sight appear but insignificant. Such was, however, far from being the case, for it must be borne in mind that now for the first time within the Christian era Cape Boyador, which had hitherto presented an impassable barrier to Europeans into the Sea of Darkness, had at length been rounded. True, claims have been set up for the honour of a prior achievement of that exploit on behalf of Genoese, and Catalans and Frenchmen, but it will be shown in the following chapter that so far as historical evidence has been adduced in support of these claims, not one of them is tenable.

CHAPTER VII.

THE SEA OF DARKNESS.

ALTHOUGH all the parts of the Infinite are finite, they will still remain infinite to a man's fancy until in some sense brought within the grasp of his intelligence, and that which, because unmeasured, is supposed to be boundless, will become endued with the awe which is inseparable from darkness and mystery. It was thus that, in the olden times, before the maritime explorations instituted by Prince Henry had led to the magnificent achievements of Columbus and Da Gama, the vast and mysterious, because as yet unexplored, Atlantic, was known by the designation of "The Sea of Darkness." Even amongst the ancients this idea was so prevalent, that we find a friend of the poet Ovid, Albinovanus, himself also a poet, putting into the mouth of Germanicus, as he came upon the ocean, the following expression of dismay,—

> " Quo ferimur? ruit ipsa dies, orbemque relictum
> Ultima perpetuis claudit natura tenebris."

The Arabs adopted the idea of the ancients, and hence we find one of their authors, Ibn Khaldún, who wrote at the close of the 14th century, immediately before the period of Prince Henry's expeditions, describing the Atlantic as "a vast and boundless ocean, on which ships dare not venture out of sight of land, for even if the sailors knew the direction of the winds, they would not know whither those winds would carry them, and, as there is no inhabited country beyond, they would run great risk of being lost in mist and vapour.

The limit of the West is the Atlantic Ocean." Such was the state of man's knowledge respecting that trackless wilderness of waters only five centuries ago.

Nevertheless we have traditions of voyages into the Atlantic earlier than Prince Henry by three thousand years, and of importance, no doubt, for the geographical history of the ancient world, but otherwise practically useless. The value of any exploration must be looked for, not only in the traces it has left behind it in the history of human knowledge, but in its influence on human action. Had any such influence for the general welfare of mankind resulted from the explorations which preceded Prince Henry's time, the Atlantic would not have been called the "Sea of Darkness."

The oldest story respecting this mysterious sea is related by Theopompus, who lived in the fourth century before the Christian era. In a fragment of his works preserved by Ælian is a conversation between Silenus and Midas, King of Phrygia, in which the former says that Europe, Asia, and Africa were surrounded by the sea, but that beyond this known world was an island of immense extent, containing huge animals, and men of twice our stature, and long-lived in proportion. There were in it many great cities whose inhabitants had laws and customs entirely different from ours. Fabulous as the story is as a whole, we cannot escape from the thought that it suggests, though vaguely, a notion of the real existence of a great Western Country. This idea is strengthened by the remarkable story related to Solon by a priest of Sais from the sacred inscriptions in the temples, and presented to us by Plato in his Timæus and Critias, wherein he speaks of an island called Atlantis, opposite the Pillars of Hercules, larger than Africa and Asia united, but which in one day and night was swallowed up by an earthquake and disappeared beneath the waters. The result was that no one had since been able to navigate or explore that sea on account of the slime which the submerged island had produced.

Many as have been the doubts and conjectures to which

this narrative has been subjected by the learned in ancient and modern times, it is a remarkable fact that Crantor, in a commentary on Plato quoted by Proclus, declares that he found this same account retained by the priests of Sais three hundred years after the period of Solon, and that he was shown the inscriptions in which it was embodied. It is also deserving of notice that precisely in that part of the ocean described in the legend we find the island groups of the Azores, Madeira, the Canaries, and a host of other rocks and sand-banks, while the great bank of varec or floating sea-weed occupying the middle portion of the basin of the North Atlantic and covering, according to Humboldt, an area about six times as large as Germany, has been reasonably regarded as explanatory of the obstacle to navigation to which the tradition refers.

It is to the jealous secrecy of the Phœnicians, who were the first, so far as we know, to brave the perils of the Atlantic, that we owe much of the darkness with which their explorations are surrounded. When Homer first sang of those " blissful plains of utmost earth " * to which he gave the name of Elysium, it was probably from Phœnician enterprise that he derived his inspiration. Civilisation in her westward course had already passed through the portals of the great Inland Sea and seated herself on the confines of the Atlantic Ocean. On the shores of Andalusia, at a point so convenient for trade that it has ever since remained the principal port of Spain, the men of Tyre established a colony whose Phœnician name of Gadir has survived three thousand years in the modern name of Cadiz. The delicious climate, the luxuriant fertility of the soil, the rich variety of products and abundance of mineral wealth, so great that even ordinary utensils were made of the precious metals, were sources ample for that " joy " or " exultation " (Alizuth) which has been supposed to have engendered " Elysium " in the fancy of the poet. Yet still it lived but as a poet's dream, for the Phœnician was jealous of his geographical knowledge,

* Odyssey, Book 4, verse 765.

and the delights which gladdened the fields of Elysium were to the Greek as mythical as the Elysian fields themselves.

Centuries had to elapse before the eye of a Greek should rest upon the waters of the Atlantic, and then not under the guidance of Phœnician mariners, nor by the light of Phœnician experience. It was in the middle of the seventh century before Christ that a trader of the island of Samos, named Colœus, availed himself of the privilege of trading with Egypt, then first granted to the Greeks by Psammiticus, the Phœnicians having been as yet the only foreigners permitted to land upon the Egyptian shore. On his way to Egypt Colœus encountered a gale of wind from the coast which lasted long enough to carry him through the Straits into the Atlantic, where he lighted upon the Phœnician colony of Gadir. In this rich and unexpectedly discovered emporium he made purchases of goods which had never before been imported directly into his own country, and by securing the profits which had hitherto been divided between the Greeks and Phœnicians, realised an extraordinary fortune. But here the results of his accidental discovery terminated, for the Greeks took no pains to continue the trade which thus favourably invited their attention.

In the reign of Pharaoh Necho, the son of Psammiticus—supposed to have lasted from 617 to 601 B.C.—a voyage of quite another character is recorded to have taken place. This king, like his father, devoted himself to the development of commerce, and being disappointed in an attempt to unite the Mediterranean with the Red Sea by a canal, established ports and built a fleet of ships on each of them. Conceiving the probability of Africa being surrounded by water, he projected an exploration for the purpose of ascertaining the truth. The aversion of his own subjects to the sea made him engage Phœnician sailors, who starting from the Red Sea made the first authenticated circumnavigation of Africa, and reached Egypt by the Mediterranean in the third year from their departure.

During their voyage it had been the practice of these

Phœnician sailors every year, as seed-time came, to land at
whatever part of Africa they might happen to be near, sow
a crop, wait for the harvest, and then again set sail. It was
reported by them as a matter of astonishment that during a
considerable part of their voyage they had the sun on their
right hand. This important fact, the most confirmatory of
the reality of the expedition, was even discredited by Herodotus to whom we are indebted for the narrative, and the
voyage itself was so unproductive of impressions on the
minds of men that no trace of it could be found in the
Alexandrian library either by Eratosthenes in the third, or
by Marinus of Tyre in the second, century before Christ,
although both of them were diligent examiners of ancient
records.

Meanwhile it would seem that even before the foundation
of Carthage in the ninth century B.C. the Phœnicians had
colonies on the West Coasts of Africa. Eratosthenes indeed
speaks of them as being exceedingly numerous, but Arte-
midorus, who lived about a hundred years before Christ,
contradicted the assertion and declared that not a vestige
of them was apparent. In any case it is scarcely probable
that such colonies existed in great numbers till after the
famous expedition of Hanno the Carthaginian.

The date of this gigantic undertaking has been the subject
of much investigation and discussion. The latest writer
on the subject, the learned and laborious geographer, M.
Vivien de St. Martin, adopts the date of 570 B.C., which
after deep research had been accepted by Bougainville. In
his elaborate work entitled "Le Nord de l'Afrique dans
l'Antiquité," which in 1860 won a prize offered in 1858 by
the "Académie des Inscriptions et Belles Lettres," I
observe that M. Vivien de St. Martin accepts in the main,
but, as I venture to think, with some improvements, the
conclusions of the learned Carl Müller. I shall here quote
these amended conclusions for the reader's enlightenment in
following the course of the ancient navigator.

The narrative is as follows :—Hanno having received

orders from the Senate to found Liby-Phœnician cities beyond the Pillars of Hercules, set sail with a fleet of sixty ships of fifty oars each, carrying thirty thousand people, men and women, with every necessary. The first city that he founded was distant two days' sail from the Pillars, and was named "Thymiaterion," near Salee, at the mouth of the Bouragray. Proceeding westward they raised an altar to Neptune on the promontory of Solœis, the present Cape Cantin, which they found covered with a thick wood. Half a day's sail towards the east brought them to a marshy coast full of large reeds, probably near Saffi, where a multitude of elephants and other wild beasts were feeding. A day's sailing carried them past this, and between the termination of this marshy country and the mouth of a river named the Lixus, *i.e.*, the Sous, they founded the following cities on the sea-coast:—Caricum Teichos or the Carian Wall (Mogadore?), Gytta (Kouleikat?), Acra, Melitta (Wad Beni Tamer?), and Arambys (Aghader?).

On the banks of the Lixus they sojourned some time, and made a treaty of peace with its natives, who were a pastoral people. Beyond these lived Ethiopian barbarians, whose country was full of wild beasts and intersected with high mountains, in which the Lixus had its rise. They took interpreters with them from among the Lixites, and coasted along the desert southwards for two days. They then sailed eastwards one day and found a small island of five or more, (probably fifteen) stadia, or about two miles in circumference, which they named Cerné (Herne, within the estuary misnamed Rio d'Ouro). Here they established a colony. At Cerné they made a reckoning of their voyage, and found that the distance from Carthage to the Pillars had taken the same time as that from the Pillars to Cerné.

After leaving Cerné they ascended the mouth of a great river named "Chretes," or rather "Chremetes," and reached a lake in which were three islands, larger than Cerné. A day's sailing brought them to the extremity of the lake, which was skirted with lofty mountains, inhabited by

savages clothed in skins, who attacked them with stones and
prevented their landing.   This river was the northern branch
of the Senegal, and the lake Panie-Foul or Lake Nguier,
which corresponds correctly with the description.   After
this, they came to another river, wide and large, full of
crocodiles and hippopotami.   This was the large branch of
the Senegal.   Here they put back and returned to Cerné. *

Recommencing their voyage southward they sailed twelve
days along the coast and found it entirely peopled by Ethio-
pians who fled at their approach.   Their language was not
understood by the Lixite interpreters.   On the twelfth day
they came near some lofty mountains thickly wooded with
sweet-scented trees of different kinds.   It took them two
days to sail round these mountains, when they found the
coast line present an immense opening, on the opposite side
of which was plain country.   At night they saw fires rising
at intervals in every direction, sometimes more, sometimes
less.   These wooded mountains clearly represent Cape Verde,
which in the time of Prince Henry received the designation
from this very peculiarity.   The inability of the interpreters
to understand the language of the natives accords with the
fact that at the Senegal commences the country of the
blacks.   The immense opening was the estuary of the
Gambia.

Five days' sail along the coast southward brought them
to a large gulf called by the interpreters the "Western
Horn."   In this gulf was a large island, and in this island
a lake of salt water, which itself contained another island.
Here they landed, and during the day saw nothing but
forests ; but, as night came on, a great number of fires were
lighted amidst frightful cries and the clang of a variety of instruments.   They were greatly terrified, and the soothsayers

---

* This is perhaps the most irreconcilable point in the whole of M. Vivien
de St. Martin's able analysis.   From Herne to the Senegal is some eight degrees,
and it is difficult to suppose that Hanno should have retraced his course for so
great a distance without any assignable motive.   It would involve sixteen
degrees traversed uselessly, a serious awkwardness in the otherwise commendable
character of this analysis.

charged them forthwith to leave the island. This description
tallies with the real character of the coast. The Western Horn
corresponds with the great gulf into which the River Jeba
debouches a little to the north of the Rio Grande. The
south side of the gulf is as it were formed by the chain of
the Bissagos Islands, the last of which, the Island of Harang,
presents the exact configuration described by Hanno.

Departing in haste they sailed along a country abounding
in fragrant exhalations, but with streams of fire running
down into the sea, so that it was inaccessible on account of
the heat. In great alarm they hastened onwards, and four
days' fast sailing brought them at night to a country which
seemed full of fires, in the midst of which arose one much
larger and higher than the rest, which seemed to touch the
sky. When day came they found that it was a very high
mountain which they named "Theón Ochema," the Chariot
of the Gods.

In the whole range of coast from the time that the mariner loses sight of Mount Atlas he will see nothing, with the
exception of the headland of Cape Verde, that could be in
any way dignified by the name of mountain. Near Cape
Verga, in 10½ latitude, the country begins to show some
elevation, and so continues till the neighbourhood of the Isles
de Los, in 9½ degrees, where, near a broad creek which
receives the waters of the Sangaria, rises a conical shaped
mountain conspicuously distinguished by its height and
form from the rest of the chain. To this the Portuguese
subsequently gave the name of Sagres, in honour of the
headland of the same name in Algarve, where Prince Henry
had taken up his abode. It was known by the name of
Souzou. This, both from its physical character and position,
and from a calculation of the distances, is presumed to be
identical with the Chariot of the Gods. It may require
some ingenuity perhaps to explain the occurrence of fires
which Pliny and Pomponius Mela have since described as
perpetual. The very fact that these fires were seen at night
and not by day disproves the assertion and makes it reason-

able that what Hanno saw was purely incidental. A story
told by Bruce of the Shangalla of Abyssinia has a suggestive value which commends it for quotation: " As soon "
he says, " as the rain subsides, the high grass which it has
brought into existence becomes suddenly dry, brown, and
parched; and being inconvenient to the Shangalla, they set
fire to it.   Flame rapidly extends over the country and fire
actually flows down ravines and gullies in which, but a few
weeks before, another element was seen rushing on its
course."

Three days' sail beyond this mountain brought them to a
gulf named the Southern Horn, at the bottom of which was an
island like that before described, containing a lake, in which
was another island peopled with savages.   The females,
more numerous than the males, had hairy bodies, and the
interpreters called them " Gorillas."   They were not able
to seize any of the males, for they fled across the precipices,
and defended themselves with stones ; but they took three
females who broke their bands, and bit and tore their captors with fury ; they therefore killed them.   Hanno brought
back two of their skins and deposited them in the temple of
Juno at Carthage.   Beyond this they could not venture on
account of their provisions beginning to run short.

There can be little doubt that the " savages " here described are the chimpanzees.   As to the position of the gulf
and island the distance traversed shows it to be a gulf
resembling a large estuary formed on one side by the continent and on the other by Sherborough Island, in which the
peculiarities of the description may be easily recognised.

In the examination of a periplus, the details of which have
for centuries been canvassed by the learned with ever varying
results, much has been gained when the positions of two or
three salient points of the coast at important intervals have
been fixed with some degree of certainty.   This happy result seems here to have been attained.   The careful measurement of the distance from the Island of Herne to the Straits,
and its approximate coincidence with the distance from the

Straits to Carthage, is a strong point made in deciding the position of Cerné in accordance with the narrative. Another most remarkable fact is that the River Chretes is, as shown by Bochart, spoken of by several ancient writers under the name of Chremetes, and that the Chremetes is described by Aristotle as "one of the most remarkable rivers of Africa, having its source in the same mountain as the Nile, whence it flows to empty itself into the Outward Sea." We have here distinctly indicated that ancient notion of the common origin of the Nile of Egypt and the Nile of the Blacks which was maintained by geographers down to the time of Prince Henry, and which assigned the latter title to the River Senegal. When in addition to this some minute items of local description are found to correspond with the real geographical formation of the mouth of that river, another point of certainty seems to be authoritatively established. Starting from such conclusions the recognition of Cape Verde as identical with the large mountain covered with trees *round* which they sailed seems unavoidable. When therefore M. Vivien de St. Martin shows as the result of his analysis that Hanno sailed further to the South in a few months than the Portuguese did in a great many years, the claim will probably be conceded by many, and the shade of the Carthaginian chief be allowed to enjoy in its plenitude the glory which, if M. St. Martin's deductions be correct, would so justly attach to his name. But when the distinguished geographer accepts upon trust, and adds the weight of his authority to, the assertion that * " a very long time before the Portuguese discovered the Rio d'Ouro and its island, that locality had been frequented by Catalan navigators," we are bound in the most emphatic manner to take exception to the statement, and, for reasons which will hereafter be adduced at length, to declare that that statement is utterly without foundation.

This remarkable voyage of Hanno, which for centuries formed the principal source of Greek and Roman informa-

* See page 383.

tion on the African coasts of the Atlantic, was described in an inscription in the Punic language in one of the temples of Carthage. Long afterwards, probably about the middle of the fourth century B.C., it was translated into Greek by some one whose name is not known, and thus this very precious document has survived to our times.

In about the year 470 B.C., another expedition was attempted under the following circumstances. Sataspes, a nephew of Darius, had been sentenced by Xerxes to be impaled for violating the maiden daughter of Zopyrus, the devoted friend of Darius, whose fidelity had secured Babylon to his master after a siege of twenty months. The mother of Sataspes, a sister of Darius, besought from the King a commutation of the sentence, engaging that her son should, if spared, circumnavigate Africa and return by the Red Sea. Her request was granted, and in an Egyptian vessel manned by Carthaginians Sataspes sailed through the Straits, doubled Cape Solœis (Cape Cantin), and after many months' voyage southward became disheartened and returned. On presenting himself before Xerxes he related that at the farthest point of his voyage he had lighted on a shore inhabited by a people of diminutive stature, clothed with garments made of the palm-tree. He assigned as a reason for his turning back that his ship was stopped [query by weed], and that it was impossible to go any further. Xerxes, believing that he lied, had him impaled in accordance with the original sentence, because he had not completed the task imposed. Now as this voyage, which doubtless was subsequent to and suggested by that of Hanno, preceded the journey of Herodotus into Egypt in the year 448 B.C., in which he derived through Carthaginians that information respecting the mode of commerce on the West Coast of Africa which they first had gathered from Hanno's voyage, it might have been expected that greater impressions would have been left behind as to the desirableness of continuing explorations along that coast.

Another presumed exploration, or rather circumnavigation of Africa, is that attributed to the geographer Eudoxus of Cyzicus in Mysia, who lived towards the end of the second century B.C. Having left his native place for Egypt he entered the service of Ptolemy Evergetes the Second and his wife Cleopatra, by whom he was employed in making voyages to India. We have two contradictory accounts of his voyages. One, taken from the writings of Cornelius Nepos as related by Pomponius Mela, supposes that Eudoxus, starting from the Arabian Gulf, arrived at Cadiz after circumnavigating Africa, but the description given of the natives beyond the desert is so full of extravagant fables that it is utterly unworthy of any consideration. The other account of the adventures of Eudoxus is by Posidonius as preserved by Strabo, which relates only a series of unsuccessful attempts, whence we may come to the conclusion that Eudoxus did not circumnavigate Africa, and that his voyages taught nothing that was not known before his time.

So barren of influence on Atlantic exploration in after times was the expedition of Hanno, that since it there has been but one admissible intimation transmitted to us of a passage by sea to the southward of Cape Boyador before the fifteenth century, and that was made by Africans of the west coast, on their own coast line and by the mere chance action of the winds, and was as fruitless in impression on the minds either of explorers or of Arab geographers as any that had preceded it. It was first brought to light in 1848, since when it has been triumphantly adduced in derogation of the glory of Prince Henry. In that year the learned French orientalist M. Reinaud published with a French translation the Geography of the Arab Abú Al-Eidá, embodying therein the "Geography" of Ibn Said, of the middle of the thirteenth century. In the latter is recorded how a Moor named Ibn Fátimah being once at Noul-Lamtha (Wad-Nun, a little north of Cape Non, see Hartmann's Edrisi), took ship and was wrecked in the midst of some shoals. The sailors lost their bearings, and had no notion where they were. They therefore deserted the ship, and put out in a sloop to reconnoitre. Sometimes the sloop got entangled amongst marine plants, but was raised off them by dint of rowing. When they reached the middle of the gulf the sailors were astonished at the great quantity of tunny fish they saw; they also observed some white birds. Before they reached the shore their provisions were entirely exhausted. Just as they came under the Glittering Mountain (Aldjebel-allamas), Cape Blanco, so named from its being of a glittering white, "some Berbers of the tribe of Godála made signs to them not to approach the mountain. The sailors did not comprehend the intention of this warning, but nevertheless they turned northwards and managed to pass the Cape. A man then came forward who knew both the Arabic and Berber languages, and asked them how they had missed their way. The sailors related what had occurred, and asked the reason of their having been warned off the mountain. The man replied 'the whole of that mountain is one mass of deadly serpents. Strangers take it for a rock of glittering colour, and deceived thereby, come near and are devoured by the serpents.' The man took pains to reassure the sailors, and some of the latter bought some camels and rode to Tegazza, the capital of the tribe of Godála, described as in 11° of longitude and twenty of latitude. They remained some time with the Berbers of that tribe, drinking camel's milk and eating dried camel's flesh. They then returned to Noul, accompanied by some of that tribe." The correctness of the description as to the seaweed and the tunny fish leaves little room to doubt the soundness of the learned editor's conclusion that the cape to which these Moorish sailors were driven was really Cape Blanco. But such an occurrence can scarcely be accepted as in any way diminishing the honour earned by the predetermined and persistent explorations instituted by Prince Henry. If it were possible at this late period to learn that, some centuries before the discovery of the West Indies by Columbus, a native of Haiti had been carried by accident

to the shores of Cuba, would the glory of the immortal
Genoese be diminished one iota by such a discovery? Surely
not, not even had the Haitian discovered that Cuba was an
island, a fact of which Columbus was ignorant to his dying
day, whereas it is not pretended that Ibn Fātĭmah made
any discovery that was not effected by the mariners of
Prince Henry. The cosmographers and cartographers who
followed Ibn Said derived not the least addition to their
store of funded information from the romantic adventure of
Ibn Fātĭmah. Whatever may have been effected in ancient
times, this is the only instance within our knowledge in
which it can be said with certainty that Cape Boyador was,
in the middle ages, passed before the time of Prince Henry,
although that honour has been claimed on behalf of Genoese
and Catalans and Frenchmen, and finally for the Norman
Jean de Bethencourt. The high reputation of the distinguished French geographer, M. d'Avezac, who has most
prominently advanced these claims, demands the most
attentive consideration to the arguments he adduces; but
after a careful investigation I feel bound in conscience, but
with the sincerest respect for him, to give in every instance
the Scotch verdict of " non proven." But the reader must
judge for himself.

The earliest claim is set up for a Genoese expedition in
1291.

For the fullest account of this expedition we are indebted
to the learned labours of Dr. Pertz,* Principal Librarian of
the Royal Library of Berlin, and editor of the Monumenta
Germaniæ Historica. It was discovered by him among the
public annals of the city of Genoa, which form a continuation of the Chronicles of Caffaro, printed, but only in
extract, by Muratori. It is as follows: " In the year 1291
Tedisio Doria and Ugolino de Vivaldo, with his brother
and certain other citizens of Genoa, commenced a voyage

* In a memoir printed in Berlin entitled " Der Ælteste Versuch zur Entdec-
kung des Seeweges nach Ostindien," presented to the Royal Academy of Sciences
of Munich on March 28th, 1859, the centenary of its foundation.

which none has ever in any way attempted till now. They
equipped in the best manner two galleys supplied with provisions, water, and other necessaries, and despatched them
in the month of May towards the strait of Ceuta, that they
might go by sea to the ports of India, and bring back useful
articles of merchandise. The two said brothers Vivaldo
went in person, as also two friars minor. It was an undertaking that astonished not only those who witnessed it,
but those who heard of it. Since they passed a place called
Gozora no certain news has been received of them, but may
God preserve them and bring them back safe and sound to
their own homes." This account was in the handwriting of
Jacopo Doria, a near relative of Tedisio Doria, one of the
originators of, though not a participator in, the expedition.
In it we have an undeniable statement not only of the
reality of this noble undertaking, but of the purpose for
which it was set on foot, and of the farthest point from
which news had at that time been received of it. The story
is confirmed by the great astrologer and physician Pietro
d'Abano, who wrote that portion of his important work
"Conciliator Differentiarum" which contains the reference to
this event about the year 1312. This narrative is eminently
interesting as showing the geographical notions which led
to the expedition, and the courage required for so dangerous
an exploit. In a chapter treating of the possibility of living
within the tropics, he says, "According to Ptolemy persons
have reached us from the equinoctial regions, for the same
man is capable of enduring, at different times, opposite
extremes of temperature, and it is said that the Indian city
Arin* lies in those regions. Others assert that all passage

---

* The sacred city of Odjein or Ougein, in Malwa, whence the Indians reckoned
their first meridian. The change of the name to Arin in Arabic is thus explained
by M. Reinaud in his Mémoire sur l'Inde, p. 373. The dj of the Indians was
sometimes rendered z by the Arabs, and thus the Arab translators wrote the
word Ozein; but as in manuscripts the vowels were often omitted, the mass of
readers to whom the name of Odjein was indifferent would pronounce it Azin,
and as the copyist would sometimes forget to insert the point which distinguished
a z from an r, Azin would be read Arin.

THE SEA OF DARKNESS.

between here and there is prevented by certain mountains which attract men to them as the loadstone attracts iron, and that men laugh while being attracted, and at last are held fast. It is also reported that Cæsar sent two centurions to seek the head of the Nile, who related that by the help of the King of Ethiopia and his recommendations to neighbouring kings, they reached some immense lakes whose outlet the inhabitants themselves did not know, nor could any one separate the grass, which was so entangled in the water that neither on foot or in a boat was there any contending against it. They further stated that they saw two stones from which the vast body of the river fell forth, but whether that was the head of the Nile or only an affluent, or whether it then first springs out on the land or only returns diverted from a previous course is unknown. Some assert that the desert is so vast and sandy, so full of serpents and venomous animals, and so deficient in fresh water, that no one can easily make the passage from thence. *Wherefore* some time since the Genoese fitted out two galleys provided with every necessary, and passed through the Pillars of Hercules at the end of Spain, but *what became of them remains now nearly thirty years unknown.* The passage, however, is now open by going northward through Great Tartary, and so winding round to the east and then to the south." The purpose of this voyage is thus set forth, vaguely it is true, but in immediate connection with the indefinite geographical knowledge of the period.

It would be difficult to speak in terms too laudatory of this noble undertaking, the result of private enterprise. It wanted but success, or, in case of failure, to be followed up with the invincible perseverance of a Prince Henry, to have neutralised by anticipation the glory of Prince Henry himself. The narrative of Jacopo Doria mentions Gozora as the last place from which tidings had been received respecting it. The map of the Venetian brothers Pizzigani, of 1367 shows us the position of this place under the name of Caput Finis Gozole, which is very clearly Cape Non.

Another cotemporaneous writer is stated by Giustiniani *
to have written of this expedition; viz., Francesco Stabili,
better known as Cecco d'Ascoli, in his Commentary on the
treatise De Sphærâ Mundi of Sacrobosco.† Repeated ex-
aminations of the various editions of that commentary have
however failed in verifying the reference. It was not till a
hundred and sixty-four years later that any allusion was
made to the expedition. A Genoese gentleman of noble
family named Antonio de Nolli, of whom we shall hereafter hear more, being overwhelmed with debt and in
desperate circumstances, had sought to mend his fortune
in explorations on the west coast of Africa, under the
auspices of Prince Henry. In 1802 a letter apparently
by his hand addressed to his creditors under date of the
12th of December, 1455, and signed Antonius Ususmaris
(the Latinized form of his adopted pseudonym "Uso di
Mare"‡), was discovered in Genoa by M. Gräberg de
Hemsö, a learned Swedish merchant resident in that city,
among a collection of papers which had been presented to
the archives of Genoa by M. Federico Federici in 1660.
In the immediate neighbourhood of the letter were some
geographical legends in an unknown hand, apparently
prepared for inscription on a globe or mappe-monde, as
was not unusual in those times. One of the legends in
all its rude Latinity is as below,§ and translated is as

* See Giustiniani, Castigatissimi Annali di Genova. Genova, 1537 (Lib. 3.
fol. iii. verso).

† The Latinized form of the name of John Holywood, who flourished at the
beginning of the thirteenth century, and whose learned treatise De Sphærâ
Mundi, the most famous book of the middle ages, was for centuries the subject
of a host of commentaries. He was named from Holywood, in Yorkshire, the
place of his birth.

‡ I observe that in the old chronicle of Jacopo Doria, a Genoese sea-captain
named Antonius Ususmaris is mentioned under date of 1284. It is not improbable that he may have been known in Genoa to have accompanied this famous
expedition in 1291, and that hence Antonio de Nolli adopted this significant
name of his compatriot in undertaking a similar exploration.

§ "Anno 1281 recesserunt de civitate Januæ duæ galeæ patronisatæ per D.
Vadinum et Guidum de Vivaldis fratres volentes ire in levante ad partes Indiarum,
quæ duæ galeæ multum navigaverunt. Sed quando fuerunt dictæ duæ galeæ

follows: "In the year 1281 two galleys left the city of Genoa, under the command of the brothers Vadinus and Guido Vivaldi, with the view of going to the East to the parts of the Indies; which two galleys sailed a great way, but when they came to this sea of Ghinoia one of them stranded, and could not proceed further, but the other sailed on and passed through that sea until they reached a city of Ethiopia named Menam. They were captured and detained by the people of that city, who are Christians of Ethiopia, subjects of Prester John. That city is on the sea-coast near the river Gihon. The aforesaid were kept in such close confinement that not one of them ever returned from those parts." As quoted by M. d'Avezac the date is altered from 1281 to 1285, and the following sentence is added, "The aforesaid was related by a Genoese nobleman named Antoniotus Ususmaris." Here we have a statement of the locality to which the expedition attained, while it should be mentioned that in the letter of Antonio de Nolli, alias Uso di Mare, we are actually brought into contact with an individual descended from the explorers. I propose to show that both the letter and the legend, which have been adduced to

in hoc mari de Ghinoia una earum se reperit in fundo sicco per modum quod non poterat ire nec ante navigare, alia vero navigavit et transivit per istud mare usque dum venirent ad civitatem unam Ethiopiæ nomine Menam, capti fuerunt et detempti ab illis de dicta civitate, qui sunt cristiani de Ethiopia submissis presbitero Joanni ut supra. Civitas ipsa est ad marinam prope flumen Gion prædicti fuerunt taliter detempti, quod nemo illorum a partibus illis umquam redivit, qui prædicta narraverat." (Annali di Geografia e di Statistica, tom. 2. pp. 290, 291.) But quoted by M. d'Avezac (Nouvelles Annales des Voyages, tom. 108, p. 47) in the following altered and enlarged form: "Anno Domini M.C.C.LXXXV recesserunt de civitate Januæ duæ galleæ patronisatæ per D. Ugolinum et Guidum de Vivaldis fratres volentes ire in Levantem, ad partes Indiarum. Quæ galleæ multum navigaverunt; sed quando fuerunt dictæ duæ galleæ in hoc mari de Ghinoiâ, una earum se reperit in fundo sicco per modum quod non poterat ire nec ante navigare, alia vero navigavit et .ansivit per istud mare usquedum venirent ad civitatem unam Ethiopiæ, nomine Menam; capti fuerunt et detenti ab illis de dicta civitate, qui sunt christiani de Ethiopiâ submissi presbytero Joanni, ut suprà. Civitas ipsa est ad marinam, prope flumen Gion. Prædicti fuerunt taliter detenti quòd nemo illorum a partibus illis unquám redivit. Quæ predicta narraverat Antoniotus Ususmaris, nobilis Januensis. (Annali di Geografica e di Statistica, tom. i. pp. 290, 291)."

prove that Prince Henry's navigators were anticipated by the
Genoese in actual African exploration, bear internal evidence
of being a farrago of nonsense and untruth. The legend is
anonymous, and consequently without intrinsic authority, but
it quotes as its authority "Antoniotus Ususmaris nobilis
Januensis," who is also the writer of the letter. The
original of the letter is given at foot,* and, as the reader

* (The Letter)—1455, die 12 Decembris. Honorandi Fratres, quantum sciatis
de me male scripti, bene illud judicare possum, qui non sufficit vestrum vobis tenere,
sed de vestris male vos visitare contingit, vere non possendo vobis scribere rem
de ullo bono, et habendo in veritate animum ad vos esse, et me ponere in manibus
vestrorum, et aliorum creditorum, voluit ista mea fortuna me transmisisse in
una caravella ad partes Ginnoie et essendo in ista verecundia, qua jam dis-
posui citius mortem sumere, quam vivere; et transivi ubi umquam aliquis
cristianus fuerat ultra miliaria octingenta, et reperto rivo de Gamba, maximo
in extremitate in eo intravi sciens quod in ipsa regione aurum et meregeta
colligitur. Illi piscatores me insultaverunt cum archibus scive sagittis avene-
natis putantes esseremus inimici, et videndo nos recipere noluerunt, fui coactus
redire, et inde prope legas septuaginta quidam nobilis Dominus niger dedit mihi
capita tringinta una et certos dentes elefantorum, papagajos cum certo pauco
zebetto, pro certa rauba sibi presentata, et intellecta voluntate mea mecum
misit ad S. Regem Portugalliae secretarium suum cum certis clavibus, qui quidam
secretarius se obligat pacem tractare cum illo Rege de Gamba. Et sic viso S.
Rex istius Secretarii fuit contentus vadam simul tantum ad illas partes. Ideo
in Dei nomine compello aduch unam caravellam, in qua vado, et habebo caricum
de illis infantis, et me expediam per totum venturum, et infra dies decem
expediam istum ambasciatorem in una caravella, ut vadat pacem tractandam;
ipse mihi dimittit totum sum, ut ipsum implicare velim cum mea. Quare Domine
me expedit, ad hoc videre ista vice quid facere vult ista mea fortuna, quae nisi
esset mihi tantum adversa vivere sub magna audiendo quid mihi narrat ipse
Secretarius, quae si vobis scriberem, vana vobis viderentur. Verum ex toto
firmae non restabant leghae trecentos ad terram presbyteri Joannis, non dico
persona sua, imo incipit ejus territorium, et si me potuissem detinere vidissem
capitaneum regis mei, qui prope nos erat jornatas sex, cum hominibus
C. et cum eo cristiani de presb. Joannis V. et locutus fui cum illis illius
exerciti'; reperui ibidem anum de natione nostra, ex illis galeis credo Vivaldae,
qui se amiserit sunt anni 170. qui mihi dixit, et sic me affirmat iste secretarius, non restabat ex ipso semine salvo ipso, et alius qui mihi dixit de elefan-
tibus, unicornibus, et aliis stranissimis, et hominibus habentibus caudas, et
commedentes filios, impossibile vobis videretur, credatis quod si navigassem
aduch diem unum amisissem tramontanam. Et causa quia me detinere non potui
fuit quia victualia mihi deficiebant, et de suis victualibus ullo modo homines
blanchi uti non possunt nisi infirmenter, et moriantur, salvo illi nigri, qui in eis
nascuntur. Aer vero optimus et pulcrior terra quae sub caelo sit, et quasi equi-
vocum videlicet in mense Julii dies de horis duodecim et nox de horis undecim.
Recito vobis haec omnia et sum certus diceretur citius veleretis vestrum et aliorum

will see, is a most ungrammatical and incoherent document. It states that Uso di Mare, being in the neighbourhood of the River Gambia, spoke with the last living descendant of those who in the legend were said to have reached
the Ethiopian city named Menam, which was on the sea-
coast, and inhabited by Christian subjects of Prester
John. The allusion to the River Gihon, which the early
Arabian geographers connected with the Nile in Eastern
Africa, would lead us to infer that this Ethiopian city was
on the east coast,—to have reached which would have been
the glory of the expedition of 1291,—in which case we
have to imagine a navigation over four thousand miles of
most perilous African ground, with no professed object, and
not even spoken of as a wonder when accomplished, which
is incredible. The only alternative is that this Christian
city, subject to Prester John, lay on the west coast; but the
letter states that the westernmost inland boundary of Prester
John's country was "three hundred leagues off," which in-
volves an absurdity. Then an incoherent and inconceivable
story is told of there being, at a distance of only six days'
journey from the point of coast where Uso di Mare was, a
captain of Uso di Mare's king, who we must suppose to be
the King of Portugal, having with him an army of a hundred men and five of Prester John's Christians, whom, if the
writer could have stayed, he would have seen. Then although
he was unable to stop for want of victuals, this space of six
days' journey is annihilated in a breath, and Uso di Mare not
only does see, but speaks with, some men of that army, and
*just there he found one of this nation whom he believed to be from
those galleys of Vivaldi which had been lost one hundred and
seventy years before.* The person thus found told him that he

quam ista varia audire, expedit habeatis patientiam sex menses, et eo post quia
faceo me assegurare, quod certe opus non esset, essendo illa maria sicut darcina
nostra de ibi.

Ista littera sit omnibus creditoribus qui credunt, et vos cum eis si habuissem
pro posse eos contentarem de pagis de 60. non posuissem me in tali ventura cum
una caravella, tantum erit forsitan per meliora, Ideo patientiam habeant amore
Dei.

VR. ANTONIUS USUSMARIS.

or she was the last of the stock, and this was confirmed by the black secretary of a neighbouring chieftain, and by another man who told him of men with tails who ate their own children, and other matters which Uso di Mare feared his readers would think impossible. But after he had in this miraculous manner lighted upon the last living remnant of the famous Genoese expedition, is it conceivable that this man, himself a keen explorer, should fail to ascertain and declare distinctly in his letter on what part of the coast of Africa this city of Mena or Menam, the terminus of the expedition, lay? Moreover, neither Antonio de Nolli himself, who was of a noble family, nor his own explorations, were so insignificant that a communication of the kind from him upon a matter so intimately connected with the glory of Genoa should be entirely disregarded by his fellow citizens, if such a communication had been reputed valid. Yet, subsequent historians of the republic, even to one hundred and thirty years later, distinctly declare that no news whatever of the expedition of which they were so justly proud had ever reached their times.* In presence of such facts, can the reader accept this preposterous letter as evidence that Prince Henry was anticipated by the Genoese in rounding Cape Boyador? I think not.

But further, the testimony of maps has been adduced in evidence of the high probability of the Cape of Good Hope having been rounded in the middle ages before the time of Prince Henry; as for example, that of the date of 1306, by the Venetian Marino Sanuto, an earnest advocate of a new crusade for the recovery of the Holy Places,† on which map the South of Africa is surrounded by the sea. The geography of Sanuto's map itself is derived from the early middle age cosmographers, and no greater proof can be given of his real ignorance of the true form of Africa

---

* See Giustiniani, Castigatissimi Annali di Genova, 1537 (Lib. 3. fol. 111 verso); and Foglieta, Historiæ Genuensium Libri xii., 1585 (Lib. 5. fol. 110 verso).

† See his "Liber secretorum fidelium crucis," published by Bongars. Hanau, 1611, being part of the second volume of the "Gesta Dei per Francos."

A F R I C A

FROM THE

LAURENTIAN PORTULANO

1351

than the following note below the Regio 7 montium, "Regio inhabitabilis propter calorem." In short, a single glance at the map is sufficient to show that nothing could well be further from a delineation of Africa based on actual knowledge.

Far more startling than the map of Marino Sanuto is a map in a Portulano of the date of 1351, in the Laurentian Library at Florence, of which the Count Baldelli Boni gave a facsimile to accompany his valuable edition of Marco Polo, published at Florence, 1827, 4°. On this map not only is a southern extremity given to Africa, but the coast-line of Guinea is drawn with so much greater an approximation to correctness than would be expected from the period, that the Count inferred that both these delineations were the result of actual discovery. (See map.) This conclusion however is rendered untenable by a further examination of the map, for while there is not a single name of river, cape, or bay, or any other local indication of actual discovery on the whole line of coast comprising the remarkable indentation supposed to represent the Gulf of Guinea and onwards round the southern extremity of the continent, the map does contain, and for the first time as far as we know, names indicating entirely new discoveries amongst the African islands. These will be spoken of more fully in their appropriate place hereafter. For the present it will be sufficient to speak of such indications as have been referred to on the west coast of the continent itself. We there find no sign of local geographical knowledge beyond two rivers corresponding with the Wad-om-er-Biyeh of modern maps, and the River Palolus, or River of Gold, as delineated on the Pizzigani map sixteen years later. And as if to supply one with the opportunity of rendering this argument the more valid, Count Baldelli Boni has inserted at the corner of the map a facsimile of the fifth map in the Portulano, and on it are entered the names of the places already known on the northern portion of the west coast as far as Cape Boyador, exactly in accordance with the information supplied by the later Pizzigani map of 1367. These realities, combined with the fact that the later Pizzigani map of 1367 and the Catalan map of 1375 make no pretensions to exhibit such astounding discoveries as those of the Gulf of Guinea and the rounding of the Cape, lead to the reasonable conclusion that the Medicean map was not, as Count Baldelli Boni infers, intended to describe any such actual explorations, but merely to propound a geographical theory based upon traditions and inferences for which we shall presently be able to show that authority existed. The only alternative is to suppose that the successful navigator who had rounded the Cape of Good Hope, had recorded his stupendous achievement simply in outline on a carefully executed map, but that his country was so insensible to the importance and honour of such a discovery that it left the name of the explorer to sink into oblivion. This is past belief, and therefore it only remains to show how the remarkable outline in question can be accounted for. There were two opinions entertained by the geographers of antiquity respecting the conformation of the Atlantic Ocean. Hipparchus the Bithynian, who lived a century and a half before our era, maintained that it had no connection with other seas, but formed one great lake. On the other hand Herodotus, Crates of Malles, Posidonius, Cleomedes, Arrian, in the Periplus attributed to him, and Strabo, admitted the possibility of the circumnavigation of Africa. Pomponius Mela, in the first century, maintained the same belief, and Julius Solinus, in the third century, distinctly states in the sixtieth chapter of his " Collectanea rerum Memorabilium," speaking on the authority of King Juba, that " all that sea from India to Gades (Cadiz) was navigable before the north-west wind." *   St. Isidore of Seville, at the beginning of the seventh century, partook the same opinion ; and in the middle ages his native country, Spain, which was greatly influenced by his opinions as well as those of Mela and Solinus, was the focus of the geographical literature of the Arabs. We

* "Omne illud mare ab Indiâ usque ad Gades voluit Juba intelligi navigabile, Cori tantum flatibus."

have a summary of the two opposing creeds in the words of
the Alexandrian philosopher, Joannes Philoponus, who
also lived in the seventh century; in his work "De Mundi
Creatione," liber 4. cap. 5, p. 153, he says : " Some persons
have suspected, following an absurd tradition, that the
Atlantic is united on the south with the Erythrean Sea.
They pretend that several navigators have been carried
by accident from that ocean to the Erythrean Sea,. which
is evidently false, for it would require that the ocean
should extend quite across Libya and even under the torrid
zone.   Now it is impossible for men to navigate them
on account of the burning heat that prevails."   In the map
of Marino Sanuto, already referred to, the southern termination of the African continent is made to turn greatly to
the east, in conformity with an idea suggested by the author
of the Periplus of the Erythrean Sea ; but in the Laurentian
map, of which we now treat, the outline is distinctly different
from any which preceded it or have followed it.

Let us now examine under what influences that outline
may have been laid down, simply as a theoretical exponent
of old ideas confirmed by more recent observations.   During
many centuries the Arabs were exclusive masters of the
commerce of the eastern coast of Africa, and had establishments in all the ports and principal islands as far as Sofala.
Further south than Cape Corrientes, however, they did not
venture, because, as Barros tells us (Dec. 1, liv. 8, cap. 4),
their vessels, being sewn with cocoa fibre and not fastened
with nails, " could not stand the shock of the rough seas off
the Cape of Good Hope,"—and "several losses of vessels had
occurred in the direction of the Western Ocean."   Now the
westward trending of the coast south of Cape Corrientes,
suddenly first and continuously afterwards, would naturally
lead to the conclusion that the termination of the African
continent there commenced.   There is no difficulty in understanding how these notions of the Arabs were communicated to Europeans, for as the former were the purveyors of
the commerce of India and the east coast of Africa by the

Red Sea to Alexandria, they would not fail to be questioned by the merchants of the Mediterranean respecting the countries whence they came. Moreover, Marco Polo, who returned from his eastern journey in 1295, speaks distinctly (book 3, chap. 36) of the prodigious velocity of that south-ward current which led to this belief, and which was so strong off the Cape alluded to as to cause it to be afterwards named by the Portuguese " El Cabo dos Corrientes."   In addition to these facts, the very singularity of the map affords presumptive proof that it was not the result of a distinct exploration of so eminently notable a character.

But Count Baldelli Boni wished also to show the high probability of the Gulf of Guinea having been already dis-covered, and if truth always floated on the surface instead of lying at the bottom of the well, I for one should forth-with subscribe to his opinion. But let us look a little deeper. The map is an extract from the first of eight sheets, and the Count naturally adduced all the evidence which the Portulano afforded of new and original discovery. Accordingly, on the corner of this map he has inserted an extract from another, the fifth in the series, really exhibiting, for the first time within our knowledge, the unmistakable proof of new discoveries in the *African islands*. But, together with these island novelties, this extract from sheet 5 contains the corresponding part of North Africa, and there, as well as on sheet 1, we find laid down the *erroneous* geographical information derived from ancient authorities and repeated on later maps, but southward thereof no local information whatever. Under such circumstances a mere outline, however striking in its form, cannot, whether applying to the Gulf of Guinea or the Cape of Good Hope, be accepted as the result of actual exploration.

Incidental allusion has been already made to the Catalan map of 1375. We have now to speak of it in a more special manner, for on the third sheet of that map is the repre-sentation of a boat-load of explorers off the coast to the south of Cape Boyador, accompanied by the following legend

in Catalan : " Partich luxer dñ Jac. Ferer, per anar al riu de
l'or, al gorn de Sen Lorens qui es à X de Agost, y fo en
l'an MCCCXLVI :" " The ship of Jaime Ferrer started to go
to the River of Gold on St. Lawrence's day, the 10th of
August, 1346." The event here recorded is corroborated by
the following legend, which occurs amongst those already
described on page 102 as having been discovered in the
Genoese archives by M. Gräberg de Hemsö. The original
is at foot.* " On St. Lawrence's day, viz., the 10th of
August, 1346, a galley belonging to the Catalan, John Ferne,
left the city of the Majorcans with the purpose of going to
Rujaura [the River of Gold], but of said galley no news has
since been received. On account of its length that river is
called Vedamel. It is also called Ruiauri, because the gold
of Pajola is collected in it. You must also know that the
majority of the inhabitants of these parts are employed in
collecting gold in this river, which is a league wide, and deep
enough for the largest ship in the world."

" This is the Cape Finisterre of West Africa."

It has been inferred by M. d'Avezac from these two
legends that this voyage must have been preceded by many
others, " because," he argues, " one does not fit out an
armament with a fixed destination without knowing approximately at least the point one has to arrive at."

I now propose to show that the contrary was the case, and
that the expedition was fitted out for the express purpose of
finding the *unknown* mouth of a river in which gold was
collected, and the existence of which had become known to
the mercantile populations in the Mediterranean through
the medium of commercial intercourse with the Arabs. The

* Recessit de civitate majorisarum Galeatia una Joannis Ferne Catalani in
festo Sancti Laurentii, quod est in decima die mensis Augusti, anno Domini
1346, causa eundi ad Rujaura, et de ipsa Galeatia nunquam postea aliquid
novum habuerunt. Istud flumen de longitudine vocatur Vedamel et similiter
vocatur Ruiauri, quia in eo recolligitur aurum de pajola. Et scire debeatis quod
major pars gentium in partibus istis habitantium sunt electi ad colligendum
aurum in ipso flumine, qui habet latitudinem unius leguo et fondum pro majori
nave mundi.
Istud est caput finis Terrarum Affricæ occidentalis, etc.

fact of the voyage having been recorded not only in the
archives of Genoa, but also on the face of a remarkably
handsome map prepared with extreme carefulness and
labour, is a proof that the expedition was one of unusual
importance and anxiety, such as the purpose I have
suggested would involve. Had it been merely an unsuccessful venture to a point already known even approximately, we should not expect to find the expedition recorded
on the face of a map at all, but we should reasonably hope to
find that point laid down with an approximation at least to
accuracy on charts of the period, and especially on the one
on which this individual expedition was recorded. Fortunately, there are maps existing on which the river indicated by the legends is laid down, and by their help, and
in conjunction with the wording of the legends themselves,
we have an opportunity of testing how far the geographical
information they convey is, either approximately or at all,
in accordance with the knowledge which would be derived
from even one antecedent maritime exploration.

The two legends manifestly refer to the same event: they
both record an expedition which started on the same day *for
the purpose of going* (in the Catalan *per anar*, and in the
Latin *causâ eundi*) to the same river. This river, the Ruiauri
or River of Gold, was so called because gold of Pajola was
collected in it, and from its length it was called *Vedamel*.
Now in the Venetian map of the brothers Pizzigani, made in
1367, twenty-one years after the expedition of Jaime Ferrer,
we find laid down *in a latitude a little south of the Canaries*
the river Palolus, rising in a large lake, on which is the following legend in Latin, " This lake proceeds from the Mountain
of the Moon, and passes through sandy deserts." In the
middle of its course the river bifurcates, and again joins,
forming an island, on which in Latin is the inscription,
" The island Palola; here gold is gathered." Into the
opposite or eastern extremity of the lake flows the Nile, the
eastern branch of which takes its northward course towards
the Mediterranean, in its well-known position. We thus

find a river exactly corresponding with the description of
that for which Jaime Ferrer started on St. Lawrence's day,
in the year 1346. Three of the four specialities indicated in
the Genoese document are here substantiated by Venetians
who, like the Genoese, had commercial relations with the
Arabs; and that on a map bearing no reference whatever
to the voyage of Jaime Ferrer. We have a river on which
gold is collected, and it is the gold of Palola or Paiola, and
we also have an explanation of the expression that from *its*
length it is called Vedamel. That length may be judged
when it is made to extend from the Nile, delineated in its
true position as falling into the Mediterranean, to another
outlet into the Atlantic a little south of the Canaries. The
fourth speciality of the river as given in the Genoese
document is the name Vedamel itself, which I think I
can show to mean River of Nile in conformity with the
old idea of the Nile having a western outlet into the
Atlantic.

The Genoese document in which the name Vedamel
occurs is so carelessly spelt that the name of "Jayme
Ferrer," a well recognisable Catalan name (see " Ferrer," in
Torres Amat's " Escritores Catalanes "), is misspelt "Joannes
Ferne;" and the Genoese form for the Rio d'Oro is in the
course of two or three lines spelt both "Rujaura" and
"Ruiauri." It is not difficult therefore to suppose that
"Vedamel" is a misspelling for "Vedanill," in which we
recognise the Arabic words "Ved" or "Wadi," a river, and
"Nill," the "Nile." Pliny had long ago declared that a
branch of the Nile debouched on the west coast of
Africa, and an ample description of it is given by the
Arab geographer Edrisi in the middle of the twelfth
century, who, after speaking of the sources of the eastern
Nile, says, "The other arm of the Nile (the Nile of the
Blacks) flows into the western districts, and, reaching from
the east to the extreme west, empties itself into the sea not
far from the island of Ulil, which is a day's sail from its
mouth; and on that Nile of the Negroes, or on another river

which mixes its waters with it, are situated the abodes of the people of Nigritia."

A reference to the map of Abul-Hassan Ali Ben Omar (1230) shows us this Western Nile, under the name of Nil Gana, falling into the Atlantic in about the latitude of the Gambia. The map of Ibn Said (1274) has it, under the name of Os Nili Ganah, a little more northward. That of Abulfeda (1331) with the same name, yet a little more northward. The retention of the belief in this river as a branch of the Nile by the Arab geographers is shown by an Arabic map, preserved to us by M. Jomard in his "Monuments de la Géographie," by a Moor named Mohammed Ebn-Aly Ebn-Ahmed al Charfy of Sfax, and bearing date 1009 of the Hegira, which corresponds with A.D. 1600. That the river itself was the Senegal is shown by Azurara, the chronicler of the conquest of Guinea in the time of Prince Henry, who speaks of it as the Ryo do Nillo, which they call the Çanega. Both in the Pizzigani map and in the Catalan map which records the voyage of Ferrer, this river, whose existence was thus learned from Arab sources, is called the River of Gold.

But while this notion of a river of gold, debouching on the west coast of Africa, was thus handed down geographically from ancient times, the mercantile cities of Italy would have the impression more immediately brought home to them by the gold brought across the desert from Guinea into the Mediterranean. We find in the treatise "Della Decima" of Balducci Pegolotti, who was a factor in the great Florentine house of the Bardi, and who wrote in the first half of the fourteenth century, that the malaguette pepper, which was the product of the Guinea coast, was then among the articles imported into Nismes and Montpellier; and De Barros expressly states (Dec. I, fol. 33) that the malaguette imported into Italy before Prince Henry's time was brought from Guinea by the Moors, who, crossing the vast empire of Mandingo and the deserts of Libya, reached the Mediterranean at a port named Mundi Barca, corrupted into Monte da Barca, and as the Italians were not acquainted with the locality whence it came, they called it "grains of Paradise." It would be unreasonable to doubt that, with the malaguette from Guinea, gold was also transported by these merchants across the desert to their port in the Mediterranean, and though the Italians were ignorant of the country whence it came, they would not fail to learn that it lay somewhere on the western coast of Africa. We have therefore but to repeat the poet's apostrophe to the "auri sacra fames," to perceive the motive which would induce an enterprising party of men to encounter extreme danger for the sake of discovering a sea-path to the mouth of such a river.

But these very maps themselves prove how utterly ignorant the bold Majorcan adventurer was of the position of that mouth. The Pizzigani map places it north of Cape Boyador, the Catalan map itself offers a *suggestion only* of where that mouth *might be*, some short distance south of that cape. But both these indications resolve themselves simply into conjectures, inasmuch as *neither north nor south of Cape Boyador is there any river at all which could by any pretence be made to correspond with the Vedamel or Rujauri till we come to the Senegal, which is at least seven hundred miles south of Cape Boyador.* Whether Ferrer himself passed Cape Boyador or not it is impossible to state and futile to conjecture, for the legend itself tells us that nothing more was heard of the expedition. That which was subsequently named the Rio d'Ouro by the Portuguese could by no possibility have anything to do with the Rio d'Oro which Ferrer went to seek, for the simple reason that the former is no river at all, but only an arm of the sea, the appearance of which deceived the Portuguese, and to which they gave the name of the Rio d'Ouro because there they first received gold in ransom for captives.

For precisely the same reason it is clear that the Rio d'Ouro of the Portuguese can in no sense be identical with the Fleuve d'Or referred to in the chronicle of Jean de Bethencourt's voyage of 1402, and in an extract therein given from the book of a Spanish mendicant friar, who

asserted that he had accompanied some Moors in a galley
to that river. It is expedient here to introduce and refute
the extract from this fable, which has also been adduced
to show that Prince Henry's explorers were anticipated on
the west coast of Africa. The words of the mendicant
friar who relates that notable expedition run thus : " They
put to sea, and steered for Cape Non, Cape Saubrun, and
Cape Boyador, and followed the whole coast southward to
the Fleuve de l'Or ; " and according to the said friar, "when
there they found on the river's bank very large ants, which
drew up the grains of gold * from under the ground. The
merchants made considerable gains in this voyage. They
then departed and proceeded along the bank, and found a
very good and very rich island named Gulpis, where they
made great profit and where the people were idolaters.
They then proceeded further and found another island named
Caable, which they left on the right hand, and then they
found on the mainland a very lofty mountain abounding in
all sorts of good things and named Alboc, from which
sprung a very great river. The mountains there are
said to be the loftiest in the world. Some call them
in their language the Mountains of the Moon, others the
Mountains of Gold. There are six, from which spring six
rivers, which all fall into the Fleuve de l'Or. There they
form a great lake, and in this lake is an island called *Palloye*,
which is peopled with negroes. Thence the friar proceeded
further till he came to a river named Euphrates, which
comes from the terrestrial Paradise. He crossed it, and
passed through many countries until he came to the city of
Melle, where dwelt Prester John. He remained there many
days, for he saw there a considerable number of marvellous
things, of which at present we make no mention in this book
that we may proceed the more rapidly, and for fear the reader
should take them for lies." The possibility of an European
thus crossing the continent of Africa and escaping to tell the

* This is but the old story from Herodotus of the Indian ants which were
smaller than a dog but larger than a fox, and which in making their subterraneous
dwellings pushed up sand charged with gold.

tale might well be doubted; but the reader has only to recognise in this language a *rechauffé* of the confused geography of Edrisi, not losing sight of the good friar's stumble over the reference to the Euphrates,* to judge whether the fear of the narrator as to his credit for veracity is a reasonable one. What then becomes of the voyage of the Moors to the Fleuve de l'Or?

Thus far it has been shown that all the claims of Genoese and Catalans to the honour of having passed Cape Boyador before the Portuguese are untenable. We have now to deal with a claim on behalf of the Dieppese, which was not set up till the seventeenth century, but which has since been repeatedly asserted. It was first put forth in a work entitled " Relation des Costes d'Afrique appellées Guinée," &c., par Villaut, escuyer, Sieur de Bellefond, Paris, 1669. For the sake of brevity, but in order that at the same time the account may be given in the words of a Frenchman, I have selected the summary of the narrative as extracted from Villaut de Bellefond's work by M. Estancelin, in his " Recherches sur les voyages et découvertes des navigateurs Normands en Afrique," &c., Paris, 1832, 8°. M. Estancelin has also made extracts from another work which followed that of M. Villaut de Bellefond half a century later, entitled "Nouvelle relation de l'Afrique occidentale," 5 tom. Paris, 1728, 12°, by the Père Labat. His summary is as follows:—

" France, so long and so cruelly the victim of the folly of her masters, began to breathe again under Charles the Fifth. This monarch knew how to appreciate the advantages of commerce, and saw the interest of encouraging that of a province which had formed his own appanage. The Dieppese took advantage of these favourable inclinations. In the month of November, 1364, they fitted out two vessels of a hundred tons each, which set sail for the Canaries. About Christmas they reached Cape Verde, and

---

* In speaking of a famous and very large city of the negroes named Kucu, Edrisi says, " Some negroes think that this city lies on the Nile itself, others on a river flowing into the Nile; but in truth the Nile passes through the city Kucu, and then diffuses itself through sandy plains into the desert, and thence merges into lakes, just as the *Euphrates* does in Mesopotamia."

anchored before Rio Fresco, in the bay that still in 1669 bore the
name of Baie de France. Passing the coast of Sierra Leone they
stopped at a place, named afterwards by the Portuguese Rio
Sestos. Struck with the resemblance which this place bore to
their native city, they named it Petit Dieppe. Their trade with
the natives procured them, for objects of little value, gold, ivory,
and pepper, from which on their return, in 1365, they gained
immense profit. Encouraged by this first success, in September
in the same year, the merchants of Rouen joined those of Dieppe,
and the company fitted out four ships, of which two were to trade
from Cape Verde to Petit Dieppe, and the other two were to go
further to explore the coast. These instructions were subject to
modifications, which proved fortunate for the owners. One of
the ships destined to pass on further, stopped at the Grand Sestre,
on the coast of Malaguette, for, finding a great quantity of pepper
in this place, it took in a cargo. The other ship traded at the
Côte des Dents, and went on as far as the Gold Coast. It returned
with a large quantity of ivory and a little gold. The people of
this coast not having welcomed the sailors so hospitably as those
of the coast of Malaguette, the company resolved thenceforth to
fix their depôts at Petit Dieppe and the Grand Sestre, which the
sailors had then named Petit Paris, in honour and memory of the
capital of their country.

" These expeditions were all made during the reign of Charles the
Fifth. Factories, which they then called ' loges,' were established to facilitate their intercourse with the natives. The ships
thus found their cargoes prepared, and on arriving had only to
unload and reload. As they were too weak to attempt to govern
the natives and to reduce them to submission, the colonists and
sailors felt the necessity of gaining their affection and confidence.
In this they succeeded without trouble ; it needed only to be
humane and just, and above all not to use the scourge of religious
proselytism, the odious and fatal pretext, of which the Spaniards
and Portuguese have made such cruel use to legitimatise the
atrocities which their thirst of gold caused them to commit in the
countries which they conquered. It does not appear that the
kindly relations which united the Africans with their guests ever
altered. They were, on the contrary, deeply rooted in the
memory of the people, who even preserved for a long time, in
their language, a number of French expressions, which Villaut de
Bellefond, from whom these details are borrowed, found in his in-

tercourse with them. 'The little of the language,' says he, 'that one can understand (in 1666) is French; they do not call pepper, as in Portuguese, sextos, but malaguette, and when one lands, if they have any, they cry, "malaguette tout plein, tout à force de malaguette," which is the little of our language which they retain.'

"The abundance of spices which the Normans brought back in their annual voyages, produced a diminution of their value. This branch of commerce no longer offering such great profits, the company sent out, in 1380, a ship of a hundred and fifty tons, called the 'Notre Dame de Bon Voyage,' which sailed from Rouen in the month of September, to trade at the Gold Coast, and if possible to form a settlement there. This ship arrived, towards the end of December, at the same landing where, fifteen years before, the second expedition had traded so advantageously. This expedition was very successful; the 'Notre Dame' returned to Dieppe, nine months after, very richly laden. 'Thus commenced,' says Bellefond, 'the prosperity of the commerce of Rouen.'

"The year following (1382) three vessels, 'La Vierge,' 'le Saint Nicholas' and 'l'Espérance,' set sail on the 28th of September. La Vierge stopped at the first place which had been discovered on the Gold Coast, which had been named La Mine, because of the quantity of gold found there. Le Saint Nicholas traded at Cape Corse and at Mouré below la Mine, and l'Espérance went as far as Akara, having traded at Fantin, Sabou and Cormentin. Ten months after the expedition returned safe and sound with rich cargoes. Their reports fixed the attention of the company, which thenceforward conceived the idea of directing all their speculations exclusively to that point. For this purpose three vessels, two large and one small were sent out in 1383. The small one was to go to Akara to discover the southern coasts. The two large ones were ballasted with building materials which were employed in constructing a station at La Mine. There they left ten or a dozen men and returned after an absence of ten months. The small vessel was retarded by the currents which in those parts present the remarkable phenomenon of two parallel streams in contact with each other, running with great velocity in opposite directions. It was only partially successful in its object, and returned with an incomplete cargo three months before the two others. When they arrived it was again sent out to carry provisions to the new colony, which soon afterwards became of sufficient importance to build a church, 'which,' says Bellefond, the Dutch now make use of, and in which may still be seen the arms of France.' The development of this prosperity was checked by the frightful calamities which burst upon France shortly after the accession of Charles the Sixth. The decay of commerce followed that of the state, and when its sovereign had lost his reason, France, delivered over to party contentions, became the prey of England. At this unhappy period the African trade began to decrease from year to year, and finally disappeared. The station of La Mine was abandoned before 1410, and from that time till after 1450 there is reason to suppose that the Normans did not attempt any maritime expedition whatever."

The work containing this astonishing pretension, thus made for the first time in 1669, without any documentary corroboration whatsoever, was addressed to Colbert by its author on his return from a voyage which he had himself made to Guinea. Its supporters assert that " there is reason to think that the elements of the accounts were derived from the registers of the Admiralty at Dieppe, subsequently destroyed in the bombardment of 1694." The claim was reasserted in 1728 by the Père Jean Baptiste Labat, who says that " the date and other circumstances which he relates are taken from the MS. annals of Dieppe, which can be seen in the cabinet of Monsieur ——, the king's advocate, in the same city," and, thus unauthenticated, the story has been over and over again repeated up to the present time. As therefore, in the one case, the presumable elements of the account were destroyed, and in the other, have never been forthcoming from the cabinet of their nameless possessor, we naturally look for some evidence, either external or internal, of the trustworthiness of such a pretension. An octavo book full of damaging analysis was published by the late learned Vicomte de Santarem in 1842, * but a few conclusive facts will render a host of minor ones needless.

And first as to internal evidence. The credibility of M. Villaut as an historian may be judged from the fact that he

* Recherches sur la priorité de la decouverte des pays situés sur la côte occidentale d'Afrique au-dela du Cap Bojador, par le Vicomte de Santarem. Paris, 1842. 8°.

makes the Island of St. Thomas, in the Gulf of Guinea, which was not discovered till 1471, when Prince Henry had been dead eleven years, to have been discovered by the Portuguese on the 23rd of December, 1405, when that Prince was eleven years old. An author capable of such circumstantial misrepresentation is surely not to be trusted.

Secondly: In the original of Villaut de Bellefond, he says (page 160) that the word Malaguette, the name of the spice imported from the west coast of Africa, was *French*, and uses this fact thus asserted as an argument in confirmation of his claim. But this assertion is entirely disproved by the fact already stated at page 114, that in the " Della Decima " of Giovanni Balducci Pegolotti (Vol. iii. p. 229), which was written about 1340,* *a quarter of a century before* the date of *the earliest pretended intercourse of the Dieppese with Guinea,* malaguette is mentioned as being imported into Nismes and Montpellier.

But again, if, as is here stated, the merchants of Rouen fitted out vessels in conjunction with those of Dieppe, and continued these expeditions to the coast of Guinea every year, whence comes it that the archives of Rouen, which were not destroyed by a bombardment like those of Dieppe, should not contain the slightest record of any such fact? Now it happens that such an association had really been formed, not in the fourteenth century, but in the year 1626, and this fact has an interesting connection with the evidence of maps in reference to the point in dispute. The Dieppese have indeed much to boast of without resorting to unauthenticated assertions to secure to themselves yet greater honour than they deserve. The finest charts that we possess of the first half of the sixteenth century are the productions of a school of hydrography established at Arques, near Dieppe, by Pierre Deschelier in the beginning of that sixteenth century. There can be no doubt that this school was the offspring of the many daring maritime explorations by

* See p. 279 of Colonel Yule's " Cathay and the Road thither," printed for the Hakluyt Society, Lond. 1866.

which the Dieppese had for many years distinguished themselves. But, as we shall presently have occasion to show, the watchful jealousy of the Portuguese over their African possessions kept for a long series of years even Dieppese daring in check. The result of this was, that although at the close of the century Dieppese perseverance conquered, yet up to the date of 1626, the beautiful maps of the Dieppese contain no mention of either Petit Dieppe or Petit Paris, whereas in 1631, five years after the establishment of the Rouen and Dieppe Company, we do *for the first time*, find the name of Petit Dieppe laid down on the MS. map of Jean Guérand of Dieppe.

In such a position of things we naturally look around for any corroborative evidence whatsoever of this extraordinary claim. Nor has such been wanting. M. Estancelin has brought to light two pieces of testimony by way of confirmation, which it is a duty to lay before the reader in full. The first of these is a statement made by Samuel Braun, a surgeon of Basle, who went out in a Dutch vessel to the Gold Coast, and resided at Fort Nassau from 1617 to 1620. His account is printed in the "Appendix Regni Congo," in De Bry's collection of Petits Voyages, published in Frankfort, 1625. In it he says, " In this Fort (Fort Nassau), as well as at Accra, I saw some people above a hundred and thirty years old, who told me that the Fort Mina had begun to be built many years before by French merchants who came to traffic there. As every year for three months there were constant rains with strong whirlwinds which the sailors call Travada, so that the goods were damaged, the French begged permission of the inhabitants to build a magazine or warehouse, which, as they were on very friendly terms, the blacks willingly conceded. Accordingly they built a tolerably large warehouse and brought their goods to land. This was a great furtherance to traffic, as the natives who had neither coin or weights exchanged their gold for merchandise without any measurement but that of the eye. When the Portuguese learned that the

French carried on this profitable trade with the negroes they fell on them unawares, took possession of the magazine, gave the merchandise to the inhabitants, and assured them that they would deal with them on better terms than the French. These poor people readily believed them and assisted in murdering those who came there afterwards. Finally the magazine was converted into a castle [misprinted ' chapel '], which is now very strong, and only serves to the great injury of the natives."

The second quotation is from Doctor Olivier Dapper's description of Africa, published in Dutch at Amsterdam in 1668 : " The Castle of La Mine " he says, " is a very old building, as is shown by different dates in various places. In a ruined battery restored by our people, some years ago, and named the Batterie Française (because it was of French construction, and because the French, according to the natives, were established in this place before the Portuguese), our people found the first two figures of the date 13—, but the following numbers could not be deciphered. In the small inner court there also exists an inscription cut in the stone, between two old pilasters, but almost entirely effaced by exposure to the weather, and consequently illegible ; while at the provision magazine one sees at once that it was built in 1484, under John II., King of Portugal, as is shown by the date placed on the door, which is still as clear and as entire as if it dated only from a few years ago, whence we must conclude that the other before-mentioned date must be very ancient."

M. d'Avezac, adverting to these quotations, comments on them as follows : " This tradition of the natives, thus repeated in 1617 to Braun by men one hundred and thirty years old, that is to say, by men born in the first years of the establishment of the Portuguese, and whose fathers had witnessed the facts recounted in the narrative, is a fact of importance not to be annulled by simple contradiction. * It is moreover con-

---

* See Recherches sur la priorité, etc., pp. 32, 33, and Bulletin de la Société de Géographique, cahier de Janvier, 1846, pp. 18—19.

firmed by material indications which are not without their
value. The old inscriptions defaced by time, especially that
which was found by the Dutch in the ruins of the old
Batterie Française, show that the first constructions of Fort
La Mine date from the fourteenth century. Besides which
this French warehouse, transformed into a chapel by the
Portuguese, retained even in 1667 the traces of its former
masters. Villaut de Bellefond, who then visited those
coasts, attests this in the most exact manner. 'The Dutch,'
he says, 'now use for their preaching the same chapel
which we built there, in which are still to be seen the arms
of France.'" It is not often that M. d'Avezac makes a
mistake, but here he has chanced to fall into two errors at
once, and between the two M. Villaut makes a fortunate
escape. Had he, as M. d'Avezac supposes, alluded in the
quoted sentence to the " French warehouse transformed into
a chapel by the Portuguese," he would have been in the
ludicrous position of making Dutchmen preach in 1669 in a
chapel which had been built in a moment by a printer's devil
in 1625. A reference to the original of Braun's " Fünff
Schiffarten," three lines beyond M. d'Avezac's extract, will
show by the words " dieses Castell " that the printer had
misprinted " Capell " for " Castell;" but M. Villaut, it will
be seen, had stated that the French built their own church,
and therefore he derives no confirmation from the extract on
that head.

But all this time the reader will naturally infer that
these several statements could not possibly have been made
unless the French had really been in those parts some
considerable time before ; and so in truth they had been.
The only question is as to the period. The French themselves are very indistinct, or rather they seem entirely at
fault, at to the date of what they would call the resumption,
we the commencement, of their intercourse with the coast of
Guinea. That the Portuguese fenced them off with jealous
perseverance during the early part of the sixteenth century
is demonstrated by the words of a Dieppese captain in 1539,

which for a more comprehensive purpose we shall presently
have to quote in full from Ramusio. But that in the latter
half of that century the French did succeed in establishing
themselves on that coast is shown by an expression which
occurs in the third volume of Ramusio, published in 1565,
page 417 verso, in the editor's " Discorso sopra la Nuova
Francia," where, in speaking of " Guinea and the Malaguette
coast of Africa," he says, " which the French constantly frequent with their ships:" and that they carried on an immense
trade in Guinea grains and ivory is shown by a letter in the
British Museum (Lansdowne xxv. art. 72), addressed to Lord
Burleigh under date of the 9th of March, 1577, by Doctor
David Lewis, Judge of the Admiralty, concerning a cargo
of Guinea grains and elephants' teeth taken by one Batts
from a ship called the Petit Margaux, coming from the river
of Cestos, and belonging to one Thomas de Verins of Dieppe.
We there find the words " by the reporte of suche as be best
acquainted with them of Roan, Deepe, and other places in
Normandye, it should seame that they have an ordinary trade
to the sayd ryver of Cesto." From the documents which
accompany this letter we learn the amount of this capture,
which consisted of twenty-four tons or butts of Guinea grains,
and seven hundred elephants' teeth. If the cargo of one out
of many vessels " having an ordinary trade " to the Guinea
coast was so rich as this, we can form some notion of the
footing which the French had gained on that coast even for
some years previous to this period.

And this leads us to another incidental proof, though
of a different nature, that it was about the beginning
of the last half of the sixteenth century that the French
established these relations with that coast. M. Vitet, one
of the most zealous advocates of the prior claims of the
Dieppese to African discovery, in his " Anciennes Villes
de France," Paris, 1833, page 244, after having, without
any evidence whatever, simply asserted that the ivory
carvings of the Dieppese were as old as the close of the
fourteenth century, lets slip the following confession:—

" Unfortunately their works of this period are only known to us by reputation; no trace of them has reached us. These delicate bijoux with difficulty escape destruction. The churches only had it in their power to procure some of them, but the pillage of the altars in the sixteenth century caused the disappearance of the paxes and crucifixes of ivory, together with the shrines and gold-embroidered chasubles. The oldest works of the kind that can be found at Dieppe were made in the seventeenth century, and they are extremely rare." Then in a note he says :—" M. Flammand, ivory uealer in the Grande Rue, is said to possess two small bas-reliefs of an earlier date than 1600, one representing the self-sacrifice of Curtius, the other Mutius Scævola before the King of Etruria, but not having seen them I cannot answer for their style or their antiquity." Bombardments and pillage seem to have been more effectual in exterminating material testimony on important historical questions during a few centuries in Dieppe, than all the accidents of thousands of years have effected in Assyria. We have Assyrian ivories of three thousand years old, as delicate bijoux as any ever made at Dieppe, and others of intermediate periods from various countries in abundance. No antiquary, assuredly, will admit so vapid an argument as that adduced by M. Vitet for the disappearance of his fourteenth and fifteenth century ivories.

Meanwhile it has been shown that the French had been connected with the Gold Coast quite long enough to admit of the existence, in Villaut de Bellefond's time, of a church then occupied by the Dutch, but containing the arms of France. And although Dapper's character for trustworthiness seems to be not much better than that of M. Villaut, for the late learned French geographer, M. Eyries, says of him (Biographie Universelle de Michaud), that " he was sometimes so undiscriminating in selecting his materials, that he has misled authors who have trusted him without making a critical examination of his statements," nevertheless a whole century was long enough to render possible,

in his time, the existence of a French battery as well as
the obliteration of a date (if date it was), exposed to the
annual corrosive action of three months of rain. At the
same time it is clear that the date of 1484 if less exposed or
cut in harder material would, even if older, survive comparatively intact. That such was either its position or the
texture of the material, or both, is made quite certain by
Dapper's own words; for, as ·the annual rains were an unfailing reality and the one hundred and eighty years between
1484 and Dapper's time were realities equally inexorable, and
yet the inscription was as sharp and unimpaired as if it had
been only made a few years before, it follows that either its
position must have been sheltered or the material uncommonly
hard. We have thus, it is hoped, cleared the field of all the
witnesses except the superannuated negroes whose memories
at the age of one hundred and thirty enabled them to supply
M. Samuel Braun with intelligence as to what had happened
two hundred years before. These shall be answered presently.

Such is the nature of the evidence adduced in corroboration
of this extraordinary claim; and M. d'Avezac, by far the most
learned of its advocates, leaves the question with the following
judicial verdict: — " The contemporary documents which
proved the authenticity of these explorations have perished,
and modern criticism takes advantage of this loss to dispute
the genuineness of the narratives which show the establishment of the French on the coast of Guinea, as far as the Gold
Coast, before the end of the fourteenth century. She is in her
right ' dans son droit,' and she avails herself of it. It would
be discourtesy in us to disregard it." *Noblesse oblige*, and if
the evidence closed here, no duty could be more imperative
or more grateful, in the presence of such language, than to
refrain from pressing a case in which an eminently distinguished *savant* finds his arguments weakened solely by an
accident. But we are not left to rest on so loose a foothold
as this. In contravention of the enlightened testimony of
the venerable negroes we are able to adduce *Norman* evidence,
before the Portuguese time, and *of the very period* when the
asserted prior discoveries of the Normans were at their height,

to show that the Normans knew nothing of the coast south of Cape Boyador; and we have further the evidence of a *Dieppese* captain, *after* the Portuguese discoveries, and *earlier* than the *earliest* statement of any kind in favour of the Dieppese, showing that the Portuguese were the first discoverers of that coast. The asserted period of Norman exploration ranges, as we have just seen, from 1364 to 1410, and we have the unquestioned Norman narrative of the voyages of Bethencourt in 1402-5, written by the chaplains who accompanied him. Referring to the Spanish friar's book already spoken of at page 116, they say:—

"The mendicant friar says in his book that the distance from Cape Bugeder to the River of Gold is only a hundred and fifty French leagues as the map shows."

The reader will remember that I have shown (page 115) that the indications on the maps were assumptions, not demonstrated facts. But to continue:—

"This would take three days' sailing in ships or barks, but longer in the galleys, which can only sail along the coast, so that it is not in our power to go there. If things be as the Spanish friar's book says, and as those say who have visited these countries [see the disproof of this on page 117], it is M. Bethencourt's intention, with the help of God and that of Christian princes and people, to *open* the way to the River of Gold. If he succeeded it would be a great honour and a great profit for the kingdom of France and all Christian kingdoms, seeing that we should get near to the country of Prester John, from which so many good things and so much wealth are derived. It cannot be doubted that much remains to be done which *might* have succeeded in times past, *if it had been undertaken.* He does not boast that he will succeed, but he will so demean himself that, if he do not succeed, both he and his crew shall be held blameless, for he will spare no pains to decide whether success is possible or absolutely impossible. But with the help of God he will conquer and convert to the Christian faith a host of men now in a state of perdition for want of enlightenment. It is a great pity, for go in any part of the world you will, you will not find handsomer or better made men and women than in these islands (the Canaries). They have great intelligence, and only want teaching. And as the said Lord de Bethencourt has a great desire to know the condition of other neighbouring parts of this country both islands and mainland, he will spare no pains to get exact information respecting all these countries."

Such was the state of knowledge and such were the intentions with reference to the west coast of Africa beyond Cape Boyador entertained by a Norman gentleman of seafaring tendencies, whose estate of Grainville la Teinturiere lay only five-and-twenty miles from Dieppe at the very time when the pretended Dieppese relations with Guinea were at their height. Surely the secrecy of those wealth-producing expeditions must have stood unexampled in the history of the world.

But to proceed. After Bethencourt's time follow the well-recognised voyages under Prince Henry. From the rounding of Cape Boyador to that of the Cape of Good Hope occupied from 1434 to 1497; and in 1539 we have a document by a *Dieppese captain,* preserved to us by Ramusio (tom. iii., p. 426 b., 1565 edition), in which occurs a passage, of which the following is a translation :--

" This land of Brazil was first discovered in part by the Portuguese about five and thirty years ago . . . . . . and as I might be asked why the Portuguese prevent the French from going to Brazil and the other places whither they have navigated, such as *Guinea* and Taprobana, I could give no other reason than their insatiable avarice, and although they are the smallest nation in the world, even this does not seem enough to satisfy their cupidity. I think they must have drunk of the dust of the heart of King Alexander to have brought about in them such unlimited greediness, and they affect to hold in their closed fists more than they could enclose in both their hands. I think it is their belief that God made the sea and the land for them alone, and that other nations are not worthy to navigate, and if it had been in their power to put limits and shut up the sea from Cape Finisterre to Ireland, the passage would have been closed long ago ; and there is just as much reason why the French should not go to those lands, in which they (the Portuguese) have not planted the Christian religion and where they are neither loved nor obeyed, as we should have reason to prevent them from going to Scotland, Denmark, or Norway, *if we had been there before them.*"

If words mean anything this is an acknowledgment on the part of a *Dieppese captain,** of Portuguese priority in the discovery of the coast of Guinea; and be it observed that this acknowledgment is made unguardedly in the midst of a bitter expression of rivalry and complaint against the latter apropos of their possessions by right of discovery, in which the power of claiming priority, had the complainant possessed it, would have been beyond all price.

But lest these more splendid pretensions should be met with an obdurate incredulity, a minor claim has been set up on behalf of Jean de Bethencourt himself, for the honour of having at least passed southward of Cape Boyador in anticipation of the Portuguese. The claim is made in respect of the following occurrence. M. de Bethencourt being in the island of Fuerteventura, set sail on the 6th October, 1405, with three galleys for the Great Canary, and the words of the chronicler are:—

"The vessels were separated at sea, and all three came near the lands of the Saracens *very near to the port of Bugeder*, ' bien près du port de Bugeder.' M. de Bethencourt and his people landed and went a good eight leagues into the country. They took some men and women whom they carried away with them, and more than three thousand camels. But they could not take all on board; they killed some and hamstrung (or potted) some, and then returned to the Great Canary."

On the strength of the word "port," in the foregoing narrative of de Bethencourt's adventure, it has been reasoned that henceforth no one can deny to M. de Bethencourt the honour of having passed Cape Boyador thirty years before the Portuguese, who boast of having been the first to double it, inasmuch as it is pointed out that the Port of Boyador is south of the Cape. Cape Boyador had been for ages the ne plus ultra of navigation along this coast. When Gil Eannes really doubled this Cape in 1434, it was, as Azurara tells us,

---

* Now known to be Jean Parmentier, who made a voyage to Sumatra in 1529, the first made to the South Seas by a Frenchman. The account was in all probability written by his friend and eulogist the poet Pierre Crignon, who accompanied him in the voyage.

" by avoiding certain shoals and rocks which are on certain
reefs exactly laid down on the charts that have been drawn
by the orders of Prince Henry." The French Admiral Baron
Roussin in his " Memoire sur la navigation aux côtes occiden-
tales d'Afrique," Paris, 1827, 8°, tells us that Cape Boyador
" when seen from the northward shows a strand of red sand
with a gradual descent towards the sea," and the " African
Pilot," published by our own Admiralty, says that " the
surf is exceedingly heavy all along this shore." Hence we
can understand the difficulties in the way of doubling this
Cape which would be offered to small craft making their
way southward along the African coast. These difficulties
were such as to baffle the efforts of Prince Henry's sailors
for a long series of years. But now for the position of the
bay. In the old editions of the " African Pilot," now very
scarce, is a passage extracted by M. d'Avezac, who has written
an article on the subject,* to the following effect: " Cape
Boyador is surrounded by a reef which extends above a
league into the sea; *to the southward* of it you may anchor in
a little bay in four or five fathoms water, but coming from
the northward you must not approach it nearer than twenty
fathoms." The position of this little bay is very minutely
described by Admiral Roussin. He says that " *the western
extremity of the Cape, which is very low, forms a small bay with
the cliff which immediately follows.*" So that the claim set up
for Bethencourt is that he anchored in a bay south of the
very tip of the nose of Cape Boyador. In early maps and
narratives of travels the words port, cape, river, &c., are
used with marvellous recklessness, and if, as is almost certain, Bethencourt landed a little *north* of Cape Boyador,
then famous as the limit of exploration, nothing would be
more natural than the expression that " they came very
near to the Port of Bugeder." The editor of Bethencourt's
voyage himself took this view of the case, for in his preface

* Entitled " Sur la véritable situation du mouillage marqué au sud du Cap de
Bugeder dans toutes les cartes nautiques." Lue à la Société de Géographie de
Paris dans la séance du 20 Mars, 1846. Paris, 1846, 8o.

he says that Bethencourt was " thrown by a storm on the
Coast of Africa, vers le Cap de Bogeador," and the French
word " vers " can by no process be made to mean " beyond."
The authors of the narrative also, who elsewhere show themselves quite alive to the importance of the subject, make
no claim for the honour of such an achievement on behalf of
their hero. It was not till four centuries and a-half had
transpired that a fellow countryman of Bethencourt's,
my good friend M. d'Avezac, discovered for him a claim
to glory of which neither himself nor his companions
were in any way conscious. But let it for a moment be
assumed that the ships had indeed arriyed at the port in
question, it will be seen that neither by skill or knowledge
or courage, but under the unwelcome action of the winds,
they were driven thither from a point in the open sea, which
entailed not one tittle of the special danger or difficulty
which constituted the glory of " *doubling* " the redoubted
Cape. Gil Eannes, in 1434, by dint of sheer courage and
perseverance *rounded* Cape Boyador, which de Bethencourt
never did. The argument seems scarcely deserving the
learned labour of a distinguished writer, for it proves neither
merit nor usefulness in the feat assumed to have been
accomplished, and I should not have thought it desirable
to lay it before the reader were it not that its author regards
it as so great a triumph over Prince Henry's navigators
that he uses the following words, " Thus twenty-nine years
before the so much vaunted enterprise of Gil Eannes we see
the French make a Ghazyah of eight days" [it should have
been leagues] " on the African lands beyond Cape Bugeder.
In the presence of this fact simply enunciated, of what
value are the resounding noise and pompous éclat of a blind
renown ?" To our view the honour that attaches to that
fact resolves itself into the very quintessence of a bagatelle,
but by all means let the invited comparison be duly made.
Let the bandage be withdrawn from the eyes of Fame,
that the goddess may hie her beyond Cape Boyador to cull a
garland for the brows of Jean de Bethencourt. It is to be

feared she will return but empty handed. Yet there are
flowers beyond that stormy cape, but they grow for those
only who shall win them by predetermined purpose and
unflagging exertion. Fame leaves untouched the roses of
St. Mary till they shall be gathered by the weather-beaten
hands of Gil Eannes, that in her name he may offer them at
the feet of his princely master. The value of the meritorious
rounding of Cape Boyador by Gil Eannes in 1434 was that
it led to the rounding of the Cape of Good Hope by
Bartholomew Dias in 1487, to the discovery of America by
Columbus in 1492, to the unfolding of the sea way to India
by Vasco da Gama in 1497, to the maritime discovery of
China in 1517, to the discovery of the Straits of Magalhaens
in 1520, and the circumnavigation of the globe in the same
voyage, and to the discovery of Australia by the Portuguese
some ten years later. In the celebration of such mighty
results what "bruit" could well be too "retentissant,"
what "éclat" could well be too "pompeux?" And now
let us turn aside to do full justice to the transaction which
precluded the merit of the Portuguese by a priority of
nine and twenty years. Driven by stress of weather, Jean
de Bethencourt, as is supposed but by no means proved,
lights upon a little bay, looking south west, on the extreme
western point of the Cape which was not rounded till nine
and twenty years later. He lands, and thirty years in
advance of the Portuguese, sets an example of that for
which the latter were afterwards reproached. He captures
men and women. In so doing he proceeds eight leagues
into the country, but whether his steps were turned northward or southward is not recorded, and, if it were, the fact
could not by any possibility have the worth of a straw. In
short, to put the value of this fortuitous expedition to the
most unequivocal test, it could not enable any cartographer
to add one iota of information to the chart of the west coast
of Africa. That the sea of darkness was visited by some
glimpses of light before the time of Prince Henry is true,
but these must be treated of in another chapter.

CHAPTER VIII.

GLIMPSES OF LIGHT.

THERE has been a belief prevailing in every religion from
the oldest times that the souls of the departed cannot enter
into bliss without first crossing a river. The doctrine
originated apparently in India, whence it passed into Assyria
and Chaldea, and so into Persia. From Asia it extended
into Greece and Egypt, thence through Ethiopia to the
country of the Gallas, and at length we find it, as Bowdich
tells us, among the natives neighbouring on the country of
the Ashantees. Even to the mythic Jordan of the Christian
the idea has still obtained. In the poems of Homer the
ocean is treated as a river beyond which, at the earth's confines, were the Elysian fields, which Hesiod and Pindar
made to be surrounded by water, so that the habitations of
the blest were transformed into islands, and hence, as it has
been supposed, originated the name of the Insulæ Fortunatæ
or Fortunate Islands.* These remained, however, no better
than islands of fable, lying remote wherever fancy suggested,
till solidity was given them by the discovery of the Canaries
in the outlying ocean, and at length the land of spirits had
assigned to it in men's minds a somewhat more definite
geographical position. It is in the highest degree probable

* I am indebted for this deduction to the learned treatise by the venerable
Joaquim da Costa de Macedo, entitled : " Memoria em que se pertende de provar
que os Arabes não conhecerão as Canarias antes dos Portuguezes." Lisboa
1844. fol.

that the Phœnicians had been the original discoverers of the Canaries. Strabo tells us, lib. 3, " The poets make mention of the Islands of the Blest, and we know that even now they are to be seen not far from the extremity of Mauritania, opposite Gades (Cadiz). Now I say that those who pointed out these things were the Phœnicians, who before the time of Homer had possession of the best part of Africa and Spain." It may therefore be reasonably presumed that the Canary Islands were known to the Carthaginians established at Gadir or Cadiz, but' that the monopolizing policy of that nation referred to by Diodorus induced them to conceal from other countries the extent of their commercial relations.

After the third Punic war the attention of the world was directed to other conquests, and it was not till about eighty-two years before our era that we find the Fortunate Islands brought afresh under notice. The Roman armies were in Spain. Sertorius fleeing from the ships of Annius had passed through the Straits and turning to the right had landed a little above the mouth of the Guadalquivir, then called the Bætis, when he fell in with some Lusitanian sea captains who had just returned from the Fortunate Islands, and their description of them is given us by Plutarch in his Life of Sertorius. It is as follows: " They are two islands separated only by a narrow strait, and distant from Africa ten thousand stadia [or five hundred leagues]. They are called the Fortunate Islands. It rains there but seldom and then softly. The winds are generally agreeable and bring with them refreshing showers which fertilise the earth, and make it not only produce anything that is planted, but supply spontaneously excellent fruits for the abundant nourishment of a happy people who pass their lives in the most delicious idleness. The changes of the season are always gentle and the air is pure and whole-some. The north and east winds which blow from our continent in traversing so great a space expend their force before they reach those islands. The winds from the west and south sometimes bring gentle rains, but for the most

part only refreshing vapours sufficient to make the ground
fertile. All these advantages have established even amongst
the barbarians the generally received opinion that these
islands contain the Elysian fields, the abode of happy souls
celebrated by Homer." M. Bory de St. Vincent fancied
that he recognised in these islands Madeira and Porto
Santo, but the fact of the latter not being inhabited
precludes that idea. It is more probable, judging by their
distance from each other and from the continent, that
Lancerote and Fuerteventura are referred to. This glowing
account inspired Sertorius with the most ardent desire to
retire from the toils of life and seek repose in these blissful
islands, but circumstances forbade the realization of his wish.

Twenty years after the death of Sertorius we have five
islands specified by distinct names in a vague itinerary
drawn up by one Statius Sebosus from the accounts of
navigators of his time and preserved to us by Pliny. He
represents the group to which he gives the name of Hespe-
rides as one day's sail from the western promontory (Cape
Non). He names them (1) Junonia, at 750 miles from
Gades (Cadiz), (2) Pluviālia and (3) Capraria, 750 miles
west of Junonia, and 250 miles beyond to the left of
Mauritania and towards the ninth hour of the sun were the
great Fortunate Islands, one called (4) Convallis and the
other (5) Planaria on account of their form; but all these
indications are too indistinct to supply us with any information beyond the fact that in the time of Sebosus five islands
of the Canary group had received individual names.

Happily we are supplied also by Pliny with information of
a far more distinct character respecting these islands. When
King Juba the Second was reinstated by Augustus on the
throne which his father had lost, on his return to Mauritania he turned to account the geographical knowledge
which he had acquired through his education in Italy, and
sent out an expedition for the express purpose of exploring
the Fortunate Islands. On the return of the navigators he
wrote a narrative of the voyage from their report, and sent it

to the emperor. A fragment only of that narrative survives, and has been transmitted to us by Pliny in the following shape: "The Fortunate Islands lie to the south-west, at 625 miles from the Purpurariæ. To reach them from the latter they first sailed 250 miles westwards and then 375 miles to the east.* The first is called Ombrios, and contains no traces of buildings. There is in it a pool in the midst of mountains, and trees like ferules, from which water may be pressed, which is bitter from the black kinds, but from the lighter ones pleasant to drink (sugar cane). The second is called Junonia, and contains a small temple built entirely of stone. Near it is another smaller island having the same name. Then comes Capraria, which is full of large lizards. Within sight of these islands is Nivaria, so called from the snow and fogs with which it is constantly covered. Not far from Navaria is Canaria, so called on account of the great number of large dogs therein, two of which were brought to King Juba. There were traces of buildings in this island. All the islands abound in apples and in birds of every kind, and in palms covered with dates, and in the pine nut. There is also plenty of honey. The papyrus grows there, and the Silurus fish is found in the rivers. (See Pliny, Nat. Hist., lib. 6, cap. 37.) In Ombrios we recognise the Pluvialia of Sebosus, the words being synonymous. Convallis becomes Nivaria, and Planaria is replaced by Canaria, which name is still borne by the large central island, and has now been given to the whole Archipelago. There is no difficulty in fixing the island named Nivaria, a name which clearly indicates the snowy peak of Teneriffe, almost constantly capped with clouds. In Ombrios or Pluvialia, with its pool in the midst of mountains, we recognise the island of Palma, with its famous Caldera or cauldron, the crater of an old volcano.

* The "three hundred" is omitted in some editions of Pliny, but that they are necessary is evident from the account of Pliny himself. It is clear that the 625 miles is reckoned in making the periplus of the whole group, the 250 tally-ing with the distance from Fuerteventura, one of the Purpurariæ, to Ombrios or Palma. The 375 would be the length of the eastern return track from Palma round the group.

The distance also of this island from Fuerteventura agrees
with that of the 250 miles indicated by Juba's navigators
as existing between Ombrios and the Purpurariæ. It has
been already seen that the latter agree with Lancerote and
Fuerteventura in respect of their distance from the continent
and from each other, as described by Plutarch. That the
Purpurariæ are not, as M. Bory de St. Vincent supposed,
the Madeira group, is not only shown by the want of inhabitants in the latter, but by the orchil, which supplies the
purple dye, being derived from and sought for specially from
the Canaries and not the Madeira group, although it is to be
found there. Junonia, the nearest to Ombrios, will be Gomera. It may be presumed that the temple found therein
was, like the island, dedicated to Juno. Capraria, which
implies the island of goats, agrees correctly with the island
of Ferro, which occurs next in the order of the itinerary, for
these animals were found there in large numbers when the
island was invaded by Jean de Bethencourt in 1402. But a
yet more striking proof of the identity of this island with
Capraria is the account of the great number of large lizards
found therein. Bethencourt's chaplains, describing their visit
to the island in 1402, state :—"There are lizards in it as big as
cats, but they are harmless although very hideous to look at."

It was probably the desire to bring these mysterious
islands within the grasp of history that induced King Juba
to send out this expedition, and although the blessedness that
was looked for formed no part of the discovery, yet as these
were the only islands that were lighted upon in the ocean where
they were sought for, they were assumed to be the genuine
Insulæ Fortunatæ, and accordingly retained the name.

For thirteen centuries from the time of which we have
been speaking, the Fortunate Islands were destined again
to be almost buried in oblivion. The destruction of the
Roman Empire re-plunged Europe into ignorance, and
although the Fortunate Islands were vaguely known to the
Moors of Spain under the designation of the Islands of
Khaledat, it has been elaborately shown by the eminent

Portuguese savant, now venerable in years, Senhor Joaquim
José da Costa de Macedo, that the Arabs had no practical
knowledge of the Canaries before the times of the Portuguese discoveries. He maintains that the only notions they
had respecting them, were such as they derived from Greek
and Latin authors, and he seems satisfactorily to have proved
his point.

It was not till the beginning of the fifteenth century, when
the Norman Jean de Béthencourt established himself in
the Canaries, that something like substantial information
respecting those islands was made accessible to Europeans.
Much earlier expeditions it is true had been attempted, but
of the navigators who visited those islands before the
fifteenth century, some only landed accidentally, and others
went for the purpose of taking slaves, or goats' flesh, or else
to gather orchil for dyeing, and dragons' blood or other
products that might be useful in commerce. That the
Canaries were visited, but visited only, by the Portuguese,
even earlier than the year 1345 is proved by a passage in a
letter from Alphonso fourth King of Portugal to Pope
Clement VI. which was written under the following circumstances. When Alphonso the eldest son of the Infant Don
Ferdinand, and grandson of King Alfonso the Wise, was
deprived by his uncle Dom Sancho of the succession to the
Crown of Castile, he retired in indignation to France to the
Court of his uncle Philippe le Bel. He there married
Marhaut or Mafalda daughter of Amery VI., Viscount of
Narbonne, by whom he had Luis of Spain, called by almost
all the Spanish historians, Luis de la Cerda, Count of
Talmond, and Admiral of France. On the death of John
III., Duke of Britany, a civil war divided the country into
two parties. England took the part of the Count de Mont-
fort, the Duke's brother, while the King of France maintained that of his nephew the Count de Blois, who had been
called to the succession by the Duke himself. In this
contest, Don Luis commanded in several engagements
against England, till at length Pope Clement VI. obtained

a truce, signed at Malestroit on the 19th January, 1343, which was to last three years, so that terms of peace might in the interval be negociated in the Pope's presence at Avignon. One of the plenipotentiaries was Luis of Spain, and as the negociations were greatly protracted by repeated delays on the part of the King of England, he remained there till the beginning of the year 1345.

During his stay at Avignon, Don Luis represented to the Pope that there were islands in the ocean, named the Fortunate Islands, some of which were inhabited and others not, and that he wished to obtain possession of these for the exaltation of the Faith and the spread of Christianity, and for this purpose he prayed his Holiness to grant him the necessary authority and the title of King of these islands. The Pope granted his request, and by a Bull dated from Avignon, November 15th, 1334, bestowed on him the lordship of the Fortunate Islands with the title of Prince of Fortune, to remain in perpetual fief to the Apostolic See, to which it should pay annually 400 florins of good and pure gold of Florentine coinage; and Don Luis gave an acknowledgment of the fief on the 28th of November of the same year. At the same time the Pope wrote letters to the Kings of France, of Sicily, of Aragon, of Castile, and Portugal, as well as to the Dauphin, and to the Doge of Genoa, desiring them to help the new king in this enterprise. The reply of the King of Portugal contains the passage to which allusion has been made. While submitting, from habitual reverence, to the desire of his Holiness, he reminded him that he had already sent out expeditions to those islands, and was only prevented from sending out a large armada by the wars in which he became involved, first with the King of Castile, and afterwards with the Saracens. The letter finished with the King's excusing himself on account of the exhausted condition of his treasury from supplying Don Luis with ships and soldiers, but expressing his willingness to furnish him to the extent of his power with provisions, and other supplies. This letter was dated

from Monte Mor, 12th of February, 1345. The war with
Spain, to which the king referred, broke out at the close of
1336, whence it follows that his assertion that he had
thereby been prevented from sending out a large armada to
those islands, either means that previously to that year the
Portuguese had sent out expeditions to the Canaries or that
expeditions which he had sent out during the war would, but
for the war, have been equipped on a grander scale. Meanwhile we have evidence to show that in 1341 a voyage was
made to the Canaries, under the auspices of the King of
Portugal, in a narrative for which we are indebted to the poet
Boccaccio, and which has been rescued from oblivion so
recently as 1827 by the learned Sebastiano Ciampi. It was
derived from letters written to Florence by certain Florentine
merchants established at Seville, under date of the 17 kalend
of December, 1341.[r]

The narrative records that " On the 1st of July of that same
year, two vessels furnished by the King of Portugal with all
the necessary provisions, and accompanied by a smaller
vessel, well armed and manned by Florentines, Genoese,
Castilians, and other Spaniards, among whom were naturally
included Portuguese, for the word Hispani included all inhabitants of the Peninsula, set sail from Lisbon, and put out
into the open sea.[*] They took with them horses, arms, and
warlike engines for storming towns and castles, in search of
those islands commonly called the ' Rediscovered.' The
wind was favourable, and on the 5th day they found land.
They did not return till the month of November, when they
brought back with them four of the natives, a large quantity
of goat skins, the fat and oil of fish, and seal skins; red
wood which dyed almost as well as the verzino (Brazil
wood), although connoisseurs pronounced it not to be the
same; the barks of trees to stain with a red colour; red
earth and other such things. Nicoloso de Recco, a Genoese,

---

[*] " The Florentine who went with these ships was Angelino del Tegghia dei
Corbizzi, a cousin of the sons of Gherardino Gianni," according to what we learn
from a marginal note by Boccaccio.

the pilot of this expedition, stated that this archipelago was nearly nine hundred miles from the city of Seville ; but that reckoning from what is now called Cape St. Vincent, the islands were much nearer to the continent, and that the first of those which they discovered [most probably Fuerteventura] was a hundred and fifty miles in circumference ; it was one mass of uncultivated stony land, but full of goats and other beasts, and inhabited by naked men and women, who were like savages in their appearance and demeanour.   He added that he and his companions obtained in this island the greater part of their cargo of skins and fat, but they did not dare to penetrate far into the country.   Passing thence into another island [Great Canary], somewhat larger than the first, a great number of natives of both sexes, all nearly naked, came down to the shore to meet them.   Some of them, who seemed superior to the rest, were covered with goats' skins coloured yellow and red, and, as far as could be seen from a distance, the skins were fine and soft, and tolerably well sewn together with the intestines of animals.   To judge from their gestures, they seemed to have a prince, to whom they showed much respect and obedience.   The islanders showed a wish to communicate with the people in the ship, but when the boats drew near the shore, the sailors who did not understand a word that they said did not dare to land. Their language however was soft, and their pronunciation rapid and animated like Italian.   Some of the islanders then swam to the boats, and four of them were taken on board and afterwards carried away.   On the northern coasts of the island, which were much better cultivated than the southern, there were a great number of little houses, fig trees and other trees, palm trees which bore no fruit, and gardens with cabbages and other vegetables.   Here twenty-five of the sailors landed, and found nearly thirty men quite naked, who took to flight when they saw their arms.   The buildings were made with much skill of square stones, covered with large and handsome pieces of wood.   Finding several of them closed the sailors broke open the doors with stones, which enraged the fugitives, who filled the air with their cries. The houses were found to contain nothing beyond some excellent dried figs, preserved in palm baskets, like those made at Cesena, corn of a much finer quality than the Italian, not only in the length and thickness of its grain but its extreme whiteness, some barley and other grains. The houses were all very handsome and covered with very fine wood, and as clean inside as if they had been whitewashed. The sailors also came upon a chapel or temple, in which there were no pictures or ornament, but only a stone statue representing a man with a ball in his hand. This idol, otherwise naked, wore an apron of palm-leaves. They took it away and carried it to Lisbon. The island seemed to be thickly peopled and well cultivated ; producing not only corn and other grain, but fruits, principally figs. The natives either ate the grain like birds, or else made it into flour, and ate it with water without kneading. On leaving this island they saw several others, at the distance of five, ten, twenty, or forty miles, and made for a third, in which they remarked nothing but an immense number of beautiful trees shooting straight up to the skies [most probably Ferro, remarkable for its magnificent pines]. Thence to another, which abounded in streams of excellent water and wood [Gomera]. They found also many wild pigeons, which they killed with sticks and stones. They were larger and of better flavour than those in Italy. Falcons and birds of prey were numerous. The sailors ventured but a very little way into the country. At length they discovered another island, the rocky mountains of which were of immense height and almost always covered with clouds, but what they could see during the clear weather seemed very agreeable, and it appeared to be inhabited [Palma]. They afterwards saw other islands, making in all thirteen, some of them inhabited and some not, and the further they went the more they saw. They remarked the smoothness of the sea which separates these islands, and found good anchorage, although there were but few harbours, but all the islands were well provided

with water. Of the thirteen islands five were inhabited,
but some were much more populous than others.* The
languages of these people were said to be so different, that
those of one island did not understand those of another, and
they had no means of communication except by swimming.
A phenomenon which they witnessed on one of these islands
[Teneriffe] deterred them from landing. On the summits
of a mountain which they reckoned to be more than thirty
thousand feet high they observed what from its whiteness
looked like a fortress. It was however nothing but a sharp
point of rock, on the top of which was a mast, as large
as a ship's mast, with a yard and a lateen sail set upon it.
This sail when blown out by the wind took the form of a
shield, and soon afterwards it would seem to be lowered,
together with the mast, as if on board a vessel, then again
it was raised and again would sink, and so alternately.

" They sailed round the island, but on all sides they saw
the same phenomenon, and thinking it the effect of some
enchantment, they did not dare to land. They saw many
other things also, which Niccoloso refused to relate. At any
rate the islands do not seem to have been very rich, for the
sailors hardly covered the expense of the voyage.

" The four men whom they carried away were young and
beardless, and had handsome faces. They wore nothing
but a sort of apron made of cord, from which they hung a
number of palm or reed fibres of a hair's-breadth and a half
or two hairs'-breadth, which formed an effectual covering.
They were uncircumcised. Their long light hair veiled their
bodies down to the waist, and they went barefooted. The
island whence they were taken was called Canary, and was
more populous than the others. These men were spoken to
in several languages, but they understood none of them.
They did not exceed their captors in stature, but they were
robust of limb, courageous, and very intelligent. When

* Thirteen is correct if the desert islands be added to the seven inhabited
ones. Those inhabited are here counted five instead of seven, doubtless from
defective exploration.

spoken to by signs, they replied in the same manner, like mutes. There were marks of deference shown from one to another; but one of them appeared more honoured than the rest. The apron of this chief was of palm leaves, while the others wore reeds painted in yellow and red. They sang very sweetly, and danced almost as well as Frenchmen. They were gay and merry, and much more civilised than many Spaniards. When they were brought on board, they ate some bread and figs, and seemed to like the bread, though they had never tasted it before. They absolutely refused wine, and only drank water. Wheat and barley they ate in plenty, as well as cheese and meat, which was abundant in the islands, and of good quality, for although there were no oxen, camels, or asses, there were plenty of goats, sheep, and wild hogs. They were shown some gold and silver money, but they were quite ignorant of the use of it; and they knew as little of any kind of spice. Rings of gold, and vases of carved work, swords and sabres were shown to them; but they seemed never to have seen such things, and did not know how to use them. They showed remarkable faithfulness and honesty, for if one of them received anything good to eat, before tasting it, he divided it into portions which he shared with the rest. Marriage was observed among them, and the married women wore aprons like the men, but the maidens went quite naked, without consciousness of shame."

Meanwhile the Prince of Fortune made but little progress towards the acquirement of the royal domain with which the Pope had endowed him. In short, the whole project proved a mere abortion, and neither the treasury of the Pope, the property of Don Luis, nor the knowledge of the geography of the Canaries, were advanced one iota thereby.

The enterprise of the Norman Jean de Bethencourt, a century and a half later, was of a far more persistent and effectual character. Having conceived the project of conquering the Canaries, which were then only frequented by merchants or Spanish pirates, he assembled a body of adventurers, among whom was a knight named Gadifer de la Salle, who joined him at Rochelle. He first made a descent on the island of Lancerote, established himself there, and undertook the conquest of the other islands; but not having enough people to effect this enterprise, he went to ask help of the King of Castile, to whom he made homage of the islands. The King conceded to him the sovereignty of the Canaries, with the right of coining money. He also gave him twenty thousand maravedos for present expenses, and a well-found ship with eighty men. By means of these reinforcements he subjugated the island of Fuerteventura. He then revisited France, and there collected a new troop of people of all classes, with their wives and children, whom he brought to his new states, and succeeded in conquering the island of Ferro. Resolving now to finish his days in France, he distributed his lands to those who had helped in his conquest, and named his nephew Maciot de Béthencourt governor-general, as his representative; enjoining him to do justice according to the customs of France and Normandy. He set out on the 15th December, 1405, first for Spain, where he renewed his homage, and obtained a bishop for the Canaries. Thence he went to Rome, where he received from the Pope the bull of installation for the Spanish bishop. He returned in 1406 to his lands in Normandy, and died in 1425.

It will have been seen in a previous chapter that on the authority of the great Portuguese historian, De Barros, the names of Porto Santo and Madeira were for three centuries accepted as having been, for the first time, given to those islands on their assumed first discovery by Zarco and Vaz, in 1418-20. But while it is to the Portuguese, under the auspices of Prince Henry, that we owe the colonization of the lovely island of Madeira, and the development of its valuable resources, there can be no doubt that its discovery, although neglected, had already been made at an earlier period. From Lisbon, nevertheless, it would seem, if an ingenious inference by the learned M. d'Avezac be correct,

the first expedition took place which was to remove, though
but with a shadowy hand, some portion of the mist which
held that island enveloped in Atlantic obscurity.

In that deluge of Mahometan invasion which, so soon
after the rise of the false prophet, overswept the surface of
civilized humanity with a force unexampled in the history of
ancient Rome, the Iberian peninsula at length became a
victim. After the death of Don Roderic and the extinction
of the kingdom of the Goths at the beginning of the eighth
century the Moslems with wonderful rapidity compelled
nearly the whole of Spain as well as Portugal to yield to
their victorious arms. Cordova and Granada became the
two principal seats of government, and Lisbon also became
essentially a Mussulman city. "From Lisbon," then,
according to Edrisi, who was the first to write an account of
the voyage, "the Maghrurins, or 'strayed ones,' set sail
with the object of learning what was on the ocean, and what
were its boundaries. They were eight in number, and all
related to each other. Having built a transport boat, they
took on board water and provisions for many months, and
started with the first east wind. After a sail of eleven days
or thereabouts, they reached a sea whose thick waters ex-
haled a fœtid odour, concealed numerous reefs, and were but
faintly lighted. Fearing for their lives, they changed their
course and steered southwards for twelve days and reached
the island of El Ghanam, so named from the numerous
flocks of sheep which pastured thereon without a shepherd
or any one to tend them. On landing they found a spring
of running water and some wild figs. They also killed some
of the sheep, but the flesh was so bitter that they could not eat
it, and they were obliged to content themselves with taking
the skins. For twelve days more they sailed southwards, and
discovered an island in which were habitations and cultivated
fields. As they approached it they were surrounded by boats,
made prisoners, and carried in their own boats to a city on
the sea-shore. They reached a house in which were men of
tall stature, dark-skinned, with short but straight hair, and

women of uncommon beauty.   In this house they were confined for three days, and on the fourth there came to them
a man who spoke Arabic, and who asked them who they
were, what they sought, and where they came from.   They
related to him their adventures, and he gave them good
encouragement and told them that he was the king's interpreter.   Two days afterwards they were presented to the
King, who put similar questions to them;. to which they
replied that they had ventured out to sea for the purpose of
making themselves acquainted with its wonders and curiosities, and of ascertaining its limits.   When the King heard
them talk in this fashion he laughed heartily, and told the
interpreter to explain to them that in former times his
father had ordered some of his slaves to venture out on
that sea, and that after sailing across the breadth of it for a
whole month they found themselves deprived of the light
of the sun, and returned without having either gained or
learned anything.   The King furthermore desired the interpreter to assure the adventurers of his friendly disposition.
They returned to their prison and there remained until a west
wind arose, when they were blindfolded and put on board a
boat and taken out to sea.   When they had been out three
days and three nights they reached land, and the wanderers
were put on shore with their hands tied behind them, and
there left.   They remained there till sunrise in a miserable
condition from the tightness of the cords with which they
were bound, but hearing some laughter and human voices
near them they began to shout.   Some of the inhabitants of
the country came to them, and seeing their wretched plight,
unfastened them and questioned them as to their adventure.
They were Berbers, and one of them asked the wanderers if
they knew how far they were from their own country.   On their
answering in the negative, he told them that it was two
months' sail.   The person who seemed to be of most consideration amongst them said repeatedly Wasafi (alas), and
accordingly they took that to be the name of the locality,
and ever since it continues to bear the name of Asafi.   They

reached Lisbon in considerable confusion at their disappointment, and from that received the name of the Maghrurins,
or 'strayed ones,' and from these adventurers a street at the
foot of the hot bath in Lisbon took the name of the street
of the Maghrurins." On this story M. d'Avezac makes the
following ingenious observations :—"Eleven days west of
Lisbon and then twelve days to the south would bring them
to Madeira, which would be the island of El Ghanam or El
Aghnam, the latter being the plural of the former word,
which means, 'small cattle.' The name El Aghnam has a
remarkable resemblance in sound to the Italian name of the
island Legname, which occurs, as will be presently seen, on
maps anterior to the Portuguese discovery, and of which
name Madeira was simply a translation. It should be observed, however, that the word Ghanam or Aghnam, which
generally implies flocks of sheep, would here rather mean
herds of goats, whose flesh is rendered bitter, according to
M. Berthelot, the author of the 'Natural History of the
Canaries,' by a plant, le coquerel, which they sometimes
browze upon." Whether M. d'Avezac's ingenious derivation
be correct or not, it is certain that the Madeira group was
discovered in the early part of the fourteenth century, and
I now propose to prove that that discovery was made in
Portuguese ships commanded by Genoese captains.

In the Portulano Mediceo of the date of 1351 in the Library
at Florence, an extract from which the reader has already seen,
the Madeira group is distinctly represented, bearing names,
in the two instances of Porto Santo and the Desertas, identical with those which they at present bear, while the island of
Madeira is called "Isola dello Legname" or "Island of Wood,"
of which the name "Madeira" is simply a translation. The
Portulano is anonymous, but Count Baldelli Boni in his
valuable edition of the "Milione of Marco Polo," published
in Florence in 1827, adduced admirable proofs to show that
it was of Genoese construction. Against the island of
Lançarote in the Canaries is inserted the shield of Genoa,
distinctly claiming the priority of discovery in favour of

that republic, and Count Baldelli with reason remarks that
no Venetian or Pisan or Catalan would be the first to lay
down, on a map so important, a fact in favour of their rivals
the Genoese. It is right however to observe that on the
later Venetian map by the brothers Pizzigani of 1367, and
in the Catalan map of 1375 this remarkable indication is
inserted. Perhaps a stronger argument is derived from the
use of the Genoese dialect in the names in preference to
that of Venice or Pisa. Now if upon a Genoese map we
find both the Madeira group and the Canary group laid
down for the first time within our knowledge, but with the
arms of Genoa inserted against the latter, but not against
the former, the legitimate inference is that in the one case
a claim was reserved for Genoa to which in the other they
could make no pretensions. It is this theory which I now
propound as a new one, and which I propose to corroborate
by well authenticated historical facts. M. d'Avezac, with
his usual untiring research, has bestowed great labour upon
the inquiry into the discovery and naming of the Island of
Lançarote. He has shown that the discoverer was of the
ancient, but now extinct, Genoese family of Malocello. In
the visit of the Norman knight Jean de Béthencourt to that
island in 1402, it is said that they stored their grain in an
old castle reputed to be built by Lancelot Maloisel. In a
Genoese map of the date of 1455, made by Bartolommeo
Pareto, are inserted against the same island the words
"Lansaroto Maroxello Januensis," and further we are led to
believe that the discovery was made as early as the thirteenth
century from a passage in Petrarch which declares that
a patrum memoriâ, i.e. a generation back, an armed fleet of
Genoese had penetrated as far as the Fortunate Islands.
Now as Petrarch was born in 1304, if, as is highly probable,
Lancelote Malocello's voyage was the one alluded to, it
will have taken place at the latest in the close of the
thirteenth century. We thus find a reason for the reserva-
tion by Genoese map makers of the claim of their country
to the Island of Lançarote; but it may be asked, if the

Genoese were the first, as it appears, to delineate the Madeira group upon a map, and thereby to show that they were the discoverers of that group, how comes it that they did not claim it for their own by the same process adopted with reference to Lançarote? There can be no doubt that, if they could have set up such a claim, they would, but meanwhile we are provided by history with what appears to be a very satisfactory answer. By a treaty concluded in 1317, Denis the Labourer, King of Portugal, secured the services of the Genoese Emmanuele Pezagno as hereditary admiral of his fleet, with a distinct understanding that he and his successors should make unfailing provision of twenty Genoese captains experienced in navigation to command the king's galleys.

In the year 1326 we find this same Emmanuele Pezagno sent by Affonso IV. as ambassador to our own King Edward III., who regarded him with such favour, that on July 24, 1332, he addressed a letter to Affonso, recommending both Emmanuele and his son Carlo to his especial patronage. Further, the document in the handwriting of Boccaccio, discovered in 1827 by Sebastiano Ciampi, informs us that in the year 1341 two Portuguese vessels commanded by Genoese captains, but manned with Italians, Spaniards of Castile, and other Spaniards, comprising doubtless Portuguese, for the word " Hispani " included both nations, made a re-discovery of the Canaries. Even so late as 1373, we find the rank of admiral of the Portuguese fleet remaining in the hands of Lancelot, son of Emmanuele Pezagno, who received it from Peter I. by letters patent dated 26th June, 1357. Thus from 1317 to 1351 we have a range of thirty-four years for the discovery of the islands laid down on this important Genoese map. The exact year of this discovery is not known, but enough has been said to demonstrate that the Genoese map of 1351 indicates the discovery of the Madeira group by Genoese navigators in a foreign service, while we have the evidence that such service was rendered by Emmanuele Pezagno to the King of Portugal. Politically the question is without importance, for if any doubt could be

thrown on the claim of Portugal to these islands on the
ground of Genoese commanding Portuguese ships in this
earlier actual discovery, nevertheless the accidental re-discovery of the group by the Portuguese in 1418-20 led to the
first colonization and fertilization of the islands, and it would
be as futile to dispute such a claim as it would be to negative
the English claims to the colonization of Australia on the
ground of those early authenticated discoveries in that vast
island by the Portuguese, which it has already been my good
fortune historically to establish. This engagement of Genoese
navigators by the Kings of Portugal in the fourteenth century
cannot diminish by one iota the transcendent glory of that
heroic little nation, to whom in truth we owe the knowledge
of one-half of the globe that we possess.   My late honoured
friend the Vicomte de Santarem, in his patriotic ardour,
endeavoured to carry back the claim of the Portuguese to
comparative maritime distinction to an earlier period than
was either just or reasonable.   It was not reasonable to expect that a people seated on the open Atlantic, that dreadful
and unmeasured ocean whose mysterious immensity had
gained for it the name of the Sea of Darkness, should so early
gain experience in navigation as the comparatively protected
occupants of an inland sea, allured by the wealth of seaports
within easy reach, and encouraged by antecedents which filled
the history of centuries.

CHAPTER IX.

TANGIER.

1437.

THE personal qualities of King João's successor, Dom Duarte, promised most favourably for the maintenance of that prosperity which had been bequeathed to the kingdom by the energy and wisdom of his father, yet was his reign destined to misfortune from its beginning to its close. On the morrow of his father's death as he was about to be proclaimed king at Lisbon, his physician Mestre Guadalha, who was held in high consideration as an astrologer, counselled him to postpone the ceremony on the ground that the stars at that time foreboded him misfortune. The king gave no heed to the superstitious words of the soothsayer, who forthwith, in the presence of a great concourse of people, prognosticated that the years of the king's reign would be few and full of troubles. The prediction and its accurate fulfilment have been consolidated in the records of history. The ceremony nevertheless took place in conformity with the usual custom.

From Lisbon the king went to Cintra, where his wife and children were, and here a noticeable novelty was introduced, for when the Princes of the royal household did homage and took the oath of allegiance, the eldest son of the King, afterwards Affonso V., but then little more than a year and a half old, received the style and title of "Prince of Portugal" instead of that of "Infant." This change had been lately adopted in the Peninsula, in imitation of the title of

" Prince of Wales," given to the eldest son of the kings of
England, and of that of "Dauphin," given to the eldest son
of the King of France. Thus the eldest son of the King of
Castile was called "Prince of Asturias," while the eldest son
of the King of Aragon received the title of "Prince of
Gerona," and so from that time forward the heir to the
crown of Portugal was styled "Prince of Portugal."

The trouble with which the king had been threatened
began to show itself betimes. The king's youngest brother
Dom Fernando was especially desirous of emulating the
prowess of his brothers in Africa. In this desire he was
greatly encouraged by Prince Henry, the aim and object of
whose life was to make discoveries and conquests in that
direction, and together they decided on attempting an
attack upon Tangier. Accordingly they besought the king
their brother to fit out an expedition for them against the
Moors. The king at first affectionately but firmly refused,
for the exchequer had been seriously reduced by many
causes, but at last their arguments and the influence of the
queen prevailed, and against his judgment he reluctantly
gave his consent.

His first measure was to meet the Cortes at Evora, and
demand of them the necessary funds for the expedition.
These were readily granted, but the grant called forth much
discontent and many complaints from the people. Dom
Pedro, Dom João, and the Count of Barcellos also remon-
strated with the king on the course he was pursuing, and, as
in his heart he acquiesced in all their arguments, he resolved
to rid himself of the responsibility by applying to the Pope.

The question having been laid before the consistory, and
duly considered, the following answers were returned: That
if the infidels in question occupied Christian territory and
turned churches into mosques, or if, though occupying their
own lands, they did injury to Christians, or even if, while
doing none of these things, they were idolaters or sinned
against nature, the princes would be justified in making war
upon them. Nevertheless they should do so with piety and

discretion, lest the people of Christ should suffer death or losses. Concerning the levying of imposts for the prosecution of the war, it was decided that war might justly be made against the infidels in two ways: 1st. Of necessity, in defence of territory; 2nd. Voluntarily, for the purpose of conquering land from the heathen. In the first case taxes might be imposed, but the voluntary war could only be carried on at the personal expense of the king. Before this decision arrived, however, the king, influenced by the queen or by the promise he made to his brothers, had brought his preparations for the enterprise to such a point as to render the answer futile.

Prince Henry has not been held entirely free from blame in the matter. True it is that the advancement of Christianity and civilization, the good of his country, the dictates of chivalry, the furtherance of his brother's wishes, and his own love of glory, all conspired to set before him in the light of duty, the enterprise which he thus warmly advocated. True it is, also, that the original invasion of Ceuta had been attended with an unlooked for success in the highest degree encouraging to the aspirations of a courageous and ardent mind, and that in that invasion his judgment, no less than his valour, had given him so high a standing in the estimation of his illustrious father, as to gain him the chief command in preference to his elder brothers, yet there can be no doubt that in this instance, as on the occasion of his proposed attack on Gibraltar, his zeal was allowed to outrun his discretion. The dictum of the consistory respecting the indiscreet sacrifice of Christian life in waging war against the infidels might, had it arrived in time, have been accepted by him as a wholesome warning, but it did not arrive in time; and it may be further urged in his extenuation, that if a hesitating cautiousness had always been allowed to repress enthusiasm, history would now be wanting in the records of full many an heroic achievement.

At length the preparations were completed, and on the 26th of August, 1437, the princes landed at Ceuta, of which

Count Pedro de Menezes was still the commander. Their arrival soon became known in all the surrounding districts, and the tribe of Ben Hamed sent messages to Prince Henry praying for peace, and offering him tribute of gold, silver, cattle, and wood, and the Prince accepted them as vassals of the king. The Prince then reviewed the force which he had brought with him, and found it to consist only of two thousand cavalry, one thousand cross-bowmen, three thousand infantry, so that of fourteen thousand which had been promised him, eight thousand were missing. This shortcoming was caused by the reluctance of the people to risk their lives and property in what they considered a rash adventure, and also by the lack of ships to convey a greater number of men to the African shore. In consequence a serious question arose among Prince Henry's counsellors as to whether Dom Duarte should not be applied to for a sufficient force before further steps were taken, but the Prince, fearing lest any delay might be fatal to the expedition, overruled their doubts, and promised them the greater honour if they conquered with so small a force.

Finding that the shortest road to Tangier across the Sierra Ximera was strongly guarded, they decided to go by the Monte Negrona through Tetuan and the valley of Angela. Dom Fernando, being ill and unequal to the journey by land, went by sea. After two days' march they came before Tetuan, which surrendered without resistance. On the 13th of September Prince Henry arrived with his army before Old Tangier, which was already deserted, and there found Dom Fernando awaiting him. He then made arrangements to encamp along the sea-coast, and while the troops were thus engaged, a report was spread that the inhabitants of Tangier had opened their gates with the intention of abandoning the place. This news proved to be so far from the truth that the Portuguese were engaged till nightfall in endeavouring to force the gates, and then withdrew, carrying off the Count de Arrayolos and Alvaro Vaz de Almada wounded. There were in the city about seven thousand fighting men, including many cross-bowmen from Granada. They were commanded by Zalá ben Zalá, the same who had twenty-two years before lost Ceuta.

On Saturday, the 14th September, Prince Henry had completed his encampments, and from that time till the following Thursday was occupied in landing the artillery and munitions. On the morning of Friday, the 20th of September, Prince Henry ordered the trumpets to sound to battle. Dom Fernando, the Count Arrayolos, and the Bishop of Evora were to scale the walls at different points, and Prince Henry took upon himself the attack on the gate of the fortress, where the greatest resistance would be made. For this purpose he took with him only two mantas or mantelets,* without any scaling-ladder.

The engagement commenced in the morning and lasted till six o'clock, when the Portuguese were obliged to retire with loss. All attempts to force the gates had been utterly useless, for they had been very strongly walled up by the Moors, with stone and mortar. The contemplated attack with the scaling-ladders proved abortive, for the ladders were too short to reach the top of the wall. Prince Henry was therefore compelled to withdraw, and on mustering his people found that he had five hundred wounded and twenty killed. He ordered that the artillery should remain in charge of the Marshal, and the Captain Alvaro Vaz de Almada, who being left close under the walls, and at a distance from the camp, received much injury from the Moors, but nevertheless valiantly stood their ground. Prince Henry now sent to Ceuta for longer scaling-ladders and also for two large pieces of cannon, together with powder and shot, for the guns which he had were too small and ineffective. During ten days there were repeated skirmishes, in which several Portuguese noblemen were slain.

At length, on the 30th of September, a body of ten thousand Moorish horsemen and ninety thousand foot came to the

---

* Mantelets were temporary and movable defences formed of planks, under cover of which the assailants advanced to the attack of fortified places.

assistance of the city, and took their stand on a hill within
sight of the camp. But when Prince Henry went out to
meet them with fifteen thousand cavalry, eight hundred
cross-bowmen, and two thousand infantry, they were seized
with a panic and took to flight. The next day the same
manœuvre was repeated, and on Thursday, the 3rd of
October, the Moors advanced in yet greater numbers and
drew near to the camp. The Prince again went out to
meet them, and drove them from their position with considerable loss. Meanwhile another attack was made upon
the camp by the Moors, but they also were repulsed by
Diogo Lopez de Souza, who had been left to defend it. This
engagement was of the highest importance, for had either of
the attacks proved successful, the Portuguese army must
inevitably have been destroyed.

On the 5th of October the scaling-ladders were replaced,
and a wooden tower moving on wheels, and containing
men supplied with missiles, was provided for the purpose of
being brought up to the level of the walls, to facilitate the
escalade by driving from the parapets those who were sta-
tioned there. The Prince then ordered a second assault to
be made upon the town, at a spot where the batteries had
made a breach in the wall. This attack was led by himself
in person, the remainder of the troops under arms being
entrusted to Dom Fernando, the Count de Arrayolos and the
Bishop of Evora, to make a stand against the Moorish army,
in the event of their attacking the lines during the action.

This assault was as unsuccessful as the former, for only one
scaling-ladder was brought to rest against the wall, and that
was burnt by the Moors, and those who were upon it were
killed. Not one of the others, nor even the wooden tower
could be brought up to the wall, for as no other attack,
either feigned or real, was made elsewhere, the whole garrison
was able to repair to the point assailed, and with firearms
and other missiles compelled the Portuguese to withdraw
with great loss.

On the 9th the Moors appeared in great multitudes, accompanied by the Kings of Fez and Marocco and the other neighbouring princes. They forthwith attacked the advanced posts of the Portuguese army, and opened communication with the fortress, at the same time taking possession of the Portuguese batteries with all the artillery and munitions for the siege. Prince Henry had his horse killed under him, and found himself fighting on foot in the midst of the enemy, from which peril he escaped at the sacrifice of the life of his chief engineer, Ferdinand Alvarez Cabral, who with devoted gallantry came to his rescue. An additional act of devotion on the part of a page of Dom Fernando provided him with another horse, mounted on which he cleft his way in safety through the enemy.

When the Prince reached the camp he found the Portuguese overwhelmed with the great odds against which they had to contend, and to add to his dismay he found that about a thousand of his men had fled to the ships. Happily Dom Pedro de Castro, who was in command of the fleet, came to his aid with reinforcements. Oppressed as he was with toil and anxiety, the Prince showed no sign of shrinking from the high requirements of his responsible position. Though surrounded by danger the most imminent, he encouraged his men by an appearance of confidence and cheerfulness, which he was far from feeling in his heart.

On the following day the Moors again attacked the trenches, but they were now more strongly fortified, and after four hours of hard fighting the Moors were repulsed with immense loss.

At length when their provisions were well-nigh all consumed, Prince Henry came to the resolution to force a passage in the night-time to the shore and withdraw with the fleet. In this plan however he was frustrated by the treachery of his chaplain, Martin Vieyra, who deserted to the Moors and informed them of the Prince's resolution. The Moors now suspended their attacks and deliberated as to the best course to pursue in the probable event of the Portuguese falling into their hands. Some were for exterminating them without mercy, others with greater wisdom suggested that such a massacre would only provoke the Christians to revenge, and that therefore the most prudent course would be to let them freely depart, upon condition that they surrendered Ceuta, and delivered up their artillery and arms and baggage, with all the Moors that had been taken prisoners. This proposal was made, and after a short deliberation accepted by the Portuguese, who in fact had no alternative. Prince Henry accordingly sent Ruy Gomez da Sylva, chief constable of the camp, a man of great prudence and courage, and Payo Rodriguez the secretary of dispatches, to conclude the treaty with the King of Fez and the other Princes of Marocco.

Meanwhile a great number of Moors, who either were ignorant of the importance of Ceuta or were very doubtful of its being surrendered, were determined to make another vigorous onset upon the Portuguese camp. They principally directed their attack upon the side which was defended by Dom Fernando, and their numbers and the ferocity of the onslaught placed the Prince in considerable danger. But the Portuguese fought with desperation, and the neighbourhood of the intrenchments was soon covered with the bodies of the dead and wounded Moors. They then endeavoured to set fire to the palisades, but the indefatigable energy of Prince Henry averted this danger also. At his side fought the Bishop of Ceuta, whose intrepidity encouraged the soldiers with a fervour of pious zeal which worked wonders in the unequal contest. The struggle having lasted for seven hours without any decisive result on either side, the Prince determined on reducing the area of the camp and bringing it nearer to the sea, a task which, in spite of the fatigues of the preceding day, was effected in one night. The Moors offered no opposition, but contented themselves with occupying the ground between the camp and the shore, and guarding the neighbouring passes.

Meanwhile the Portuguese were obliged to kill their horses for food, and to use their saddles for fuel to cook them. In addition they were tormented with thirst, for within the lines there was but one well, which was not sufficient to supply a hundred men with water, so that if some rain had not fallen, they must all have perished. Many of these disasters would have been averted, had Prince Henry in the first instance kept his camp near to the sea-shore, in accordance with the wise instructions which had been given him by the King, his elder brother. Before leaving Lisbon he had received an autograph dispatch from the King, containing a special injunction so to fix his camp before Tangier that he should touch the shore at two points, and if, from a deficiency of numbers, that could not be effected, he was by no means to neglect retaining a communication with the sea at least at one point.* This recommendation was accompanied with an urgent request that it might often be read and never infringed, and Prince Henry had promised to observe it to the fullest possible extent. Nor does there appear to have been reason for deviating from these precautionary instructions, and men of calm judgment attributed much of the disastrous result of the expedition to this want of implicit attention to the King's instructions. To establish a communication with the fleet had now become a matter of great difficulty and danger, if not of impossibility.

Fortunately for the Portuguese the enormous losses suffered by the Moors gave them an inclination to subscribe to terms of peace. Hence happily it followed that on the 15th of October a treaty was concluded, by virtue of which the Portuguese were at liberty to embark, but simply in their clothes as they stood, delivering up their arms, their horses, and their baggage. Ceuta, with all the prisoners therein, was to be surrendered, and a pledge given by the King of Portugal, on behalf of his country, that peace should be maintained with all Barbary for a hundred years, both by sea and land. Dom Fernando, with twelve other nobles, was given over as an hostage until the surrender of Ceuta and the prisoners, while on the side of the Moors the eldest

---

* Sousa, Provas, tom. i. p. 533, et seq. Pina, cap. xxi. p. 138.

son of Zalá ben Zalá, the Lord of Tangier and Arzilla, and
one of the most powerful vassals of the King of Fez, was
delivered as a pledge of their security.

When the delegates returned to the Prince, they informed
him that the Moors had conceived the treacherous plan of
taking all the Portuguese prisoners, if they availed themselves of the conditions of the capitulation to enter the town
with the view of embarking. Prince Henry consequently
gave orders for every preparation to be made for embarking
as quickly as possible. In the attempt, however, to reach
the boats, about sixty men of the rear-guard were slain.

On Sunday, the 20th of October, the fleet set sail. Out
of thirty-seven days that they had been under the walls of
Tangier, twenty-five had been occupied in besieging the
Moors, but during the remaining twelve they had been themselves besieged. Their losses, however, they reckoned at
only five hundred men, while the Moors must have counted
at least four thousand slain and many thousands wounded.
To the latter this loss was insignificant when compared with
the extent of their population, whereas Portugal, with its
limited range of territory, had no superfluity of men to
spare; but, worst of all, the Portuguese had failed in their
object.

Such was the disastrous termination of this imprudent
enterprise, and however much we may admire the distinguished heroism of Prince Henry, or honour the nobility of
the motives which overruled his judgment, it must be confessed that to him the blame of the disaster must be mainly
attributed. The foresight and wisdom which he had so often
exhibited in matters of detail were wanting in his consideration of the requisites for an enterprise which was dictated to his feelings and his fancy by the prevailing instinct
of his nature, viz., a chivalrous devotion to what he conceived to be religious duty to God and to his country. It
was, in the first place, unjustifiable to imperil on a foreign
shore the lives of a courageous little army so inadequate in
their numbers to the work set before them, and, in the

second place, it was an imperative duty to secure, as far as possible, the safety of such courageous followers by every prudent precaution; and proportionately culpable was the dereliction from that duty when enforced by the most emphatic injunctions, even in the handwriting of the sovereign. Of the indomitable energy and valour of the Prince we have already witnessed proofs of an extraordinary kind, yet even these, supported by efforts to which they proved a most encouraging example, were insufficient to avert the melancholy result which we have had to describe. But this was not the end of the tragedy. We have now to recount the sad story of the sufferings and death of the devoted but hapless Dom Fernando, who was left behind as a hostage in Barbary.

After the departure of the army, the prince and his companions were conducted by Zalá ben Zalá, on the 22nd of October, 1437, to Arzilla. On their road they were treated with every insult by the Moors, who were still smarting from the losses they had suffered from the Portuguese. Meanwhile Prince Henry, having dispatched the Bishop of Evora and the Count of Arrayolos to Portugal, retired to Ceuta to await his brother's release, but on his arrival there his fatigues and grief induced an illness which entirely prostrated him. About this time he was joined by his brother, D. João, who agreed to negociate with Zalá ben Zalá the exchange of the Moorish prince, his son, for Dom Fernando, and if the terms were rejected to release his brother by force of arms. He set sail on the 20th October, but his project was frustrated by a violent tempest, which forced him, after many perils, to take refuge in the Algarves.

The King, in great grief at the sad fate of his brother, and desiring to save him, even at the cost of Ceuta, convoked the Cortes in the beginning of 1438, that he might have their consent and counsel on the subject. The members were desired to give their votes separately and by writing, and after much deliberation they finally resolved that Ceuta should not be abandoned, but that every other possible step should be taken for the release of the prince. King Duarte, in despair at this decision, applied to the Pope, the King of France, and other friendly powers, for active assistance, and received from them nothing but condolence and words of consolation. His attempt to ransom his brother was also fruitless.

After seven months of suffering and illness Dom Fernando and his followers were transferred by Zalá ben Zalá to the King of Fez, May 25th, 1438. The journey to Fez lasted six days, and was accompanied by even greater insults than had been offered on the road to Arzilla. Arrived at their destination, they were confined in the *Darsena*, a species of castle, in rooms from which every ray of light had been carefully excluded.

The unhappy prince and his companions were now in the power of the ferocious Lazurac, an unscrupulous monster, who, in the name of Abdallah the young King of Fez, exercised unlimited authority over the State.

After three months' captivity, during which they owed the very food they ate to a Majorcan merchant, they were set to work loaded with chains in the royal gardens. The only food the prisoners were allowed was two loaves daily without meat or wine. Their bed was composed of two sheepskins, their pillow a bundle of hay, and they had no covering but an old cloak. The prince slept with eleven persons in a room only large enough for eight, and they suffered much from filth, vermin, and hunger. Dom Fernando, however, suffered greater grief at the news of the King's death, than had been caused by any of his own misfortunes.

In the May of 1439, the King of Portugal offered Ceuta in exchange for the Infant, but Lazurac, hoping for a large ransom, contrived to protract the negociations. Meanwhile the unhappy Prince was treated with even greater cruelty, to which Lazurac was excited by the ulemas or holy men of the country. On one occasion letters directed to him from Portugal were intercepted, and the unhappy Moor who was the bearer of them was scourged and stoned. Some of Dom Fernando's companions narrowly escaped the same fate, and he was separated from them and placed in a more wretched dungeon than before. In this miserable hole he languished during the remaining fifteen months of his existence.

At length he was attacked with dysentery, and his enfeebled frame being unable to struggle against the malady, the Constant Prince, for such was the title which his pious resignation has won for him, breathed his last on the evening of the 5th of June, 1443.

Even the ferocious Lazurac was forced to offer tardy homage to his virtues, and to declare that, had he been a Mahometan, he would have been a saint, and that the Christians were much to blame in leaving him thus to die.

The doctor and chaplain watched over the remains till the next evening, when they were conveyed into the common prison, that his followers might remove his chains. But so overwhelmed were they by their grief that they were unable to perform this office. Lazurac had the body embalmed, that it might be preserved till he saw what the Portuguese would do to regain the body of their Prince. But his companions carefully preserved the heart, and kept it in a secret place till an opportunity should occur of conveying it in safety to Portugal. The corpse was hung up at the gate of the city head downwards, and exposed to the brutal insults and mockeries of the people for four days. It was then placed in a wooden coffin fixed in the same place on two stakes fastened into the wall; where it remained for a long time.

His faithful servants, with the exception of five who died soon after him, returned to Portugal on the death of Lazurac, and brought with them the heart of their dear master on the 1st June, 1451. By order of the King it was conveyed with great solemnity to Batalha, and placed in the tomb destined for the prince by his father. The melancholy procession was met at Thomar by prince Henry, who was about to undertake a journey. When he saw them he dismissed his equipages and joined with them in rendering the last tribute of love and respect to his devoted brother. Two-and-twenty years afterwards, the corpse of the Prince was recovered from the Moors, and brought with much pomp to Batalha, and laid in the tomb which already contained his heart.

The thought of the hapless condition of his unfortunate brother had weighed so heavily on the mind of Dom Duarte that it shortened his life. The recollection that, in spite of his own convictions and the counsels of Dom Pedro and the wisest of his grandees, he had sanctioned the attempt on Tangier, was an unceasing torment to him. Nor was his brotherly affection, wounded as it was by the pitiable sufferings to which Dom Fernando was exposed, the only cause of his distress, for he was contravened in his desires to rescue his brother from captivity by the expressed wish of the Pope, the clergy, and his ministers of state. A weak and sickly prince was by them regarded as of little worth in comparison with the retention of Ceuta, the key to the extension of Portuguese conquest on the continent of Africa, the portal, already in their possession, to the introduction of Christianity amongst the infidels, and the brightest jewel in the crown of Portugal. To him these considerations, while not without their weight, were ineffectual in removing remorse for what he regarded as an unpardonable weakness in himself, and he would thankfully have resigned his crown if he could thereby have secured the restitution of his unfortunate brother. Prince Henry, when appealed to for advice, brought no relief to the mind of the embarrassed King, for with that firm adherence to the course of duty which marked his character, great as was his love for his brother, he set aside every personal consideration when weighed in the balance with the advancement of Christianity and the welfare of his country. The surrender of Ceuta therefore was not to be thought of as the means of delivering Dom Fernando.

Two courses alone remained open for accomplishing that object; ransom, or a crusade against the Moors. The former was impracticable, and the latter by no means promised success. The deep chagrin experienced by the King at length completely undermined his health. It has been generally believed that he was struck with the plague by means of an infected letter, and that his frame, enfeebled by mental trouble, was unable to contend against the attacks of so serious a malady.* In his last will, however, he left injunctions to his successor that the freedom of Dom Fernando should be secured at all costs, and, if every other means failed, even by the surrender of Ceuta.†

This good but unfortunate King died on the 9th of September, 1438, after a reign of five years, a reign remarkable for well-intentioned effort, and as remarkable for unvarying misfortune and disappointment. Active and powerful of frame, he was unsurpassed by any of his day in feats of arms and horsemanship, yet kindliness and grace were far more noticeable in his appearance than the power and energy which he really possessed. This effect may have been in some degree increased by his habit of wearing his hair long and floating, and by his round and almost beardless face. His love of justice and of truthfulness was so great that "the King's word" became a proverbial expression for that which could be implicitly relied upon. The love of study had been inculcated and cultivated in him betimes by his excellent mother. With a mind well stored with information and manners graceful in the extreme, the keenness of his intelligence and correctness of his judgment gave to him a power of expression which won all hearts, and thus he obtained the cognomen of "the Eloquent." Nor did he content himself with communicating pleasure and instruction to his cotemporaries; as an author he has left a valuable legacy to posterity in a variety of treatises on ethics and philosophy, not so much distinguished by any profoundly scientific investigation into the principles and bases of these sciences as the expression of a warm and noble nature whose instincts were directed by integrity and clearness of judgment. They embody views upon the right conduct of life and maxims for good government, derived not only from his own thoughts

* Ruy de Pina, cap. 43, p. 187.
† Ib., cap. 44, p. 189.

and experience, but from maxims and opinions received from his father, King João I.*

One great anxiety of King Duarte was to replace the royal revenues in the same position that they had been in before reduced by the excessive liberality of his father King João. This was an undertaking of the greatest difficulty. Donations made by his father had to be revoked, and it was not an easy matter to manage this with any appearance of equity. King João, being illegitimate, had been compelled when Regent to buy the influence of the grandees, whose votes were indispensable, by large concessions of land, which were held to be irrevocable. In this dilemma Don João das Regras, whose subtle intellect had turned the scale when it was a question of raising King João to the throne, lighted on an expedient for saving the honour of the late king as well as that of his successor. He counselled the latter to make known the declaration, made by King João on his death-bed, that it was his intention when he alienated such large estates from the Crown that they should descend to the male heirs only, born in the direct line from the original grantees, but that, such male line failing, the estates were to revert to the Crown. But in order the more fully to make known this intention of King João, which he had always kept secret, and had only declared immediately before his death, João das Regras recommended the king to proclaim a new law which should be named the Lei Mental or Mental Law.

This celebrated law became established in Portugal, and by the plan thenceforth adopted life donations made to individuals for special services would from time to time fall back into the possession of the Crown. João das Regras was the first to feel the effect of this law. He had but one child, and that a daughter, and all his fortune had been derived

* Among these the obligations of a monarch are thus compactly expressed : " The fear of ruling amiss; justice combined with love and moderation; the reconciliation of divided affections; the achievement of great deeds with small means," &c. "Temor de mal reger ; justiça com amor e temperança ; contentar coraçoes desvairados ; acabar grandes feitos com pouca riqueza," &c.

from the King's liberality, so that he was compelled to ask for a dispensation to insure to his daughter her right of succession. He appears to have been the only one however who sued for this favour, which was conceded, and the law was accepted without remonstrance.

After the disastrous affair of Tangier, the Prince retired to Sagres, and continued there until September of 1438, when the King Dom Duarte fell ill at Thomar. So soon as Prince Henry heard of his brother's illness, he hastened to his side, and after the King's death was charged by the Queen, his widow, to consult with Dom Pedro and the grandees of the kingdom as to the best means of meeting the difficulties into which the state was thrown by this unhappy event. This was done, and it was resolved that the Cortes should be convened to take such measures as should be deemed convenient.

It was Prince Henry's opinion that the letters convoking the Cortes should be signed by Dom Pedro, but as the latter refused to do this, all the papers were signed by the Queen, but with an intimation that she would continue to sign, until the assembly of the States General should adopt a regulation on the subject. Meanwhile, Prince Henry, on account of his habitual prudence, was selected as interlocutor between the Queen and Dom Pedro. It accordingly resulted, from the propositions made by Prince Henry and discussed at different conferences, that the Queen was charged with the education of her children and the management of their property, and that Dom Pedro was to undertake the administration of the government of the kingdom, with the title of " Defender of the Kingdom for the King." As, however, there was a considerable party who would not consent to this arrangement, and much discord arose, Prince Henry again endeavoured to conciliate the opposing factions, by obtaining the consent of the Council and Deputies of the people to the following resolutions, which were proclaimed on the 9th of November, 1438, viz. :—

1. That the education of the King, while a minor, and of

his brothers, as well as the power of nominating to places in the Court, should be left in the hands of the widowed Queen Leonora, and that a suitable sum should be assigned to her for the payment of the expenses of the royal household.

2. That the royal Council should consist of six members, who should—each in turn at certain fixed periods—have charge of the affairs of the State within their powers, which should be regulated by the order of the Cortes.

3. Besides this Council, there should be elected a permanent deputation from the States to reside in the Court. This was to consist of one prelate, one fidalgo, or gentleman of family, and one citizen, each elected for one year by his respective chamber.

4. All the business of the Council should be dealt with by the six councillors and by the deputation from the three estates, under the presidency of the Queen, and with the approval and consent of Dom Pedro. If the votes were equal upon any question of business, it should be referred to the Princes, and Counts, and to the Archbishop, and the majority should decide. If the Queen and Dom Pedro should be of the same opinion, their vote would be decisive, even if the whole Council thought differently.

5. All matters of revenue, except such as fell within the administration of the Cortes, should be dealt with by the Queen and Dom Pedro, and decrees and orders should be signed by both, and the Comptrollers of the revenue should be charged with its execution.

6. It was finally determined that the Cortes should meet every year to resolve such doubts as could not be settled by the Council alone, as for example—" The death of grandees, the deprivation of high offices, the loss of lands, the correction or forming of laws and ordinances, and that in future Cortes any defect or error that might have existed in past Cortes might be corrected or amended."

The Queen, instigated by a violent party, refused to accept or sanction these resolutions in spite of the earnest persuasions of Prince Henry. This refusal produced great excitement in the public mind, and finally in the Cortes themselves, so that they began to contemplate investing Dom Pedro with the authority of Regent.

It is to be observed that Prince Henry invariably expressed his disapprobation of the deliberations of the Chambers at Lisbon, and other meetings, and publicly declared that such assemblies were illegal, in assuming a power which belonged only to the Cortes. Guided by an enlightened policy, and by prudence resulting from experience, this wise Prince showed equal indignation when he learned that the Queen had fortified herself in Alemquer, and had sought for help from the Princes of Aragon. But this did not prevent him from going to Alemquer to persuade the Queen to return to Lisbon, to present the young King to the Cortes (1439), and so great was the respect entertained for his opinion, that the Queen, who had obstinately resisted the persuasions of all others, yielded to those of the Prince.

In the following year the troubles which existed in the kingdom obliged the Prince to occupy himself with public affairs and the reconciliation of parties, in order to avert a civil war. Such were the events that interrupted the course of the expeditions and discoveries in the interval from 1437 to 1440.

CHAPTER X.

THE AZANEGUES.

1441-1444.

AFTER the voyage of Affonso Gonsalves, recounted in Chapter VI., nothing noteworthy occurred for three or four years. Two ships set out for those parts, but one returned on account of bad weather, and the other went only to the Rio d'Ouro for the skins and oil of sea-calves, and, having completed their cargo, returned home. In this year, as has been seen, Prince Henry went over to Tangier, and therefore was too fully occupied to send any more ships to the west coast. In the year 1438 the disturbances consequent on the death of Dom Duarte (on the 9th of September at Thomar) called imperatively for the Prince's presence, and he lost sight of everything else in his efforts to remedy the dangers and troubles in which the country had become involved. In the year 1440 two caravels were fitted out for the west coasts, but the voyage was an utter failure.

In 1441 the affairs of the kingdom becoming somewhat tranquillised, Antam Gonsalves, the Prince's Master of the Wardrobe, was sent out in command of a small ship, but solely with the order to bring home skins and oil of *sea-calves* as before, for as he was but young, the Prince put less charge upon him than upon his predecessors. When he had taken in his cargo, Gonsalves proposed to continue the voyage, in the hope that some of the natives might come to the sea-side for traffic, and so he might be the first to take captives to present to the Prince. Accordingly he selected

nine of the most active of his crew, and proceeded with
them inland. He succeeded in taking two, and as he was
about to set sail on the following day, there arrived an
armed caravel, commanded by Nuño Tristam, a young
knight who had been brought up from his boyhood in the
Prince's household, and was full of zeal in his master's
service. He had come out with a special command from
the Prince to pass as far as he could beyond the port of
Galé, and to endeavour by all means to make some
captures.

Nuño Tristam had brought with him a Moor, a servant of
the Prince's, to act as interpreter. It proved, however,
that the language of the captives was entirely different.
The small capture made by Gonsalves by no means contented Nuño Tristam, and after some discussion he agreed
with Gonsalves to set out in search of natives, with men
selected from their respective crews, and the result was the
capture, after a sharp contest, of ten natives, one of whom
was a chief. When the conflict was over, at the unanimous
request of his companions, Gonsalves was knighted by
Tristam, in spite of his modestly disclaiming his right to
such honour. Hence the place was named the " Porto do
Cavalleiro."

The chief alone among the captives understood the
Moorish language, and was able to converse with the interpreter. The rest spoke the Azanegue language. Hoping
to treat for the ransom of some of the prisoners, the interpreter went on shore with one of the female captives, but
he was detained prisoner, after having in vain tried to
negociate with the natives.

Gonsalves now returned to Portugal, but Tristam, having
orders to proceed farther, and finding that his caravel needed
repairs, put into land and careened her, keeping his tides as if
he were in Lisbon roads, a bold feat which astonished many
of his crew. He then pursued his voyage, and passing the
port of Galé, came to a cape to which from its whiteness he
gave the name of Cabo Branco. Here they found tracks of

men and some nets, but gained speech of no one. And as Tristam observed that the coast took the form of a bay, in which the currents seemed likely to impede their progress beyond the time that their provisions would last, he resolved to return to Portugal.

Prince Henry was in the highest degree gratified by the prospect thus opened of bringing these barbarous natives under the influence of Christianity, and extending the honour and prosperity of his country, and rewarded the two captains commensurately with the value which he set upon this successful issue of their labours.

Although the language of the captives was unintelligible to other Moors in that country, the Prince was nevertheless able to gather from the chieftain whom Gonsalves had captured considerable information respecting the country where he dwelt. Foreseeing that he would have to send out many expeditions to contend with the infidel natives of that coast, he sent to the Pope the news of this discovery as the first fruits of his long-continued exertions, and prayed for a concession in perpetuity to the Crown of Portugal of whatever lands might be discovered beyond Cape Boyador to the Indies inclusive, especially submitting to His Holiness that the salvation of these people was the principal object of his labours in that conquest. In addition to these important requests the ambassador Fernando Lopez d'Azevedo was charged to beg of the Pontiff indulgences for the Church of Santa Maria da Africa which the Prince had founded in Ceuta. The news of this discovery was considered so valuable by the Pope and the College of Cardinals that the Holy Father readily complied and issued a Bull to that effect, which was subsequently confirmed by the Popes Nicholas V. and Sixtus IV. The Regent Dom Pedro also granted to his brother Prince Henry a charter, authorizing him to receive the entire fifth of the produce of the expeditions appertaining to the King, and in consideration of the great labour and expense which the Prince undertook at his own sole cost, issued a mandate that none should go on these expeditions without Prince Henry's license and especial command.

The captive chieftain, although treated with all gentleness, chafed under his servitude much more than those of lower condition, and repeatedly begged Gonsalves to take him back to his country, where he engaged to give as ransom five or six negroes. He also said that there were two boys among the captives, for whom a liberal ransom would be given. This and the hope of gaining further information induced Gonsalves to ask permission to return to Africa.

He was accompanied in his voyage by a nobleman named Balthazar, of the household of Frederick III., Emperor of Austria, the husband of the Infanta Leonora of Portugal. This Balthazar had joined the household of Prince Henry with the intention of winning his spurs at Ceuta, and gallantly he won them. He had often expressed a desire to witness a storm off the coast of Africa, for he had been told that storms on that coast were very different from those on the coasts of Europe. In this wish he was gratified to his heart's content, for they encountered so severe a tempest that Gonsalves and his crew narrowly escaped with their lives, and were compelled to put back to Lisbon. Once more, however, they set forth on their expedition, and when they reached the point where the ransom was to be effected, they landed the chief, and Gonsalves agreed with him where they should meet after he had made his arrangements. The chief was handsomely dressed in clothes which the Prince had given him, for Prince Henry hoped thereby to induce the natives to enter into commercial relations with him. Gonsalves was blamed for the trust he placed in the chief's faith, and a detention of seven days at the appointed place, four leagues up the Rio d'Ouro, seemed to justify the blame. At the end of the week, however, a Moor on a white camel appeared with full a hundred slaves, out of which number ten negroes of both sexes were given up in exchange for the two boys. Martin Fernandes, the Prince's messenger, acted as interpreter, and proved himself an excellent linguist.

Besides the negroes, Gonsalves received in that ransom a small quantity of gold dust, a leathern buckler, and a great number of ostrich eggs, three dishes of which rarity were one day served at the Prince's table perfectly fresh and good.

The natives stated that there were merchants in those parts who trafficked in gold, which the chronicler Azurara evidently supposed was found in their own country. He was, however, not aware that gold was brought thither from the interior by the caravans which for many years had carried on that trade across the desert, and principally since the invasion of the Arabs. During the sovereignty of the Caliphs this commerce with the interior of Africa extended not only to the western boundaries of that continent, but even as far as Spain. The caravans crossed the valleys and plains of Sus, of Darah, and of Tafilet to the south of Marocco. Thibr, the Arabic name for gold, was brought from Wangara. The Rio d'Ouro, or River of Gold, received its name from the fact that gold was there first received in barter by the Portuguese. It has retained that name ever since, although it is in fact no river at all, but simply an estuary occupying an indentation in the coast of about six leagues in depth. Gonsalves now returned to the Prince, and met with a grateful reception, as did also the German knight, who afterwards returned to his country with much honour to himself and large reward from the Prince's bounty.

In the year 1443, the Prince fitted out another caravel, the command of which he gave to Nuño Tristam, the crew consisting principally of people of his own household. They reached to twenty-five miles beyond Cape Branco, and found a small island, to which they gave the name of Gete.* Here twenty-five canoes put out from shore, containing a host of natives entirely naked. This was not on account of

---

* The Island of Arguim. Barros says (Decade i. c. 7), Nuno Tristam in this voyage reached an island which the natives called Adeget, but which is one of those that we now call Arguim. The Arabs call it Ghir, which Azurara changed into Gete, and Barros into Adeget or Adeger.

their being in the water, but it was their habitual custom. Each canoe held three or four who hung their legs over in the water and paddled with them as with oars. The Portuguese at first took them for birds of monstrous size, but when they found their mistake they pursued them to the island and captured fifteen of them. They would have taken more but for the smallness of their boat. The discovery of this point was of great importance to the Portuguese. It facilitated their obtaining information and establishing intercourse with the negro states on the Senegal and Gambia. The Prince subsequently had a fort built there, the foundations of which were laid in 1448.*

Near the island of Gete they found another, on which was an infinite number of herons which came there to breed, and many other birds which afforded them a good supply of provisions. They gave this island the name of Ilha das Garças, or Heron Island. Nuño Tristam returned the same year with his booty, which was a greater source of satisfaction to him than on his former adventure, for not only had he taken more, but he had reached to a greater distance, and moreover had not to divide his gains with any one.

When the Prince began to colonize the islands which he had discovered, and to open a road to the people to turn the discoveries to profit, those who had been loudest in their censure were the first to turn their blame into praise. After the return from Tangier the Prince was almost always at his own town, which he then had built in the kingdom of Algarve near to Lagos, where vessels discharged the prizes which they brought; and the first to beg permission to make a voyage at his own cost to the newly-discovered country

* As will hereafter be seen, Cadamosto gives us considerable information concerning the state of the commercial relations which the Portuguese had in the course of seven years established with the inhabitants of the interior. In 1638, this fort was taken from the Portuguese by the Dutch. In 1665 the English took it, but again lost it. In 1678 the French gained possession of it, and destroyed the old fortress built by the Portuguese. The Dutch recovered the place in 1685, and retained it till 1721, when the French took it by surprise, but were once again driven out by the Dutch in the following year.

was Lançarote, an esquire who had been educated from his childhood in the Prince's household, but who was now married and held the post of King's almoxarife or receiver of customs in that city of Lagos.

Having fitted out six caravels, he sailed from thence in 1444, taking with him as commanders Gil Eannes, the same who had first passed Cape Boyador, Stevam Affonso, Rodrigo Alvares, João Dias, and João Bernaldes. After a successful expedition he returned with about two hundred captives, chiefly taken from the Islands of Naar and Tider in the Bay of Arguin. The Prince received him with great honour, and knighted him at the instance of the companions of his exploit. The captives, who presented every variety of colour from nearly white to the deepest black, very soon became Christians, and were treated with great kindness by their Portuguese masters. Some of the young girls were adopted by noble ladies, and brought up as their own children.[*]

* The island of Arguin, as well as those just mentioned under the respective names of Garças or Heron Island, Naar, and Tider, all lie on the great Arguin Bank comprised between Cape Blanco and Cape Mirik, but these latter lie in a group some five-and-forty miles to the south of Arguin Island, which is in 20° 28′. Cadamosto later speaks of this group as being named by the Portuguese thus: "The first Ilha Branca, the second Garza or Heron Island, and the third Cuori."

CHAPTER XI.

THE SLAVE TRADE.

THE old chronicler, Azurara, in a chapter full of eloquent pathos, gives expression to his feelings of commiseration for the poor captives on the occasion of their being distributed amongst their several owners or purchasers, and thereby separated from those most closely bound to them by the ties of nature. He thus describes the scene :—" On the 8th of August, 1444, early in the morning on account of the heat, the sailors landed the captives. When they were all mustered in the field outside the town they presented a remarkable spectacle. Some among them were tolerably light in colour, handsome, and well-proportioned ; some slightly darker; others a degree lighter than mulattoes, while several were as black as moles, and so hideous both in face and form as to suggest the idea that they were come from the lower regions. But what heart so hard as not to be touched with compassion at the sight of them ! Some with downcast heads and faces bathed in tears as they looked at each other ; others moaning sorrowfully, and fixing their eyes on heaven, uttered plaintive cries as if appealing for help to the Father of Nature. Others struck their faces with their hands, and threw themselves flat upon the ground. Others uttered a wailing chant, after the fashion of their country, and although their words were unintelligible, they spoke plainly enough the excess of their sorrow. But their anguish was at its height when the moment of distribution came, when of necessity children were separated from their parents, wives from their husbands, and brothers from brothers. Each was compelled to go wherever fate might send him. It was impossible to effect this separation without extreme pain. Fathers and sons, who had been ranged in opposite sides, would rush forward again towards each other with all their might. Mothers would clasp their infants in their arms, and throw themselves on the ground to cover them with their bodies, disregarding any injury to their own persons, so that they could prevent their children from being separated from them. Besides the trouble thus caused by the captives, the crowds that had assembled to witness the distribution added to the confusion and distress of those who were charged with the separation of that weeping and wailing multitude. The Prince was there on a powerful horse, surrounded by his suite, and distributing his favours with the bearing of one who cared but little for amassing booty for himself. In fact he gave away on the spot the forty-six souls which fell to him as his fifth. It was evident that his principal booty lay in the accomplishment of his wish. To him in reality it was an unspeakable satisfaction to contemplate the salvation of those souls, which but for him would have been for ever lost. And certainly that thought of his was not a vain one, for as soon as those strangers learned our language they readily became Christians, and I have myself seen in the town of Lagos young men and women, the children and grandchildren of these captives, born in this country, as good and true Christians as those who had descended generation by generation from those who had been baptized in the commencement of the Christian dispensation. Nevertheless there was abundant tear-shedding when the final separation came, and each proprietor took possession of his lot. A father remained at Lagos, while the mother was taken to Lisbon and the child elsewhere. This second separation doubled their despair. However, they were not long in becoming acquainted with the country, and in finding in it great abundance. They were far less obstinate in their creed than the other Moors, and readily adopted Christianity. They were treated with kindness, and no difference was made between them and the free-born servants of Portugal. Still more: those of tender age were taught trades, and such as showed aptitude for managing their property were set free and married to women of the country, receiving a good dower just as if their masters had been their parents, or at least felt themselves bound to show this liberality in recognition of the good services they had received. Widow-ladies would treat the young captives that they had bought like their own daughters, and leave them legacies in their wills, so that they might afterwards marry well and be regarded absolutely as free women. Suffice it to say that I have never known one of these captives put in irons like other slaves, nor have I ever known one who did not become a Christian, or who was not treated with great kindness. I have often been invited by masters to the baptism or marriage of these strangers, and quite as much ceremony has been observed as if it were on behalf of a child or relation."

It is impossible to read this eloquent expression of sympathy with the sufferings of the negro captives at the time of their partition without deep compassion for the disruption of natural ties which then of necessity took place. The scene then described was the consequence of the explorations instituted by Prince Henry. He was present thereat; and the first result in the mind of an Englishman hating, and righteously hating, the very name of slavery and the sale of human beings, would be that of reprobation of the Prince and of the people sent out under his auspices, by whom these slaves were thus brought in large numbers to Portugal from the African coast. There are many, however, who will see, in the conclusion of the chapter just recited, ample reason for withdrawing that reprobation, when they consider the motives, full of beneficence, which influenced the Prince in these transactions. The comprehensive purposes which he had in view, in the matter of exploration alone, made the capture

of natives of the west coast necessary, in the first instance,
for the sake of acquiring local information. The mere process of capture is in itself in the highest degree offensive to
us, as we sit in our easy chairs, free from the necessity of
making any exertion in subduing the evils of barbarism
beyond a little loosening of the strings either of the heart
or of the purse. But no sooner do we take a survey of the
active processes which, through all history even up to the
present time, have been brought to bear in the extension
of civilisation by encroachment on barbarian soil, than we
find that violence, the details of which, if presented to us
equally closely, would be equally offensive, has invariably
had to be resorted to. It will, however, be observed that
this violence was highly repugnant to the Prince's nature.
In Azurara, we find that, so soon as he found himself
in a position to do so with a fair hope of safety to his
mariners, he charged them to resort to peaceful means with
the natives, and to refrain from doing them injury. We
have the same testimony from Diogo Gomez de Cintra,
and the same from Cadamosto. It must be acknowledged
that three independent cotemporary witnesses are sufficient
to clear the Prince from the imputation of cruelty as to the
mode of deportation of these negroes.

As to the *object* of the Prince, in allowing his sailors in
the first instance to capture and afterwards to purchase
slaves, there can be no question that his first motive was to
rescue them from their original condition of spiritual, moral,
and physical degradation, and his second to add to the wealth
of his own country by an accession of valuable labour cheaply
paid for by the real advantages bestowed upon these captured
negroes. But there are some who, in their justifiable hatred
of the slave trade as they know it in connection with
America and the West Indies, will not patiently listen to any
such reasonings, but simply object that since the Western slave
trade originated in the deportation of negroes from Africa,
if such negroes were for the first time brought wholesale
from the African coast by the sanction of Prince Henry,

upon his head must undeniably rest the odium and the gravamen of what we abominate in that slave trade. But here I must demand a pause. The opprobrium thus attributed must consist either in the intrinsic immorality of the transaction, or in priority in introducing that which, even if in any way excusable at the time, has since become detestable. As regards its intrinsic morality, I think enough has been said to demonstrate the integrity of Prince Henry's intention, and, where integrity of intention exists, I conceive it is impossible to bring the charge of immorality. That there were co-existent evils, and that such evils have in later days been aggravated to the most painful extreme, no one can doubt; but in what phase of human life do not such appear? And if they awaken our sympathy or our regret, we have assuredly no reason to question the existence of such sympathy or such regret in the character of one so eminent, not only for the benevolence of his natural disposition, but for his exalted views of Christian duty.

It therefore only remains for us to inquire—

1st. Respecting the origin of the traffic in slaves;

2ndly. Respecting the first deportation of slaves from the west coast of Africa; and

3rdly. Respecting the originators of what we now know as the slave trade to the western world.

And in each of these we shall find that Prince Henry had no share.

1. And first respecting the origin of the traffic in slaves.

History from the remotest ages tells us of men being bought and sold as slaves and often reduced to a condition more wretched even than that of the brutes. The Pentateuch, which, apart from its divine origin, is the most ancient and venerable record of history and legislation, makes frequent mention of slaves among the Hebrew people; some who under the pressure of their necessities made themselves such; others who were sold by their own fathers; others captives of war, &c. It refers also to several laws, given by God for the same people, some for the regulation of the

rights and obligations of masters and slaves, others to miti-
gate by all means the miserable condition of the latter.
(See Gen. xxxvii. 26-28 ; xlvii. 18-22 ; Exodus xxi. 2-7 ;
Levit. xxv. 39-54 ; Deut. xv. 12-18, &c.)   That these
unfortunate people were bought and sold for fixed prices like
any other marketable commodity, and that in such traffic
existed what we reprobate in the "'slave trade," there can be
no doubt.   We are supplied with a notable instance of this
in the case of Joseph, who was sold by his brothers to the
Ishmaelitish merchants, and by them again sold into Egypt
(Gen. xxxvii.)

The Greeks and Romans also, the most civilised and refined
of the nations of antiquity, not only practised slavery to an
almost incredible extent, but treated their slaves with a
barbarity and ferocity hardly to be equalled among modern
nations.   Every schoolboy knows the unmitigated cruelty of
the Spartans to the unfortunate Helots. And as to the Romans,
Lucius Florus in his *De totâ historiâ Titi Livii Epitome*,
lib. 3, cap. 19, attributes the revolt of the slaves in Sicily
headed by Eurus Syrus, to the barbarous treatment of these
poor wretches, who were forced to plough the earth, tethered
together (*catenati cultores*) like brute beasts.   Seneca, in
his treatise *De irâ*, lib. 3, cap. 40, tells us of one Vedius
Pollion, who ordered a slave to be thrown into the tank in
which his lampreys were fattening, because he had broken a
crystal vase.   The *virtuous* Cato was not ashamed of being
a slave merchant; and Trajan, that *admirable prince, who
only had the weakness of great hearts, an excessive love of
glory*, gave public games at which ten thousand gladiators
and eleven thousand animals slaughtered each other for the
amusement of a cruel people who dared to stigmatise other
nations with the name of barbarians, &c., &c (Diodor.,
lib. 48.)

When Paulus Æmilius conquered Macedonia, says Pliny,
he decreed in one day the ruin of seventy-two cities.
(Historia Naturalis, lib. 4, c. 10.)   A hundred and fifty
thousand Epirotes and Macedonians were sold at that time

in Rome, by auction, in the same place where afterwards were exposed to public sale the no less unfortunate remnant of the Hebrew people ; and Seneca tells us that in his time there were in Rome slave-warehouses in which the slaves were kept by the dealers in that kind of stock, and from which they were taken to the public markets to be sold like brute beasts.

On the dismemberment of the Roman Empire, the nations which took possession of its provinces continued the traffic in men which they found established, and to which they themselves were not strangers. Christianity, it is true, tended greatly to ameliorate the condition of the slaves, and was for many centuries, by the tendency of its eminently humanizing principles, one of the most powerful causes of the diminution and decadence of this inhuman commerce. It did not, however, altogether abrogate the universally adopted practice, nor did its Divine Author alter, or appear to wish to alter directly, the established order of human society, or the different gradations and respective civil conditions which the laws and customs of the people had adopted. (See Eph. vi. 5, 9 ; Colos. iv. 1 ; Philemon 1 ; 2 Pet. ii. 18.)

The churches and monasteries had slaves. The old grants mention them constantly among the donations. The councils of different centuries and nations were full of canons relative to slaves, taking for granted the existence of slavery without reproof or condemnation. Some forbade that slaves should be admitted to holy orders or any ecclesiastical ministration. Others made provisions for the giving up to their masters of slaves who might have sought the protection of the Church with the view of obtaining their liberty. Others ordered that Jews should not have Christian slaves. Others established rules for the manumission of slaves belonging to Churches, &c. The Venetian merchants in the eighth century, traded in *Christian slaves*. History has left us an account of the pious zeal of Pope Zacharias, who in the year 748, knowing that these merchants had bought in Rome many slaves of both sexes, that they might sell them, after mutilating them, to the infidels of Africa, redeemed a great number of these unfortunates and restored them to liberty. In the year 785 Charlemagne expelled from his territory the Greeks who came to buy Christian slaves to sell them to the Mussulmans of Spain and of the East. In the year 820 and in all the following century, the merchants of Verdun also applied themselves to this traffic, selling their own fellow-citizens, after mutilating them, to the Kaliphs and Moors, to be employed in guarding their seraglios.*

At a council at London in 1102 it was determined, that no one *should sell men like brute beasts,* as had been done formerly in England, says Fleury.† Nevertheless this prohibition was still in force in the year 1171, when Henry II., having conquered Ireland, convened the bishops of that island to a council, in which it was ordered that all English slaves should be set at liberty, so that (says a cotemporary writer) the Fathers were persuaded that Divine Justice had subjected them to the English as a punishment for their crimes, and especially *because they used to sell Englishmen as slaves to merchants and pirates.*‡

In the thirteenth century, in the midst of Europe, among Christian nations, were sold not only prisoners of war, but also at times the peaceful and unarmed inhabitants of conquered cities or places. The letter which Pope Gregory IX. wrote to the Archbishop of Estrigonia in 1231, blaming certain unworthy things practised in Hungary, as showing contempt of religion, mentions, that the Saracens went there to buy Christian slaves, which was an affront to his authority; and that the Christians, forced by want and poverty, sold their own children as slaves to the infidels. Pope Gregory IX., in his apostolical letters of the 20th of April, 1376, directed against the Florentines, added to other grave penalties and censures which he fulminated against them, that

---

* See "Mémoire historique et diplomatique sur le commerce et les établissements français au levant," read at the public session of the Institute in 1827.

† See his Histoire Ecclesiastique, lib. 65, § 22.

‡ Fleury, Histoire Eccles., lib. 72, § 38.

for as many as they had seized, so many of themselves should be reduced to slavery, and says the historian, many of these then in England were in fact made slaves to the king, and had their goods confiscated. These facts, and many others abundantly furnished by history, are enough to show that the Portuguese were not the originators of the slave trade, unless it be maintained that there is any substantial difference between the traffic in negro slaves and in white ones, or between buying them in Africa to take them to America, and buying them not only in Africa, but in France, Venice, and Rome, to take them to Spain, Africa, or the East.

Let it only once be granted on both sides that the sufferings incident to slavery take the full range from the regretable to the execrable, and it may be fairly asserted that, *cæteris paribus*, the introduction of negroes into the benefits of Christianity and civilization was an act very far from blamable, although the concomitant miseries were to be regretted; while the sale of Christians into eastern slavery was nothing short of execrable.

2. So much for our first question respecting the origin of the traffic in slaves : we will now proceed to the second, respecting the first deportation of slaves from the west coast of Africa. The enslaving of and trafficking in negroes in general may safely be said to be as old as the knowledge that there were negro nations, since the trading in men has been a custom in all countries and in all ages, and there is no reason why the negroes should have been exempt from the common fate, as soon as there was an opportunity of taking, selling, and buying them. The facts of history confirm this thought.

One of the relievi which Mr. Champollion, jun., a few years ago observed in the temple of Isambul in Nubia, represents the triumphal car of one of the Pharaohs accompanied by troops of negro *prisoners from Nubia*, which proves that the *negroes from Africa* were from the remotest ages subject to the same laws of slavery which were practised among the white nations in all the world. Josephus in his

work on Jewish Antiquities, lib. 8, speaking of the merchant
ships of Solomon, says, that among the objects which they
imported were gold, silver, ivory, Ethiopian slaves, and
apes. " Pro rebus exportatis aurum, argentum, regi refere-
bant, multumque eboris, et *mancipia æthiopica*, et simias."
So old, so inveterate, and tenacious is this practice, which
derives its origin, its continuance, and inveteracy to the
inspirations of avarice and barbarity.

But, not to dwell on so remote a period, it is known that, in
the time of the Crusades, the use of negro slaves spread
much in Europe, and became a fashion among the great
lords who were engaged in those romantic expeditions. The
reader may also be reminded that long before the Portuguese
discoveries, as stated in page 114, the malaguette pepper of
Guinea was known in Italy, and consequently must have
been brought from Guinea by the Moors, who crossed the
country of the Mandingoes, and the deserts of Libya, to the
port of Barca, on the Mediterranean ; and we have the
distinct evidence of Azurara that captured negroes were
among the commodities brought to Barca for sale by the
Moors.   Azurara, it is true, was a cotemporary of Prince
Henry's ; but it is manifest that he spoke of a practice of
long standing, and which could by no possibility have anything to do with the Prince's expeditions to the west coast of
Africa.   Cadamosto, also, whose explorations in the service
of the Prince will hereafter be related, mentions the same
fact.   It will further be recollected that (as shown on page
175), Antam Gonsalves received from the Moors negroes in
ransom of the Moors which he had himself captured, which
shows that among the Moors was already practised not only
the *enslaving of negroes*, but also the *traffic in them*, since they
promised and gave them as ransom for their own persons
and property in the same way as they gave gold or ivory, or
any other merchandise of their country.   And should any
objector amuse himself with the frivolous argument that at
least the ships of Prince Henry were the first to deport
negroes by sea from the west coast of Africa, as if deporta-

tion from that coast in ships were more criminal than the conveyance of slaves across the desert for sale in the Gulf of Tunis, even this fanciful stigma will be found not to attach to the Prince's name. Although not preceded by Jean de Bethencourt in the rounding of Cape Boyador, he was preceded by him in the capture of natives from the west coast (see ante page 133). So much for the second question.

3rdly. That the importation of negroes into the West Indies and America is not due either directly or indirectly to Prince Henry is indisputable. The very time when that importation commenced is not known, but the earliest date that any one has ever ventured to suggest was half a century after the death of the Prince. The country was Spanish, and jealously exclusive of Portuguese encroachment of any kind. The most probable conclusion to be formed on a point not already settled by history is, that when Lisbon was full of negro slaves from Africa, the Portuguese exported them to Seville for sale, and that at a later period the Spaniards, who were interested in the exploration and working of the mines in the New World, sent their slaves thither, at first in small numbers, to be employed in these works. As the number of slaves brought from Africa increased, the transport of them to America would become more general, till at length the public authorities would find themselves obliged to legalize it, and to control its practice by laws and established regulations.

CHAPTER XII.

1445-1448.

In the year 1445, some time after the return of Lançarote, the Prince gave the command of a caravel to Gonsalo de Cintra, an esquire of his household, with strict orders to go straight to Guinea without putting in anywhere on the road. He, however, allowed himself to be deceived by the natives and his own ambition into disobedience of these orders, and landing on the island of Naar for the purpose of taking captives, was slain in a fight on the shore, not being able to swim back to his boat. The unwieldy name of Angra de Gonsalo de Cintra has been given to a bay some forty miles south of the Rio d'Ouro as commemorative of the death of the unfortunate commander, but the island of Naar is in the Bay of Arguin.

In the same year, 1445, Prince Henry again sent out Antam Gonsalves in a caravel to the Rio d'Ouro with one of his own servants, Diogo Affonso, in another. They were accompanied by Gomes Pires, who was sent out by the Regent, Dom Pedro, in a third caravel. The express purpose of the voyage was to treat with the natives and endeavour to make converts to Christianity, but they returned without effecting anything worth notice. João Fernandes went out with this expedition, and remained seven months alone in the wilds of the interior, in order to gain information for the Prince respecting the language and manners of the people.

An old Moor returned voluntarily with Gonsalves, wishing to see Prince Henry, who received him with great kindness, and afterwards sent him back to his own country.

About this same time Nuño Tristam made another voyage, in which he went straight to the Islands of Garças in the great bank of Arguin. These were now left desolate, for the natives had withdrawn for fear of the invaders. The Portuguese therefore went further on and came to a country very different from the sandy wilderness they had left, for it abounded in palms and other trees of great beauty. The roughness of the sea prevented their reaching the shore, and they were driven further south, where having effected a landing, they came upon a village and took one-and-twenty captives. The old chronicler, Azurara, who set a higher value upon such details than the reader is likely to do, remarks that the circumstances of this capture will never be known, because Nuño Tristam was dead at the time that King Affonso ordered his history to be written. The expression is of value as showing that Azurara did not confine himself to written documents in compiling his history, but consulted the discoverers themselves. The Vicomte de Santarem calls attention to a comparison which he made of the description of this voyage with certain early manuscript maps which he had the opportunity of consulting. He shows that after visiting the islands of Arguin, Tristam sailed southwards past places which bear on those maps the following names :—Ilha Branca, R. de S. João, G. de Santa Anna, Moutas, Praias, Furna, C. d'Arca, Resgate and *Palmar*, which last is doubtless the spot where he found the many palms. It is also particularly worth notice that Azurara, at the commencement of his very vague description of this voyage, historically asserts that *Nuño Tristam was the first who saw the Country of the Blacks*, whereas later in the book, at page 237, he assigned that honour to Diniz Dias,* the account of whose voyage immediately follows.

This bold adventurer, who had already distinguished himself in the service of King João, was the next to beg

---

* Barros calls him Diniz Fernandez, and as he wrote before the publication of Azurara, his original authority, he has been copied by all succeeding historians and geographers.

permission to make explorations in the service of Prince
Henry, who fitted out a caravel for him. Diniz had made
up his mind to sail further than any of his predecessors, and
this resolution he carried into effect, for he never struck sail
till he reached the land of the Negroes. It was not till now
that the mouth of the Senegal was passed, which separates
the Azanegues or Tawny Moors from the Jaloffs, the first
real Blacks.

The Portuguese looked upon the Senegal as identical with
the Niger, and afterwards, when they found that the Mandingoes gave to the Upper Senegal the name of Bafing or
Black River, they unhesitatingly concluded that it was the
Niger of Herodotus, Pliny, and Ptolemy, although Barros
was surprised that the Çanaga or Senegal should have so
little water. It was not till 1698 that the old error respecting
the western course of the Niger received its rectification,
when Brué, the Director of the French African Company,
became informed that that river flowed eastward and passed
near the city of Timbuctoo. It was not, however, till 1714
that this rectification was geographically laid down. This
was on the mappemonde of Delisle, on which we see the
Niger and the Senegal for the first time represented as
issuing from two neighbouring lakes, the one flowing westward, the other to the east. The name of Niger is itself
entirely unknown in Africa. It was introduced by Edrisi,
and afterwards employed by Leo Africanus. The Arab
geographers still regard it as the same as the Nile, but to
distinguish it from the Nil-as-Massr, or Egyptian Nile, they
call it the Nil-as-Soudan, and also the Nil-el-Kebir, or Great
Nile, the latter being regarded as the greater of the two.
Mollien in his vocabulary gives Baleo, as the Poula word for
Black, and in that language the Senegal is called Baleo or
the Black River, while in Mandingo it is called Bafing, which
has the same signification, and it seems to have been an
error in Mungo Park to ascribe to the Joliba or Quorra the
name of the Niger, which had always been declared by the
ancient and Arabic geographers to flow westward into the

Atlantic, while it contained, as shown by M. Golberry, "in the shallower parts of the river hippopotami and crocodiles, or rather the caymans of prodigious size" (see page 98 of "Modern Traveller," vol. 22), attributed to it by the ancient geographers. The largest seen by Adanson, the celebrated naturalist, were from fifteen to eighteen feet in length.

The Portuguese gave the name of Guinea to the western country of the Senegal or Senegambia, whereas it is now confined to the southern coast. In fact, originally Guinea was supposed to commence at Cape Non. Even so late as the beginning of the sixteenth century, in a treaty between Spain and Portugal about the boundaries of their respective conquests in Africa, the opinion was held that the borders of Guinea began between Capes Non and Boyador. Azurara was the first to make the Senegal the northern boundary of Guinea. He says the negroes were called Guineus, showing that if he sometimes called the country of the Azanegues Guinea, it was in obedience to custom, and not because he did not recognise the difference between the two countries. The town of Jinnie, on the river Genna or Niger, was founded in 1043-4, and soon became wealthy, owing to the trade in salt from Tegazza and gold from Bitu, and continues to be an emporium for the commerce carried on by the Mandingoes between Soudan and the west coast near Senegal and Arguin. The negroes call the country lying on the Niger Genna, Ghenea, Ginea, as well as Jennii, Gennii, and Jinne, but it is not certain whether the name originated with the country or the town. The Guinea Coast, as now understood, began to be known by that name after the construction of the Fort da Mina by the Portuguese in 1481, when the King of Portugal assumed the title of Lord of Guinea. But we must return from our digression.

As Diniz Dias coasted along this newly-visited shore, the caravel caused great astonishment among the natives, till at length four of the latter, being unable to decide whether it was a fish, a bird, or a phantom, took courage and approached it in a canoe, but when they found it contained men they fled with such speed that the Portuguese could not overtake them. As, however, it was far more Dias's purpose to discover land for the service of the Prince than to take slaves for his own profit, he proceeded still south till he reached a remarkable headland, to which he gave the name of Cabo Verde. Little more is known respecting this voyage, but as the Prince set very great value on this new discovery of the negro country, he largely rewarded Diniz Dias and his companions.

Seven months had now passed since João Fernandes had been left by his own desire among the Moors at the Rio d'Ouro. Antam Gonsalves therefore reminded the Prince of the circumstance, and volunteered, if the Prince should think fit to provide him with ships, to do his best not only to bring Fernandes back, but to make the voyage repay its expenses. These ships were promptly provided, and the principal command given to Antam Gonsalves. The other two captains were Garcia Homem and Diogo Affonso, of the Prince's household. Being separated by a violent storm, the first that reached Cape Blanco was Diogo Affonso, who set up a large wooden cross as a notice to the rest, if they should arrive after him, that he had gone on in advance. This cross was fixed so firmly that it lasted there for many years, and as Azurara was informed, was still there at the time that he wrote. "Well," he says, "might it astonish any one of another nation that might chance to pass by that coast to see such a signal among the Moors, if he should happen to be entirely ignorant of the Portuguese navigations along that coast."* The other vessels soon joined him, Gonsalves being the last to reach the Cape. As the first vessels had had no success, in consequence of the natives fleeing from

---

* This sentence of Azurara's has been adduced by M. d'Avezac as a proof that vessels of other nations were in the habit of passing that way. I submit that it demonstrates the exact contrary. Had the vessels of any other Christian nation been in the habit of passing that coast, a similar elevation of a cross would have been a matter of perfectly easy occurrence, and would be far from the matter of extreme astonishment pictured by Azurara. Nor could such frequenters of the coast by any possibility be ignorant, as Azurara in fancy suggests, of the navigations of the Portuguese.

them, Gonsalves proposed that they should leave the ships in charge of lieutenants, and go in their boats to the island of Arguin, but they found no one there, except one native and his daughter, whom they captured. In consequence of information received from this man, they took twenty-five more, and here Azurara remarks :—" It was a marvellous thing that as soon as one of these people was taken, he took refuge in pointing out to the enemy not only other natives, but his friends, and even his wife and children." They then returned along the coast in search of the caravels, from which they had been absent three days. The caravels, meantime, according to orders, sailed for the island of Arguin, but, not knowing its position, passed beyond it to the country farther south. Here they cast anchor, and in little more than an hour observed a man on the shore opposite. This proved to be Fernandes, who had been watching with anxiety on the coast to see if any vessel were coming to fetch him away. As the caravels could not come close to the land, he ran along the shore till he met the boats returning, and was received with great rejoicing.

It appeared that he had engaged the affections of the natives during his sojourn amongst them, and he told Gonsalves of a chief named Ahude Maymom, who wished to barter with him some negroes whom he had taken captive. Gonsalves received the offer gladly, and exchanged articles of trifling value for negroes and gold. The place was called Cabo do Resgate, or Cape of the Ransom. Here Gonsalves knighted an old man of noble family from Madeira, named Fernando Tavares, who considered it an especial honour to be dubbed knight on the newly discovered land. The caravels then proceeded to the island of Tider, where they had an encounter with the natives, and narrowly escaped great danger from an ambush. In his homeward passage Gonsalves put in at Cape Branco, and made a capture of about sixty natives, after which he made his way to Portugal and reached Lisbon in safety.

Prince Henry was at the time in his duchy of Viseu,

whence he sent to claim his fifth, and the rest of the slaves were disposed of in the city by the captains, to the great benefit of all concerned.

However pleased the Prince may have been with the general success of this voyage, his principal satisfaction was in seeing João Fernandes back safe and sound, and able to give him information respecting the country and the people. Fernandes related that the first thing the natives did was to strip him of his clothes, and give him a mantle such as the rest of them wore. The people among whom he lived were shepherds, who wandered with their cattle wherever they could find pasture. The fodder was scanty, the land desert and sandy, with no trees except small ones, such as *figuieras do inferno* (*Palma Christi*), thorn trees, and a few palms. There were very few flowers. All the water was from wells, except a very few running streams.

The people were called Alarves, Azanegues, and Berbers. They were Mohammedans. Their language, written and spoken, differed from those of other Moors.* They had neither law nor lordship, and waged war with the negroes, who were stronger than they, more by craft than strength. Some of these negroes they would sell to the Moors, who came to their country for that purpose. Others they would take to Barca, beyond Tunis, to sell them to the Christian merchants who resorted thither, receiving in exchange bread and other commodities, just as they did at the Rio d'Ouro. The people had negro prisoners in their possession when Fernandes was among them, and some gold which was obtained from the land of the negroes. Their camels were very numerous, and could travel fifty leagues in a day, and they had plenty of cattle in spite of the thinness of the pasture. There were a great number of emus, antas, and gazelles, partridges, and hares. The swallows leave in the spring and return to winter on the sands; the storks go to the land of the negroes to winter.

* This would seem to indicate that the Berbers had not at this time adopted the Arabic character.

This country extended from Tagaoz or Tagazza to the land of the negroes, in one direction, and to the Mediterranean at the end of the kingdom of Tunis, at Barca, in the other.

Fernandes further related that one day two horsemen came up to him, who were going to join the before-mentioned chief, Ahude Maymom, and asked him to accompany them. He accepted their invitation with pleasure, and they mounted him on a camel and went their way. On the road their water failed them, and for three days they had nothing to drink. There was no certain road except by the sea-shore, and they guided themselves by the signs of the sky and the flight of birds. At length, after bearing their thirst as best they could, they came up to Ahude Maymom and his family, which, with their retinue, were about one hundred and fifty in number. Fernandes made his obeisance, and was welcomed by the Moor, who ordered milk to be given him, and treated him so well that when he was received by the caravels he had recovered his good looks and was in his usual health, though he had suffered much from the heat of the country and the sand of the desert.

Azurara gives further particulars respecting the Azanegues among whom Fernandes dwelt. Their food was chiefly milk, and sometimes a little meat with seeds of wild herbs gathered on the mountains. Wheat was considered a luxury. For many months they and their horses and dogs lived entirely on milk. Those on the sea-shore ate nothing but fish, mostly raw or dried. Their garments were vests and breeches of leather, the better classes wore mantles. They had a few good horses, with saddles and stirrups, and some few of the chiefs kept brood mares. The women wore mantles over their faces, but the rest of the body they left uncovered. The women of the chiefs wore rings of gold on their ancles, and other jewels. Their merchandise, besides the slaves and gold which they get from the negro country, consists of wool, butter, cheese, dates, which they imported, amber, civet, gum anime, oil and skins of sea-wolves, which are abundant at the Rio d'Ouro.

The success of Antam Gonsalves' expedition induced a gentleman of Lisbon named Gonsalo Pacheco, who belonged to the household of the Prince, to request permission to make the voyage. He obtained leave to equip a caravel which he had built for himself, and two others which he wished to accompany him. He took with him as captains Diniz Eannes da Grãa, his wife's nephew, an esquire of the Regent, Alvaro Gil, assayer of the mint, and Mafaldo of Setuval. When they reached Cabo Branco, they found an inscription left by Antam Gonsalves, warning them not to go to the neighbouring village, as it was deserted. They then went to the island of Arguin, in which, and on the mainland near, Mafaldo, guided by the pilot Goncalves Gallego, who had been there with Antam Gonsalves, made a capture of fifty-three natives in one night. They then went south, and at a place thirty-five leagues south of Tider, Alvaro Vasques took seven captives, and the next day Luis Affonso Cayado took ten. They coasted along some distance and came to a cape which they named "Cabo de Santa Anna," where Alvaro Vasques and Diego Gil took thirty-five more. Finding they could make no more captures, as the natives were aware of their presence and fled from them, they sailed eighty leagues yet further south, and would have landed, in spite of the hostile appearance of the people, but were prevented by the roughness of the sea. From the distance they could see that the land was very verdant, with a large population, and abundance of domestic cattle. They would have proceeded further south, but a storm which lasted three days drove them back, and when the weather set in fair they found themselves at the place where Alvaro Vasques had taken his seven captives. Encouraged by their former success, the boats were sent on shore; and they took twelve prisoners.

Between Cabo Branco and Cabo Tira they saw a small sandy island, where they found traces of men, fishing nets, and abundance of turtles.

The next day they returned and found the nets had been

removed, but there were some turtles with ropes round them just as they had been caught. Observing another island near, they went to it, little suspecting an ambush. They were attacked by a large body of natives, and compelled to retreat with the loss of seven men killed, and one of the boats, which was taken to Tider and broken up for the sake of the nails. The ships then proceeded to Arguin to take in water. The Portuguese were afterwards told by some captives taken from that place that the natives ate the Portuguese whom they had killed, but others denied that such an enormity had been committed. Azurara declares that it was certainly their custom, when they revenged the death of a relative, to eat the liver and drink the blood of the murderer—but only as an act of vengeance.

Meanwhile the recollection of the death of Gonsalo da Cintra caused the inhabitants of Lagos to appeal to the Prince for his permission that an expedition should be sent out, of sufficient strength to intimidate the natives, who were in such great numbers at the island of Tider and the neighbourhood, and so to quell their force that Portuguese vessels might henceforth pass along any part of that coast without jeopardy. To this Prince Henry gave his approval, and fourteen vessels were forthwith equipped for that object.

At this time (1445), Prince Henry was summoned to Coimbra by his brother, the Regent Dom Pedro, to invest with knighthood his eldest son Pedro, who was Constable of the Kingdom and under orders for Castile; for in such profound esteem did the Regent hold Prince Henry, that he regarded it as the greatest honour that could be conferred upon his son to receive knighthood from such hands.

Before Prince Henry started from Lagos, he entrusted the chief command of the fourteen caravels to Lançarote, who had already proved himself so able and successful a navigator on the African coast. This was a great distinction, for the other commanders were men of great eminence—Soeiro da Costa,

Alcaide of Lagos, Lançarote's father-in-law, a fine old soldier; Alvaro de Freitas; Gomes Pires, captain of the King's caravel; Rodrigueannes de Travaços, of the Regent's household; Pallenço, who had distinguished himself in the wars against the Moors; Gil Eannes, who first passed Cape Boyador; Stevam Affonso, and other distinguished natives of Lagos. Besides these fourteen ships, there were sent out from Madeira three others, the captains of which were Tristam Vaz, commander of Machico, and Alvaro Dornellas, each in his own caravel, but these were driven back by the weather before they reached Cape Branco. Alvaro Fernandes also came out in a caravel, belonging to his uncle João Gonsalves Zarco, commander of Funchal. From Lisbon Diniz Dias (who first reached the land of the negroes) went out in a caravel of D. Alvaro de Castro, chief chamberlain of King Affonso, and João de Castilha in another belonging to Alvaro Gonsalves d'Ataide, the King's tutor. Altogether there were six-and-twenty caravels, besides the pinnace in which Pallenço went out.

The fourteen from Lagos set sail in company, on the 10th of August, 1445, having agreed, if they were separated, to meet at Cape Branco, and as wind and tide were favourable they tried which had the advantage in speed, and Lourenço Dias, one of the captains, soon began to take the lead. He first reached the island of Arguin, where he found the three caravels of Pacheco's expedition cn the eve of returning to Lisbon after their discomfiture. From him they heard of the fleet that was on its way thither, and of the purpose with which it had been sent out, and they promised themselves ample vengeance for the loss they had sustained. They had already made a fair capture for the one voyage, and their provisions were running short, but rather than not accompany the new expedition, they preferred to live on short rations for a time. Accordingly, they proceeded in company with Lourenço Dias to the Ilha das Garças, where they remained for three days in expectation of the other caravels.

They there found birds in great number, which helped

their stock of provisions, and some peculiar to that part, called crooes, entirely white and larger than swans. Their beaks were more than a cubit long and three fingers in breadth, and looked as highly polished as a pacha's scabbard. The mouth and gullet were large enough to take in a man's leg of the largest size up to the knee. There is evident exaggeration in the description, but the bird appears to be a kind of Marabou Jabiru.

When some of the vessels had arrived, and among them those of Lançarote, Soeiro da Costa, Alvaro de Freitas, Gil Eannes, and Gomes Pires, two hundred and seventy-eight men were selected for the attack and sent on shore in three boats, steered by pilots who had been there before and knew the locality. They had intended to take the natives by surprise, but everything went against them. The pilots proved unequal to their work, the night was dark, the water was low. The boats stranded and were obliged to wait for the tide, and the sun was well up before they reached the island. They proceeded for three leagues along the shore till they reached Tider, near which they perceived a host of natives showing every readiness to fight. A conflict ensued, in which eight natives were killed, and four taken. They then took to flight, leaped into the water and swam to the mainland, having already sent away their wives and children. Before returning to the ships the Portuguese went to the village which the natives had deserted, and to their great delight discovered water, for they were nearly perishing of thirst. They also found a few cotton trees.

Here Soeiro da Costa and Diniz Eannes de Grãa received knighthood from the hands of Alvaro da Freitas, and De Grãa then returned, with the three caravels of Pacheco's expedition to Lisbon.

On the following day the natives returned to about a stone's throw from the caravels, and danced on the shore as if in defiance. A number of Portuguese, headed by a brave lad of the Prince's household, named Diogo Gonsalves, and Pero Alleman, of Lagos, swam ashore and soon put them to flight. Fifty-seven were captured. They pursued the fugitives as far as a village called Tira, on the sea-coast, at about eight leagues distance, but found that and two other villages deserted.

On the next day, the commanders of the fleet being all assembled, Lançerote announced to them that as the object for which they had come out was now accomplished, inasmuch as the island of Tider had been conquered, and its inhabitants dispersed, his duty as captain-general ceased; for the Prince's orders were, that after that island was taken, each of the captains should be free to take his course in any direction that might seem to promise best. Then, after making a fair distribution of the captives, he inquired of the different captains what they proposed to do. Soeiro da Costa, Vicente Dias (the outfitter), Gil Eannes, Martim Vicente, and João Dias, decided upon returning to Portugal, as their caravels were small, and the winter was coming on.

On their way back, they determined to explore to its extremity that arm of the sea which is formed by Cape Blanco. They anchored near the entrance of this estuary, and after pulling four leagues in, the boats reached the head of it, and there landed. They found a few huts, from which they captured eight natives, who told them there were no other inhabitants near. Soeiro da Costa now altered his mind, and went back to Tider with the view of obtaining a ransom for a woman and a chief's son of that place. He had cause to repent his determination, for after he had handed over a Moor and a Jew as a guarantee of his good faith, the woman leaped overboard, and swam to land, and the natives would only surrender the three on condition that three others were given in exchange. The alternative was hard, but Da Costa was obliged to put up with the loss, and return to Portugal.

Gomes Pires, who was captain of the King's caravel, in answer to Lançerote's inquiry, declared his intention to proceed to the land of the negroes, and especially to the river Nile,* about which the Prince was very anxious to gain information. In this purpose he was joined by Lançerote himself; and Alvaro de Freitas not only declared his hearty good-will to join them, but to follow them if possible to the Terrestrial Paradise. It must not be supposed that this was mere empty talk. In some of the maps of the middle ages the Terrestrial Paradise was laid down in the most eastern part of Asia, but if the reader recalls the connection of the rivers Gihon and Euphrates, two of the four rivers issuing from Eden, with the mediæval notions of the Nile, he will perceive the meaning of the words of De Freitas. · Of the same opinion were Rodrigueannes de Travaços, Lourenço Dias, and Vicente Dias (the merchant), and forthwith they set out on their voyage. Two of the other caravels now parted company with them, one from Tavila, and another called the *Picanço,* or the *Wren,* belonging to a man of Lagos, but as they did not reach the land of the negroes, they will be spoken of hereafter.

The six caravels sailed along the coast till they found the land of Zaara, or desert, the country of the Azanegues, and came to the two palm trees which Diniz Dias had found, and by which they knew that they were very near the land of the negroes. They would have landed, but the surf on the coast prevented them. The smell from the shore was so fragrant that it was as if some delightful fruit garden had been placed there for their especial delectation. The Prince had told them from information he had received from the Azanegue prisoners, that twenty leagues beyond the palms they would find the western outlet of the river Nile, called by the natives Çanaga. As they proceeded along the coast,

---

* This expression shows how full of purpose these explorations were, and that the Prince did not seek simply to add to his knowledge of the West African coast, but to compare the information which could be gathered from the natives themselves with the scientific, historical, and geographical notions of ancient and mediæval times, so as ultimately to reach the east. The reader will have already seen that the river here spoken of is the Nile of the Negroes, or the Senegal, to which river the name of Niger adhered even so late as to the close of last century.

keeping a look-out for the river, they observed before them at about two leagues distance from the land a colour in the water different from the rest, which was mud coloured. This proved to be the fresh water from the river, and they soon came to the river's mouth, where they anchored. Eight of the sailors of Vicente Dias' caravel pulled ashore, and among them Stevam Affonso, who had partly fitted out the caravel. One of them pointed out a cabin near the mouth of the river, and proposed that they should try to take the inhabitants by surprise. Stevam Affonso and five others landed, and hiding near the cabin, saw issue from it a negro boy quite naked, who was immediately taken; and when they went up to the cabin, they found his sister, a girl of about eight years old.

The Prince afterwards had this boy educated, and it was supposed that he intended him for the priesthood, that he might go and preach Christianity to his people; but the youth died before he came of age.

When the Portuguese went into the cabin, they found a shield made of leather from the ear of the elephant, quite round, somewhat larger than the ordinary size, with a boss in the middle made of the same leather. They afterwards learned from the natives that all their shields were made from the hide of this animal. When the skin is thicker than they require, they stretch it more than half its original size by means of contrivances which they have for that purpose. They made no use of the ivory, but exported it. "I learned," says Azurara, "that in the Levant one of these bones was worth on an average a thousand dobras"—a remark which shows that his knowledge of the ivory trade extended only to the ports of the Mediterranean, and not to any exportation of that commodity from the coasts of Guinea. They presently came upon the father of the children, who was busy carpentering, and did not perceive them. Stevam Affonso approached him stealthily, and springing on him clutched him by the hair, and as he himself was a little man and the African very tall, when the latter stood upright he lifted his assailant off his feet. Powerful as the negro was, he could not rid himself of his antagonist, but tossed himself about like a bull that some fierce dog had seized by the ear. The Portuguese then came up and held his arms, intending to bind him, and Stevam Affonso, imagining that he was secured by the others, loosed his hold, but no sooner did the African find his head free than he shook off the others from his arms, and fled. He was much swifter of foot than the Portuguese, and soon plunged into a forest of underwood, and while the rest were trying to find him, he made his way to his hut in search of his children and of his weapon which he had left with them. The bereaved father was furious when he could not find them, and as he looked along the shore in search of them, he saw Vicente Dias walking towards him with nothing but a gaff in his hand. The enraged African fell upon him with his azagay and inflicted a severe wound on his face; after which they closed in a deadly struggle. A negro youth came to the assistance of his friend, and obliged Dias to loose his hold; but at the approach of the other Portuguese, the two negroes made their escape.

The caravels now made their way to Cape Verde, and all reached it, excepting that of Rodrigueannes de Travaços, who lost company, and whose adventures will be related hereafter.

Off the Cape they saw an island, on which they landed to see if it were inhabited, but found only a great number of long-eared goats, some of which they took for food. Having taken in water, they went on and found another island * in which they saw fresh goat skins and other things lying about, which showed that other caravels were in advance of them, a fact which was confirmed by their seeing carved on the trees the arms of Prince Henry and the words of his motto, " Talent de bien faire." They afterwards learned that the caravel which had preceded them was that of João Gonsalves Zarco, commander of the island

* These were the Madeleine Islands off Cape Verde.

of Madeira. The number of the natives on shore was so
great that they had no chance of landing either by day or
night; but Gomes Pires, by way of trying to bring about
some intercourse, placed on shore a ball, a mirror, and a
sheet of paper on which was drawn a crucifix. These the
negroes broke and destroyed as soon as they found them.
The Portuguese now drew their bows on them, but they
returned the compliment both with arrows and azagays.
The arrows were not feathered and had no notch for the string.
They were short and made of reeds or canes with long iron
heads, and some of the shafts were of charred wood. All
their arrows, without exception, were tipped with vegetable
poison. Each azagay had seven or eight barbed points.
The poison used was very deadly.

In that island in which Prince Henry's arms were found
cut on the trees, they found many large trees of a very
peculiar kind, and among them one which measured a
hundred and eight palms round in the stem. The stem of
this tree is not higher than a walnut tree, and with its fibre
they make very good thread for sewing with, and it burns
like flax, its fruit is like a gourd, and its kernels like chest-
nuts. These they dried, and as they were in great quantity,
Azurara supposed that it was used by them for food, when
the season for the fresh fruit was over. Doubtless the tree
was the baobab.

From this point they made sail for Lagos, but Gomes
Pires became separated from the other caravels, and on
his way homewards, after taking water at Arguin, put in
at the Rio d'Ouro, where some of the natives came to him
and sold him a negro for five doubloons. They also gave
him water from the camels, and meat, and in other respects
gave him a good reception. Indeed, they were so confiding,
that they came without hesitation on board his caravel,
which he had rather they had not done. At length he
managed to have them put on shore without the occurrence
of any unpleasantness, and promised them that in July of
the following year, he would return and treat further with

them. He also laid in a good cargo of phoca skins, and then made his way home.

After the others had departed, Lançarote, Alvaro de Freitas, and Vicente Dias put in at the Bay of Arguin for water, and captured more than sixty natives. Contented with this success, they returned together to Portugal.

Pallenço, in company with Diniz Dias, after taking in water at the isle of Arguin, made for the land of the negroes. Having passed a good distance beyond the point of Santa Anna, one day when it was calm, Pallenço proposed to send some of the men on shore to make a capture. After some discussion with Dias, who wished to go straight to Guinea, the men were sent on shore. There was a heavy surf in spite of the calm, but twelve who could swim well succeeded in landing through it and capturing six of the natives. When they recommenced their voyage, the wind freshened, and Pallenço's pinnace beginning to leak, they were obliged to desert it.

After Rodrigueannes de Travaços lost company of the other caravels on their way to Cape Verde, he joined Diniz Dias. They reached Cape Verde and went to the islands (the Madeleine Islands) to take in water, and found by the same traces that Lançarote had found, that other vessels had been there before them. He observed on the island among the cattle there, two that seemed different from the rest, larger and not so tame, which he took to be buffaloes. They had a severe encounter with the natives, who far outnumbered them, but were routed at last, leaving one of their chiefs dead on the field. A young man of the Prince's household, named Martim Pereira, distinguished himself greatly. His shield was stuck full of arrows like a porcupine's back. After this encounter they returned to Portugal, stopping only at the Cabo de Tira, where they captured one man.

Hitherto it has been seen how almost all these explorers had been intent on their own gains in addition to the Prince's service; but João Gonsalves Zarco was an exception to this rule. He had fitted out a splendid caravel, and gave the command of it to his nephew Alvaro Fernandes, who had been brought up in the Prince's household, with injunctions to forego all thought of gain, and not to land in the country of the Azanegues, but to proceed straight to the negro country, and make his way as far as he could with the view of bringing back some new information that should give pleasure to the Prince. The caravel was well victualled, had a crew well disposed for work, and Alvaro Fernandes was young and zealous. They proceeded as far as the Senegal, where they filled two pipes with water, one of which they afterwards took to Lisbon. They then passed Cape Verde and came to an island, where they landed and found some tame goats, without any one tending them, of which· they took some for food. It was they who left those indications of the arms of the Prince, and his device and motto cut on the trees which were seen by Lançarote and his companions, for this Alvaro Fernandes was the first who came there.

They anchored about a third of a league from the Cape, hoping to communicate with the natives, though only by signs, for they had no interpreter. Two boats containing ten negroes put off from the shore and made straight for the ship, as if with peaceful intentions. As they approached they made signs asking for assurance of safety, which was given, and immediately five of them entered the caravel. Alvaro Fernandes received them with all possible kindness, gave them plenty to eat and drink, and showed them every attention in his power. They left with every sign of being greatly pleased. When, however, they reached the shore they encouraged other natives to make an attack, and six boats put out with thirty-five or forty men in them prepared for fighting ; but they did not venture to come close to the caravel, but remained at a little distance. When Alvaro Fernandes saw this, he launched his boat on the opposite side of the caravel so as not to be seen by the negroes, ordered eight men into it, and waited for the negroes to come nearer. At length, one of the boats containing five powerful negroes, took courage to approach. When Fernandes observed that it was in such a position that his own boat could reach it before the others could bring help, he ordered his men to sally forth suddenly and row down upon them. From the great advantage they possessed in their mode of rowing, the Portuguese were speedily on the enemy, who being thus taken by surprise and having no hope of defending themselves, threw themselves into the water, and the other boats pulled for the land. The Portuguese had great difficulty in catching them as they were swimming, for they dived like cormorants, so that they could get no hold of them. However, they took two and brought them on board.

Alvaro Fernandes saw clearly that, after this, no advantage was to be gained by staying there. He therefore proceeded further south and reached a cape where there were many dry palm-trees without any branches, and to which he gave the name of Cabo dos Mastos (the Cape of the Masts). As they proceeded, Alvaro Fernandes sent out a boat with seven men to go along the shore, and, as they went, they lighted on four negroes sitting on the beach, who were out on a hunting expedition, and armed with bows. When these saw the Portuguese, they rose quickly and fled, not giving themselves time to adjust their bows, and as they were naked, and had their hair short, the Portuguese could not catch them, but they took the bows and arrows, together with some wild boars which they had taken. Among the larger animals found there was the antelope, which on account of its tameness they would not kill. They now returned to the ship and sailed back to Madeira, and thence to Lisbon. There they found the Prince, who received them with very great favour, and showed especial honour to Zarco, who had thus at his own expense set on foot an exploration which went further than any of the others that made the voyage to Guinea that year.

It has already been stated that after the six caravels had sailed for Guinea, two separated from them and turned northward, viz., that from Tavila, and the Picanço.

On their way they met with the caravel of Alvaro
Gonsalves de Atayde, the captain of which was one João
de Castilha, going to Guinea, whom they dissuaded from
that voyage, and induced him to join them in an expedition
to the island of Palma. On reaching Gomera they were
well received, and two chieftains of the island, named
Bruco and Piste, after announcing themselves as grateful
servants of Prince Henry, from whom they had received the
most generous hospitality, declared their readiness to do
anything to serve him. The Portuguese told them they
were bound to the island of Palma for the purpose of
capturing some of the natives, and a few of the chieftain's
subjects would be of great use as guides and assistants,
where both the country and the people's mode of fighting
were alike unknown. Piste immediately offered to accompany them, and to take as many Canarians as they
pleased, and with this help they set sail for Palma, which
they reached a little before daybreak. Unsuitable as the
hour might seem, they immediately landed, and presently
saw some of the natives fleeing, but, as they were starting
in pursuit, one of the men suggested that they would have a
better chance of taking some shepherds, chiefly boys and
women, whom they saw keeping their sheep and goats
among the rocks. These drove their flocks into a valley
that was so deep and dangerous that it was a wonder that
they could make their way at all. The islanders were
naturally sure-footed to a wonderful degree, but several of
them fell from the crags and were killed. The page Diogo
Gonsalves, who had been the first to swim to the shore in
the encounter near Tider, again distinguished himself. It
was hard work for the Portuguese, for the Canarians hurled
stones and lances with sharp horn points at them with great
strength and precision. The contest ended in the capture
of seventeen Canarians, men and women. One of the
latter was of extraordinary size for a woman, and they said
that she was the queen of a part of the island. In retiring
to the boats with their capture they were closely followed

by the Canarians, and were obliged to leave the greater part
of the cattle that they had had so much trouble in taking.

On their return to Gomera they thanked the island
chieftain for the good service he had rendered them, and
afterwards, when Piste, with some of the islanders, went to
Portugal, they were so well received by the Prince that he
and some of his followers remained for the rest of their lives.

As João de Castilha, the captain of the caravel of Alvaro
Gilianez Datayde, had not reached Guinea as the others had
done, and consequently had less booty than they to carry back
to Portugal, he conceived the dastardly idea of capturing
some of the Gomerans, in spite of the pledge of security.
As it seemed too hideous a piece of treachery to seize any of
those who had helped them so well, he removed to another
port, where some twenty-one of the natives, trusting to the
Portuguese, came on board the caravel and were straight-
way carried to Portugal. When the Prince heard of it he
was extremely angry, and had the Canarians brought to his
house, and with rich presents sent them back to their own
country.

Alvaro Dornellas, after an unsuccessful attempt to make
a capture in the Canary Islands, which resulted in his only
taking two captives, remained at the islands, not caring to
return to Lisbon without more booty. He sent Affonso Marta
to Madeira to procure stores by the sale of the two Canarians.
The weather prevented Marta making the island, and he
was obliged to put in at Lisbon, where at that time was
João Dornellas, squire to the King, and cousin to Alvaro.
João had a joint interest in the caravel, and hearing of his
cousin's difficulties, hastened to his assistance. Together
they made a descent upon the island of Palma, having
obtained help from the people of Gomera in the name of
Prince Henry, and in a night attack, after a fierce encounter,
took twenty captives. They returned to Gomera, where
Alvaro had to remain, and his cousin left for Portugal. In the
homeward passage such a dearth of victuals supervened that
they were well-nigh compelled to eat some of their captives,

but happily, before they were driven to that extremity, they reached the port of Tavila, in the kingdom of Algarve.

It has been seen in a former chapter that the Norman, Jean de Bethencourt, retiring to France in 1406, had left his nephew, Maciot de Bethencourt, as governor-general of his conquests in the Canaries, comprising Lançarote, Forteventura, and Ferro. Azurara gives the Christian population of these three islands in his time as follows :—In Lançarote sixty men, in Forteventura eighty, and in Ferro twelve. They had their churches and priests.

In the Pagan islands the numbers were, in Gomera* about seven hundred men, in Palma five hundred, in Teneriffe six thousand bearing arms, and in the Great Canary five thousand fighting men. These had never been conquered, but some of their people had been taken, who gave information respecting their customs.

The Great Canary was ruled by two Kings and a Duke, who were elected, but the real governors of the island were an assembly of Knights, who were not to be less than one hundred and ninety, nor so many as two hundred, and whose numbers were filled up by election from the sons of their own class. The people were intelligent, but little worthy of trust; they were very active and powerful. Their only weapons were a short club and the stones with which their country abounded, and which supplied them also with building materials. Most of them went entirely naked, but some wore petticoats of palm leaves. They made no account of the precious metals, but set a high value on iron, which they worked with stones and made into fishing-hooks; they even used stones for shaving. They had abundance of sheep, pigs, and goats, and their infants were generally suckled by the latter. They had wheat, but had not the skill to make bread, and ate the meal with meat and butter. They had plenty of figs, dragon's blood, and dates, but not of a good quality, and

---

* Maciot attempted, with the assistance of some Castilians, to subdue the island of Gomera, but without success.

some useful herbs. They held it an abomination to kill animals, and employed Christian captives as butchers when they could get them. They kindled fire by rubbing one stick against another. They believed in a God who would reward and punish, and some of them called themselves Christians.

The people of Gomera were less civilized. They had no clothing, no houses. Their women were regarded almost as common property, for it was a breach of hospitality for a man not to offer his wife to a visitor by way of welcome. They made their sisters' sons their heirs. They had a few pigs and goats, but lived chiefly on milk, herbs, and roots, like the beasts; they also ate filthy things, such as rats and vermin. They spent their time chiefly in singing and dancing, for they had to make no exertion to gain their livelihood. They believed in a God, but were not taught obedience to any law. The fighting men were seven hundred in number, over whom was a captain with certain other officers.

In Teneriffe the people were much better off, and more civilized. They had plenty of wheat and vegetables, and abundance of pigs, sheep, and goats, and were dressed in skins. They had, however, no houses, but passed their lives in huts and caves. Their chief occupation was war, and they fought with lances of pine-wood, made like great darts, very sharp, and hardened in the fire. There were eight or nine tribes, each of which had two kings, one dead and one living, for they had the strange custom of keeping the dead king unburied till his successor died and took his place: the body was then thrown into a pit. They were strong and active men, and had their own wives, and lived more like men than some of the other islanders. They believed in the existence of a God.

The people of Palma had neither bread nor vegetables, but lived on mutton, milk, and herbs; they did not even take the trouble to catch fish like the other islanders. They fought with spears like the men of Teneriffe, but pointed them with sharp horn instead of iron, and at the other end they also put another piece of horn, but not so sharp as that at the point. They had some chiefs who were called Kings. They had no knowledge of God, nor any faith whatever.

In 1414 the exactions and tyranny of Maciot de Bethencourt had caused Queen Catherine of Castile to send out three war caravels under the command of Pedro Barba de Campos, Lord of Castro Forte, to control him. Maciot, although only Regent, for Jean de Bethencourt was still alive, ceded the islands to Barba and then sailed to Madeira, where he sold to Prince Henry these very islands of which he had just made cession to another, together with those which still remained to be conquered. Maciot subsequently sold them to the Spanish Count de Niebla. Pedro Barba de Campos sold them to Fernando Perez of Seville, and the latter again to the aforesaid Count de Niebla, who disposed of them to Guillem de las Casas, and the latter to his son-in-law Fernam Peraza. Meanwhile the legitimate proprietor, Jean de Bethencourt, left them by will to his brother Reynaud. But as yet there still remained unconquered the Great Canary, Palma, Teneriffe, and the small islands about Lançarote, and, in 1424, Prince Henry sent out a fleet under the command of Fernando de Castro, with two thousand five hundred infantry and a hundred and twenty horse, to effect the conquest of the whole of the islands; but the expense entailed thereby, combined with the expostulations of the King of Castile, caused him to withdraw for a time from the undertaking.

Subsequently, in the year 1446, he resumed his efforts at this conquest, but before taking any step he applied to his brother, Dom Pedro, who was then Regent, to give him a charter prohibiting all Portuguese subjects from going to the Canary Islands, either for purposes of war or commerce, except by his orders. This charter was conceded, with a further grant of a fifth of all imports from those islands. The concession was made in consideration of the great expenses which the Prince had incurred. In the following year, 1447, the Prince conferred the chief captaincy of the island of Lançarote on Antam Gonsalves, who went out to enforce his claim ; but unfortunately, Azurara, from whom we derive this date, and who, as it was very near the period of his writing, would be little likely to be in error, fails to tell us the result of Gonsalves' expedition. If we were to follow Barros and the Spanish historians, the date of this expedition would be much earlier. Be this as it may, when in 1455 King Henry IV., of Castile, was married to Joanna, the youngest daughter of Dom Duarte, King of Portugal, Dom Martinho de Atayde, Count d'Atouguia, who escorted the Princess to Castile, received from King Henry the Canary Islands as an honorary donation. De Atayde sold them to the Marques de Menesco, who again sold them to Dom Fernando, Prince Henry's nephew and adopted son. In 1466 Dom Fernando sent out a new expedition under Diogo da Silva, but if we are to believe Viera y Clavijo, it was as unfortunate as its predecessors. But meanwhile, at the death of Fernam Peraza, his daughter Iñez, who had married Diogo Garcia de Herrera, inherited her father's rights in the Canaries, and one of her daughters married Diogo da Silva. Still Spain maintained its claims, and it was not till 1479, when, on the 4th of September, the treaty of peace was signed at Alcaçova, between Affonso V., of Portugal, and Ferdinand and Isabella of Castile, that the disputes of the two nations on this point were settled. The sixth article of that treaty (Torre do Tombo, Gav. 17, Maç. 6, n. 16) provided that " the conquests from Cape Non to the Indies, with the seas and islands adjacent, should remain in possession of the Portuguese, but the Canaries and Granada should belong to the Castilians."

Hitherto both the gains and the losses of the Portuguese in these various expeditions had been but small. Dangers had been surmounted and captures had been made, but it may be questioned whether the greed of gain alone would have kept alive the spirit of exploration, in the face of dangers which greatly outbalanced the profit secured to individual adventurers. To the far-sighted vision of Prince Henry, the results, though small and slowly conquered, were far more promising than to those whose object was immediate profit, and hence his resolution never wavered, his zeal in the prosecution of his purpose never flagged. It needed all that zeal, supported by his princely position, and the great weight of his personal authority, to induce men to prosecute yet further search through unknown seas for lands which, with no certainty of profit, might so easily offer dangers entirely unanticipated. Such dangers were now to be encountered, and with disastrous result.

In the year 1446, Nuño Tristam set sail in a caravel, by the Prince's command, to make explorations beyond the Cabo dos Mastos, which had been discovered by Alvaro Fernandes, and, being a resolute man, he passed some sixty leagues beyond Cape Verde, and reached what is now called the Rio Grande. Anchoring at its mouth, he took two small boats with two-and-twenty men, intending to pull up the river in search of a village. The tide soon carried him up a considerable way beyond the bar, when he encountered twelve canoes containing some seventy or eighty negroes with bows in their hands, who, having seen the boat when it first entered the river, had assembled to meet him. As the tide rose, one of the native boats passed him, and landed its crew, who began discharging their arrows at the Portuguese. The others who remained in the boats came near, and also discharged their poisoned arrows at the new-comers. The Portuguese hastened back to reach the caravel, but before they got on board, four men were dead from the effect of the poison. They then made all haste to get out to sea, and were obliged to cut their cables and leave their anchors and boats behind, so fierce was the shower of arrows with which they were assailed. Of the two-and-twenty that had set out, two only escaped, one named Andre Dias, and the other Alvaro da Costa, both squires of Prince Henry, and natives of Evora. The other nineteen died, the poison being so subtle that the slightest wound touching the blood caused death.

So perished the brave knight, Nuño Tristam, who would have coveted a more glorious death, and with him another knight named João Correa, and three other gentlemen of the Prince's household, named Duarte d'Olanda, Estevam d'Almeida, and Diogo Machado. In all one-and-twenty were killed, for two were struck in endeavouring to raise the anchors. Five only remained in the ship: a common sailor who knew little enough of the art of navigation; a lad named Aires Tinoco, one of Prince Henry's grooms of the chamber, and who went out as scribe to the expedition; an African boy, one of the earliest captures in that country; and two little fellows who had been attached to the persons of some of the deceased adventurers. The pitiful position of this feeble crew on that inhospitable shore may be imagined. They naturally turned their hopes to the sailor, as the best navigator amongst them, but he freely confessed his want of skill. Aires Tinoco, however, had the good judgment to direct him to steer to the north with a little bearing to the east. For two months they knocked about without seeing land, at the end of which time they caught sight of an armed vessel, which terrified them considerably, for they feared it was a Moorish ship. It proved, however, to belong to a Gallician corsair, named Pero Falcom, who, to their great delight, told them that they were off a place called Sines, on the coast of Portugal. They then lost no time in making for Lagos. The grief of the Prince at the melancholy story related by the boys was enhanced by the fact that nearly all that had perished had been brought up from childhood in his own household. He therefore made it a duty to take the wives and children of all of them under his especial care and protection.

In proportion as Nuño Tristam had been unfortunate, good fortune seemed to await on Alvaro Fernandes, the nephew of João Gonsalves Zarco, commander of Madeira,

for in that same year he returned to the coast of Guinea,
and passed a hundred leagues beyond Cape Verde. At some
distance beyond the Cabo dos Mastos they landed and came
to a village, the inhabitants of which showed a great inclination to fight, and one of them came forward armed
with an azagay. Seeing this Fernandes hurled his lance
at him and struck him dead, upon which the rest took to
flight.

On the next day they came to an estuary, where they saw
some women. They took one about thirty years old with
her child of about two, and a girl of fourteen. The woman
was so strong that they could not force her to the boat,
and they were afraid that the delay would bring the natives
down upon them. At last they thought of taking the child
to the boat, when the affection of the mother caused her to
follow without any difficulty. They next came to a river,*
which they entered in a boat, and meeting four or five canoes
full of negroes, had an encounter with them, in which Alvaro
Fernandes received a wound from an arrow in his leg. As
he was aware of the poison, he drew the arrow out instantly
and bathed the wound with acid and oil, and afterwards
anointed it well with theriack† as an antidote, and by dint
of great care he recovered, but for some days was in great
peril of his life.

In spite of their captain being thus wounded, the caravel
pushed on to the south and reached a point of sand in front
of a great bay, where a boat was sent out to explore. As it
approached the shore, some hundred and twenty negroes
made their appearance, some armed with shields and azagays
and others with bows, and when they reached the water-side

---

* Barros says this river is now called Tabite. The Vicomte de Santarem
identifies it with that laid down on the maps of Juan de la Cosa, and João Freire
as Rio de Lagos.

† This now disused antidote, the name of which means treacle (Græce),
was a compound of a great number of drugs with a basis of viper's flesh. It
was held to be sovereign against the bites of venomous beasts. The name, which
was given by Andromachus, Nero's physician, doubtless arose from the preserva-
tive nature of treacle against putrid air and other deleterious agents.

began to play and dance in the merriest fashion, but the boat's crew not feeling any particular wish, under the circumstances, to share in their jollity, thought best to return to the ship. This was a hundred and ten leagues south of Cape Verde. But for the wound of Alvaro Fernandes they would have gone further.

On their return they put in at Arguin, and afterwards at the Cabo do Resgate, where they fell in with that same Ahude Meimom who had kindly treated João Fernandes. Unfortunately they had no interpreter, but by signs they negotiated with him the exchange of a negress for some cloths, and if they had had a greater quantity, the Moors would gladly have made a larger traffic with them of the same kind. This caravel made more way to the south than any of its predecessors, and received as a reward for so doing two hundred doubloons, one hundred from the Regent Dom Pedro and another hundred from Prince Henry.

These rewards encouraged many who would otherwise have been deterred from these explorations by the sad fate of Nuño Tristam, and accordingly in this same year nine caravels were fitted out, the captains of which were Gil Eannes, who first passed Cape Boyador; Fernando Valarinho, who had distinguished himself at Ceuta; Stevam Affonso, Lourenço Dias, Lourenço d'Elvas and João Bernaldes, an esquire of the Bishop of Algarve, commanding a ship belonging to the bishop; and three others, residents of Lagos. They first proceeded to Madeira to victual, where they were joined by two caravels, one belonging to Tristam Vaz, the commander of Machico, the other to Garcia Homem, son-in-law of João Gonsalves Zarco, commander of Funchal. Thence they sailed for Gomera, where they landed the Canary men who had been taken off by João de Castilha, and who returned very pleased with the treatment and presents which they had received from Prince Henry. But first they proposed to these Canary men to help them in making a capture at the island of Palma, and for the Prince's sake they would gladly have done so, but the plan was frustrated by the Palmarenes

having been put on the alert by the arrival of the caravel of
Lourenço Dias some days before. The Madeira vessels
accordingly returned, as also did Gil Eannes, but the rest
made their way to the Guinea coast, and passed sixty leagues
beyond Cape Verde.

Here they came to a river of great size [Rio Grande],
which they entered with their caravels; but the bishop's
vessel stranded on a sand-bank and was lost, although the
crew and contents were saved. While some of them were
engaged on the salvage, Stevam Affonso and his brother
followed some tracks that they lighted on, and found some
plantations of cotton-trees and rice, and other trees of various
kinds. The land around was hilly and had the appearance of
loaves. They presently entered a thick wood, from which issued
some natives armed with azagays and bows. Seven of the
foremost of those who went to meet them were wounded, and
of these five fell dead, two Portuguese and three foreigners.
When Stevam Affonso and the others saw the peril of their
position, they retreated and escaped with great difficulty, for
the natives were there in numbers. On reaching their ships,
they determined to return. They therefore proceeded to the
island of Arguin to take water, and thence to the Cabo do
Resgate, where, finding tracks of natives, and being unwilling to return without a capture, they followed them and
succeeded in taking eight-and-forty, with whom they made
their way to Portugal.

Stevam Affonso only went to the island of Palma, where
he took two women, a transaction which had like to have cost
the lives of the whole party had not Diogo Gonsalves boldly
snatched a crossbow from the hands of one of the Canary
men, and quickly shot seven of them. One of these was a
chieftain, as was known by his carrying a palm branch in his
hand. The rest, seeing their leader fall, surrendered. The
party then returned in safety to Portugal.

In this year (1446) Gomes Pires did not forget his promise
to the Moors in the year before, that he would return to
the Rio d'Ouro, and on his petition the Prince gave him

two caravels, with twenty men, among whom was a youth
of the Prince's household named João Gorizo, who had
charge of the accounts of the receipts and expenditure
which occurred in the Moorish traffic. It was now the
custom for all the vessels bound to the west coast of
Africa to go first to Madeira to victual, and on their arrival
Gomes Pires desired Gorizo to remain and take in the stores,
while he proceeded straight in the smaller vessel to the
Rio d'Ouro. The Moors not appearing near the entrance,
he made for the head of the estuary, and anchored in a
harbour named Porto da Caldeira, a name which does not
survive in any existing maps, but appears to have been
given by other Portuguese who had previously visited the
estuary. Although he burned fires night and day on a hill
near the harbour, it was three days before any Moors made
their appearance. When they came, he proposed to them
by his interpreters to barter cloth with them for Guinea
slaves. They answered that they were not merchants nor
were there any thereabouts, though inland there were
traffickers in merchandise who had abundance of gold and
Guinea slaves, but to reach the spot where they were would
involve a very laborious journey. Gomes Pires requested
the Moors to fetch these merchants, and gave them in
advance a remuneration for their trouble. They pretended
to go, but although he waited for them one-and-twenty days
they never returned. Meanwhile Gorizo arrived with the
other vessel, and they then set sail, and landing at different
points within a range of only eleven leagues of coast
with considerable toil and fatigue, contrived to capture
seventy-nine Moors. As they had brought out a large
quantity of salt for the purpose of salting the phoca skins in
the event of their failing to make a better capture, they
were compelled to discharge the salt in order to make
stowage room for their captives, and so they returned to
Lagos.

"Up to that period, 1446," says Azurara, "there had
been fifty-one caravels to these parts. These caravels went

four hundred and fifty leagues beyond the Cape. It was found that that coast ran southward with many points, which the Prince caused to be added to the sailing chart, and it is to be observed that what was known for certain of the coast of the great sea was six* hundred leagues, which have been increased by these four hundred, and what had been shown upon the mappemonde with respect to this coast was not truth, for it had been only delineated at haphazard, but that which is now laid down on the charts is from ocular observation, as has been already shown."

In the following year, 1447, in consequence of the failure in establishing friendly relations with the Moors at the Rio d'Ouro, where Gomes Pires made the capture just recorded, Prince Henry resolved to try if better success might be met with at Messa, a town in the province of Sus, in the empire of Marocco.† With this object he fitted out a caravel which he put under the command of Diogo Gil, a man who had already done good service against the Moors, both by sea and land. And at this time it happened to come to the Prince's knowledge that a Spanish merchant named Marcos Cisfontes had in his possession twenty-six Moors from that same place, the bargain for whose ransom had been already stipulated for in exchange for some negroes of Guinea. To turn the outward voyage to advantage, the Prince caused a proposal to be made to the merchant to carry out those Moors to Messa in the vessel which had been fitted out with that destination, with the understanding that he should receive in return a certain number of the negroes that were to be given in ransom.

As may readily be supposed, this proposal was not made by the Prince for the sake of the trifling profit that would result from the transaction. He had a double object in view of a far higher kind. He not only wished to gain information respecting the mode of traffic in that country,

---

* It should be two hundred, evidently an error into which the penman was led by adding mentally the two to the four hundred.

† Leo Africanus says, liv. 2, that it was built by the early inhabitants of Africa.

but his great anxiety, in accordance with the earnest piety which distinguished his whole life, was to rescue these negroes from heathenism, and confer on them the blessings of Christianity. The proposal was readily accepted, and João Fernandes, the same who had lived seven months among the Moors at Arguin, accompanied the party, and on arriving volunteered to negotiate the ransom. He was so successful that he procured fifty-one negroes in exchange for eighteen of the Moors.* It so happened that while he was yet on shore there came on so strong a wind from the south that they were compelled to trip their anchor and sail for Portugal. They brought back with them for the Prince a lion, which he afterwards sent to Galway by way of a present as a curiosity to an Englishman who lived there, and who had been formerly in his service. João Fernandes remained till another ship returned for him.

In this same year also Antam Gonsalves returned to the Rio d'Ouro to try if it were possible to bring the Moors of that part to terms. He anchored at some distance within the estuary, and a number of Moors came to the shore, among whom was one who was evidently a chief. This man spoke assuringly to Gonsalves, but warned him not to trust the rest unless he were present. It happened once that while he was at a distance, the other Moors made a show of friendliness to the Portuguese, and Gonsalves, thinking the chief was among them, was about to land, but the boats no sooner neared the shore than the Moors attacked the Portuguese with their azagays, and, but for the promptitude of Gonsalves, they would all have been slain. They managed, however, to effect their escape, but with one of their men so seriously wounded that he died in a few days. Another expedition to the Rio d'Ouro under the command of Jorge Gonsalves, in which he brought back a large quantity of oil and skins of sea-wolves, completed the list of voyages in the year 1447.

* This proves the influence that João Fernandes had acquired over the natives, no doubt from his knowledge of Arabic.

In January, 1448, the fame of these expeditions brought
out to Portugal a nobleman of the household of the King of
Denmark,* named Vallarte, who begged the Prince to grant
him a caravel to go to the land of the negroes. It was the
kind of request that Prince Henry was always ready to listen
to, and accordingly he had a caravel quickly fitted out to go
to Cape Verde. To Vallarte he gave the principal command,
but as he was a foreigner he sent with him one Fernando
d'Affonso, not only to aid in the command of the vessel, but
as a sort of ambassador to the king of the country. With
them also went two natives of that country as interpreters,
by whose means the Prince hoped that something might be
done towards the conversion of the people. The weather
was so exceedingly adverse that it took them six months
from the time they left Lisbon to reach the island of Palma,
near Cape Verde. But as that was not their destination
they proceeded further, and anchored at a place called by
the natives Abram, where Vallarte went on shore with some
others, and found a considerable number of negroes assembled. To these Vallarte proposed that as a guarantee for
friendly intercourse they should give him one of their people
in exchange for one of his. This was agreed to by the permission of Guitanye, the governor of the country. As soon
as they had one of the negroes on board the caravel, Fernando d'Affonso told him that their object was to instruct
him to inform his master that the Portuguese were servants
of a great and mighty Prince of the west of Spain,† and were
come by his command to treat for him with the great king
of that country. The negro told them in reply, that the
residence of their King Boor was six or seven days' journey
off, and that the king was then at a great distance fighting
against a rebel.

---

* The king in question is called by Azurara King of Denmark, Sweden, and
Norway, and as on the death of King Christopher on the 6th of January, 1448,
these three crowns were separated, we have proof both of the name of the
sovereign referred to, and of the date of the occurrence.

† The word Spain was frequently used in those times for the whole peninsula.

Fernando still desiring to treat with the king himself, the governor Guitanye, who seems to have been very friendly with the Portuguese, promised, after some delay, to send the message. During the absence of the governor, Vallarte ventured on shore with a boat's crew, and fell into an ambush of the natives, who attacked them with their azagays to such effect that, of the whole number, only one saved himself by swimming and returned to the ship. Of the end of the rest no news survived, except that the man who swam away declared that he only saw one killed, and the three or four times that he looked round as he was swimming, he always saw Vallarte sitting at the stern of the boat. But at the time that Azurara was writing his chronicle in 1448, some natives of that part came into Prince Henry's possession, who stated that in a fortress far in the interior, there had been four Christian prisoners, one of whom had died, but three were still alive, and these were supposed to be the remnants of the boat's crew. After this miserable adventure Fernando d'Affonso, not having even a boat remaining, returned to Portugal.

The people of Lagos had gained too much experience of the west coast of Africa to be insensible of the value of its fisheries, and they obtained permission from the Prince, on payment of a royalty, to turn their knowledge to account in that respect. Off the Cabo dos Ruyvos they found a large abundance. After they had been there some days, and had taken a great quantity of fish, some of which they had dried, and the rest were drying, the Moors came down upon them, and they only narrowly escaped with two men wounded.

In the course of the above explorations, to the period of Azurara's completing his chronicle, nine hundred and twenty-seven souls had been taken to Portugal.

CHAPTER XIII.

1439—1449.

WE left Dom Pedro on his return to Portugal in 1428. On the 13th of September in the same year he married Isabel, the eldest daughter of Don Jaime, second Count of Urgel, and of Isabel, Infanta of Aragon. After the death of King João, his brother, King Duarte, by a charter dated from Santarem, Nov. 6th, 1443, appointed him, conjointly with Prince Henry, guardian to his son the Infant Affonso, heir-apparent to the throne.

During the lifetime of Dom Duarte, Dom Pedro had had the misfortune not only of incurring the ill-will of Queen Leonora, but also of having to oppose, from a sense of duty to his country, the strong desire of his brothers, Prince Henry and Dom Fernando, to make the attack upon Tangier which ended so disastrously. But worse misfortunes than these were in store for him. If Dom Duarte by the virtues of his life had been unable to save his kingdom from misfortune, he left behind him a legacy which entailed upon it yet greater troubles. By his will he appointed his widow, Leonora, Regent during the minority of his son, Affonso, who at the time of his father's death was but six years old. This arrangement was obnoxious to the people, not only because the Queen was a Castilian, and had moreover instigated the disastrous expedition to Tangier, but because her appointment set aside the claims of Dom Pedro and Prince Henry, who were in every respect far better fitted for so responsible a post. The more devoted and prudent adherents of the Queen dissuaded her from assuming the Regency ; but their advice was overruled by those who insinuated that if the government were to fall into the hands of Dom Pedro, the king's life would be very insecure, inasmuch as the Prince was powerful and popular, and had sons on whose behalf he might aim at the crown.

Meanwhile the funeral of King Duarte was to be solemnized in the monastery of Batalha, and while awaiting at Thomar the arrival of those who were to be present at the ceremony, Dom Pedro availed himself of the presence of so large a number of grandees to suggest that, in consideration of the king's youth, and to remove all doubt as to the succession, Affonso's brother Fernando should be sworn hereditary prince of the kingdom, until the King should have a son. The whole of the council approved of the proposition, and Dom Fernando was sworn heir-apparent, and thence forward bore the style and title of Prince of Portugal.

On the 1st of November, 1439, Dom Pedro was nominated Regent of the Kingdom by the States General of the Realm, to whom, and not to the King, the prerogative really belonged. In this high position he showed great prudence and justice, and materially assisted the labours of Prince Henry, but without lending his name to them. He was a great patron of literature, and instilled this taste into the young King and his own family. In spite, however, of all these excellencies, he was not destined to enjoy his dignities in peace. A disagreement with the Duke of Braganza about the office of High Constable, was the prelude to a long series of disputes, in which the latter, who was a great favourite with the people, on account of his having married the daughter of the Holy Constable, was frequently enabled to make his malice triumphant.

On the 18th of February, 1445, Queen Leonora died, and, as has been supposed, by poison. In the following year at the age of fourteen Affonso attained his political majority, and the Regent accordingly resigned to him the sceptre at a convocation of the states of the kingdom, held at Lisbon for that purpose. The King requested his uncle still to direct the affairs of the kingdom in his name, feeling it to be a charge too responsible for his yet unpractised youth, and a proclamation to that effect enjoined entire obedience to the command of Dom Pedro. The King further declared that his betrothal with his uncle's daughter, which had taken place at Obidos during his minority, was entered upon at his own earnest desire, and he called upon the nobles and deputies present to confirm the marriage.

Thus far well; but the enemies of Dom Pedro, who did their best to frustrate his re-appointment as Regent, were the more embittered by their want of success. · His bastard brother, the Count de Barcellos, now Duke of Braganza, with his son Affonso, Count d'Ourem, spared no pains to prejudice the young King against his uncle, and at length they persuaded him to assume the reins of government himself. This Pedro willingly conceded, at the same time urging on the King the propriety of carrying into effect the contemplated marriage with his daughter. The King agreed to this without hesitation, but before the preparations for the marriage could be completed, at the instigation of the Prince's enemies, he demanded of him the surrender of the Regency in advance of the stipulated period. To prevent ill-consequences the Prince again yielded, and in May, 1447, Affonso established his Court at Santarem, and celebrated his marriage with Dom Pedro's daughter, though without the magnificence and rejoicings which the father of· his bride would naturally have desired. The Duke of Braganza now sought to prejudice the people against Dom Pedro. With this object he expelled with insult all the adherents of the Prince from the offices which had been given them by the King, and placed guards in their vacant castles and houses, as though the King had already declared war against his uncle. Meanwhile the Prince was forbidden access to the King, and thus debarred from the only means of defence open to him. The most dangerous of his enemies was a young

Portuguese of noble birth, named Berredes, who had retired from Rome with the rank of Pontifical Protonotary, by whose machinations the King was induced to resolve upon the dismissal of his uncle from the Court. His design, however, reached the ears of Dom Pedro, who wisely resolved to do voluntarily that which was about to be forced upon him. He therefore presented himself without any show of agitation before the King, and requested that, in consideration of his long and faithful services, he might be permitted to seek repose on his own estate, adding that, in any emergency, those services were at the King's disposal, as they had always been throughout his Regency. The King received with delight this proposition, which spared him the pain of giving his uncle a dismissal. He not only acceded to the Prince's request, but parted from him with expressions of affection and regret, and promised him faithfully to approve and confirm all that he had done during his Regency.

Towards the end of July, Dom Pedro retired to Coimbra, but was pursued by his enemies with untiring malignity. And he soon found himself accused of nothing less than having caused the death of King Duarte, his sister-in-law Leonora, and his brother João. The apathy of those who might have befriended the Prince was as remarkable as the malignity of his enemies. There was one noble exception, however, in Alvaro Vaz de Almada, the Prince's sworn brother-in-arms. This gallant nobleman had, like Dom Pedro himself, travelled much in Europe, and had everywhere been treated with great distinction. In England he had received the high honour of the Knighthood of the Garter. In Germany the Emperor had shown him especial marks of favour, and Charles VII., King of France, had conferred upon him the title of Count d'Avranches. In spite of the coldness shown him on account of his friendship for Dom Pedro, he became his warm and persevering advocate, and so powerful was his influence, that the King's evil counsellors thought it necessary to withdraw him to Cintra.

A series of bitter persecutions followed, the Prince only

opposing remonstrance to the injustice of his enemies. The
strongholds which were in possession of his adherents were
taken from them. His eldest son was deprived of the
dignity of Constable, and the arms in the arsenal of Coimbra
were demanded in the name of the King. Dom Pedro,
whose only hope of obtaining justice lay in his nephew's
learning the truth undistorted by his evil advisers, wrote
several letters to the King by his confessor. It is supposed
that the answers he received were not those dictated by
the King, for they were written as a King would write to a
rebellious subject, and he had always formerly written as a
son to a father.

Soon after this Dom Pedro received intelligence that the
Duke of Braganza intended to pass through his domains
without his permission, so that, if he submitted to this
indignity, the stigma of cowardice would fall upon him;
while, on the other hand, if he resented it, he would be
accused of breaking the peace. The Prince resolved to go
to Penella to stop the Duke's march, and as soon as the
news reached Santarem, he was joined by many nobles of the
court. Meanwhile, a message from the King reached the
Prince, commanding him to return to Coimbra, and to offer
no hindrance to the march of the Duke of Braganza. Dom
Pedro replied, that if the Duke came peaceably, he would
welcome him cordially, but if otherwise, his honour as a
prince forbade him to submit to so gross an indignity. He
therefore ranged his troops in order of battle at Penella, and
awaited the Duke. The latter, who by this time had reached
Villarinho, did not believe that the Prince would dare to
oppose the King's orders with so few troops, but when he
found that such was the case, and that many of the knights
of his suite, being secretly attached to Dom Pedro, were
unwilling for the encounter, he resolved to secure his own
safety, and escaped in the night accompanied by only two
persons. The next morning, when the soldiers found that
they were deserted by their leaders, they were seized with
panic, and fled. The Duke rallied his troops with difficulty

at Covilham, and found he had sustained considerable loss. He now persuaded the King to declare war personally against his uncle.

On his return to Coimbra, Dom Pedro received from Doña Isabel, his daughter, an intimation of the King's intention, which induced him to call a council to deliberate on his future plans. The course which he finally adopted was that proposed by the Count d'Avranches, viz., that they should march peaceably towards the King, and solicit an audience, and if that were denied, to die as gentlemen and soldiers. A few days after this, Dom Pedro and the Count d'Avranches, being sworn brothers in arms, made a vow to be together in death as they had been in life, and solemnly sealed the compact by a reception of the blessed sacrament.

Meanwhile, Doña Isabel had made a last despairing effort to preserve the peace between her husband and her father, to both of whom she was devotedly attached. By her tears and entreaties she had extorted from the King a promise to pardon her father, if he would consent to ask forgiveness, and had also succeeded in wringing that concession from the Prince. But the first burst of generosity having passed, Affonso contrived, by taking exceptions to certain portions of Dom Pedro's letter, to find a pretext for breaking his promise to the queen, and preparations for the war were continued.

On the 5th of May, 1449, the Infant left Coimbra with an army of one thousand horsemen and five thousand foot soldiers. At the convent of Batalha, Pedro visited the royal tomb, and gazed sadly at the open sepulchre prepared for him by his father, little thinking that the malice of his enemies would deprive him for a time of even this last resting-place. At Rio Mayor, five leagues from Santarem, he took counsel on his future conduct, and, contrary to the advice of most of his followers, decided to march against Lisbon, in the hope that his enemies would attack him before the arrival of the King, with whom he dreaded an

hostile encounter.   If they did not attack him he resolved
to return to Coimbra.

The King hearing of Dom Pedro's intentions, sent a
division of troops to secure Lisbon, and set out in person from
Santarem with an army of thirty thousand men, the largest
army till then raised in Portugal.   Dom Pedro selected a
good position near Alfarrobeira, above the town of Alverca;
and awaited the King's army, which came up with him on
the 20th of May.   During the sharpshooting preliminary to
the attack, a badly aimed missile from the Infant's camp
struck the King's tent.   The report spread that the King
was hurt.   This so roused the indignation of his soldiers,
that they rushed with headlong fury on the enemy, who,
unprepared for such an attack, yielded on all sides.   The
Prince, to inspire his soldiers with courage, leaped from his
horse, and pressing upon the enemy dealt terror and death
among them, till he fell, pierced by an arrow, and died
shortly after.   When the Count d'Avranches heard of his
death, he threw himself into the conflict, and fought with
desperation till, as it is reported of him, worn out with
fighting, he sank down among the heaps of slain more conquered by his own exertions than by the enemy.   A soldier
immediately cut off his head and carried it to the King,
hoping thereby to merit the order of knighthood.   The
natural brother of the hero with difficulty obtained permission to bury the mutilated trunk on the field of battle.
The body of the Infant remained all that day on the field,
and in the night some common soldiers carried it on a
shield to a hut, where it remained for three days unburied.*

After his death the enemies of the Infant sought in vain
among his papers for proofs of guilt.   His daughter Doña
Isabel now became the object of his enemies' persecution,
and they sought to persuade the King to put her away, but
the prudence of her conduct at this trying juncture, and

* In this battle perished Sir William Arnold, an English knight, who had
been Major Domo to Queen Philippa, and after her death had passed into the
service of Dom Pedro.

the affection which the King bore to her, prevented their wicked counsels from taking effect.

The courts of Europe were unanimous in reprobation of the King's conduct to his uncle. The Duke of Burgundy, and the Duchess Isabel, the sister of the unhappy Prince, sent an ecclesiastic of high standing to Portugal, who, in their name, severely reproved the King, and demanded that the body of Dom Pedro should either be buried in the tomb at Batalha destined for him by Dom João I., or delivered to him to convey to Burgundy, where it should receive honourable sepulture. Not wishing that his uncle's remains should be taken out of the kingdom, the King had them exhumed from the church of Alverca, where some common people had buried them under a staircase, and had them conveyed to the castle of Abrantes. The Burgundian priest insisted on the restoration of the children of Dom Pedro to their property and dignities. For some time the King refused out of consideration for the Duke of Braganza and the Count d'Ourem, but finally acceded to the demand, and further at the queen's request pardoned nearly all those who had fought for the Infant.

In the same year, the Queen having given birth to a son, took advantage of the favourable disposition of the King at this event, to obtain his permission for the honourable burial of her father. Accordingly the remains of the Prince were accompanied to the tomb by a long train of ecclesiastics and nobles, headed by his brother Prince Henry, and were solemnly deposited in his tomb at Batalha.

This was the last satisfaction enjoyed by Doña Isabel, and probably cost her her life, for immediately on her return to Evora with her husband from the funeral, she fell ill and died. Her sudden death was attributed by the people to poison administered by her father's enemies, who feared her influence over the King, which had just proved so powerful.

The death of the Prince having removed the great obstacle to their ambition, his enemies persuaded the King to grant them the domains which, in order that the crown lands might not be alienated, the Regent had always steadfastly refused them. Thus the Duke of Braganza obtained the town of Guimaraens, and would have had Oporto but for the determined opposition of the inhabitants.

So much was Dom Pedro honoured in Portugal, that in spite of the powerful faction raised up against him, there were yet many writers of his own time who dared to use their pens in his defence, while in the fifteenth and sixteenth centuries there were not wanting men like Camoens and Luis da Souza to sing his praises. And well did he deserve their advocacy, for not only was he a liberal patron to men of letters, but himself was an author and a poet of no mean capacity. It has been said that the first book ever printed in Portugal was a collection of " Coplas," or couplets, by him. There is no date or place of imprint in the earliest edition known to exist, but the learned academician Antonio Ribeiro dos Santos has conjectured the date of 1478. It was, however, brought out by a Spaniard, Antonio d'Urrea, and there is nothing to show that it was printed in Portugal.

CHAPTER XIV.

THE AZORES.

1431—1466.

THAT the middle and eastern groups of the Azores, like the Madeira group, were discovered by Portuguese vessels, under Genoese pilotage, at the beginning of the fourteenth century, is proved by the simple fact, that they also appear in the Laurentian map of 1351. The same facts which have enabled us to unfold the history of the discovery of the Madeira group, apply with equal force to the others.

Although on the Laurentian map each island of the Azores has not a separate name, the group is laid down with considerable accuracy for the period, the orientation only being at fault, and has collective designations, thus :—the two islands of San Miguel and Santa Maria are called *" Insule de Cabrera ;"* the islands of San Jorge, Fayal, and Pico, are called, *" Insule de Ventura sive de Colombis,"* while Terceira is called, *" Insula de Brazi."* Subsequent charts, but anterior to the effective discovery of these islands in Prince Henry's time, show their names in detail, some being identical with those which they at present bear, and others remarkably interesting as showing the observations or impressions which influenced the first discoverers in naming each island. At the same time they illustrate the application of the names on the Laurentian map. Thus, in the Catalan map of 1375, we have San Zorzo, the Catalan form of the present name of the island San Jorge, doubtless indicative of the discovery having been made on St. George's day. Fayal, which the

Portuguese afterwards so called from its forests of beech trees, is here called Insula de la Ventura, implying its discovery by an accident. Terceira received the name of Insula de Brazil from the dyewood with which it abounded, thus preceding its famous namesake in South America by nearly two hundred years. The island of Pico would seem to have been frequented by wild pigeons at the time of its discovery, since, on the Catalan map its bears the name of Li Columbi. Cabrera, or the island of goats, is a name which, on the map of Andrea Bianco of 1436, we find given to the island of St. Miguel. We thus get full explanation of what islands were indicated by the collective names which occur on the Laurentian map. The islands of Graciosa, Corvo, and Flores, are not laid down on that map, if we are to trust to the extract supplied to us by Count Baldelli Boni. Corvo had its present name given to it, on the Catalan map, under the form of Insula de Corvi Marini, and Flores was called Li Conigi, from which we must presume that it abounded in rabbits.

It has been already seen that the discovery of the Madeira group, as exhibited on the Laurentian map, produced no beneficial results in the way of colonization. We have incidental evidence that the same, as might be expected, was the case with the Azores. Father Antonio Cordeyro, a native of Terceira, who, in his "Historia Insulana," supplies us with information derived from the early and still unpublished MS. History of the Discovery of the Azores, by Father Gaspar Fructuoso, a native of San Miguel, records a tradition that, about the year 1370, a Greek was driven on the latter island by a tempest, and resolved on colonizing it, but failed in his first experiment. He brought back to the island a considerable quantity of cattle, but they soon died, and he gave up his project.

There can be no doubt that the knowledge of the existence of the Azores, as laid down in the Laurentian map of 1351, was preserved on the map brought from Venice in 1428, by Dom Pedro, and enabled Prince Henry to give directions to his navigators for the re-discovery of these islands. Thus,

we find on the Catalan map of Gabriel de Valseca, dated
1439, the entire group laid down, accompanied by the following significant legend : " These islands were *found* [not
discovered] by Diego de Sevill, pilot of the King of Portugal,
in the year MCCCCXXVII.," or MCCCCXXXII., according as one
may read the last figure but two as a v or an x.

As in 1439, the island of Santa Maria and the Formigas
were all that had been re-discovered in Prince Henry's time,
Valseca's word " found" would imply the *lighting on* the
group, which he was able geographically to depict from other
sources. Of the two readings of the date of that discovery,
I incline to think that the latter, 1432, is correct, inasmuch
as, in 1431, Prince Henry had sent out Gonzalo Velho
Cabral, a gentleman of illustrious family, in search of these
islands. He then discovered the Formigas only, but in the
following year, on the 15th of August, the Feast of the
Assumption, he fell in with the island which, on the Italian
maps, had been named Uovo, or the Egg, and named it
accordingly Santa Maria. In all probability Diego de Sevill,
the King's pilot, mentioned in Valseca's map, was the pilot
in this expedition. Prince Henry resolved to colonize the
island, and gave Cabral the rank of Captain Donatary, with
full powers to collect, even from his own household, as many
volunteers as he could, with all the requisites for that object.
Cabral devoted three years to recruiting, and finally succeeded in taking out to the island a great number of men of
rank and fortune.

Many years afterwards, a runaway negro slave, who had
escaped to the highest mountain on the north of the island,
perceived in the distance, on a clear day, another island, and
he returned to his master with the news, which he hoped
would secure him his pardon. After the fact had been
verified, intimation thereof was transmitted to Prince Henry,
and as it tallied with the information afforded by the ancient
maps which he possessed, he commissioned Cabral, who
happened at that time to be with him, to go in search of the
new island. His first essay was fruitless, but the Prince

showed him that he had passed between Santa Maria and the island he was in search of, and sent him out again. This time he was successful. Cordeiro states that he reached the island he sought on the 8th of May, 1444, which being the day of the apparition of St. Michael, he named the island San Miguel. But Azurara tells us that Dom Pedro, who, with Prince Henry's acquiescence, interested himself much in the colonization of this island, gave it the name of San Miguel from his own peculiar devotion to that saint.

The Prince gave Cabral the command of this second island, also with instructions to colonize it, and a year having been spent in the needful preparations, the explorer returned thither on the 29th of September, 1445. In the previous voyage he had taken out with him some Moors belonging to the Prince, for the purpose of tilling the soil, but on his return he found them in such a state of alarm from the earthquakes that were taking place in the island, that if only they had had a boat to escape in, they would certainly not have awaited his return. Moreover, the ship's pilot, who had accompanied him in both voyages, remarked, that, whereas in the former voyage he had seen a very lofty peak at the east end of the island and another at the west, that at the east only now remained.

It was at this time that the name of Azores was first given to these two islands of Santa Maria and San Miguel, from the circumstance that the explorers had found azores, or hawks, there, or, what is more probable, kites, which they may have taken for hawks. Prince Henry subsequently bestowed on the Order of Christ the tithes of the island, and one-half of the sugar revenues.

The third island discovered in the Archipelago of the Azores, and on that account named Terceira or "the third," would seem to have been sighted by some sailors, probably returning from Cape Verde to Portugal, whose names were not deemed of importance enough to be attached to the discovery. Nor is the date of the discovery known, but it occurred between the years 1444 and 1450, and on some

festival especially dedicated to our Blessed Lord, since it at
first received the name of the Island of Jesu Christo, and
bore for its arms the Saviour on the Cross.

The Flemings, however, claim for themselves the exclusive
discovery, to which they give the date of 1445 as made by
Josué van den Berge, a native of Bruges. This pretension
is not corroborated, but rather disproved by contemporary
evidence. Cordeiro has given us a copy of Prince Henry's
grant of the Captaincy of the island on the 2nd of March,
1450, to Jacques de Bruges, whom he describes as his
servant, who had rendered him some services, but says
nothing of the island having been discovered either by him
or any one of his countrymen. The sole reason given by the
Prince for making the grant was that Jacques came to him
and stated, that as in the memory of man the Azores had
been under the aggressive lordship of no one except himself,
and as the Island of Jesu, the third of these islands, was
entirely uninhabited, he begged permission to colonize it.
As he had no legitimate sons and only two daughters, the
Prince allowed the inheritance to descend to the female line.
This unusual grant is readily explained by the fact that the
new Captain Donatary was very rich, fitted out the armament and requisites for this rather distant colonization at
his own expense, was a good Catholic, had married a noble
Portuguese lady, and in all probability had entered the
Prince's service under the recommendation of his sister, the
Duchess of Burgundy.

The islands of San Jorge and Graciosa, being within sight
of Terceira, soon became participators in the colonization
which had been brought to the latter. The first colonizer of
Graciosa was Vasco Gil Sodrè, a Portuguese gentleman,
who, while on service in Africa, hearing of the newly
colonized island of Terceira, went thither with all his family,
but soon passed over to Graciosa, where he was joined by
one Duarte Barreto, who had married his sister, and who
had come out with the rank of Captain Donatary of half of
the island. This Barreto being afterwards carried off by

some Spanish pirates, the Captaincy of the entire island fell
into the hands of one Pedro Correa da Cunha, who had been
Governor of the Island of Porto Santo during the minority
of his wife's brother, the son of that Bartholomew Perestrello,
whom Prince Henry had originally appointed Captain of that
island.

One of the companions of Jacques de Bruges, a wealthy
and noble Fleming, named Willem van der Haagen, whose
northern name sounded so harsh to Portuguese ears, that
they translated it into Da Silveira, which means the same
thing, viz., "Hedges" or "Underwood," took from Flanders,
at his own cost, two ships, full of people and artisans of
different kinds, to make a trial of the island of San Jorge.
Selecting a point of the island which they called the Topo,
he founded the city which afterwards bore that name, but
the sterility of the island at a later period made him remove
to Fayal, which had been discovered in the interval.

On what day or in what year the island of Fayal was
discovered, or who was the discoverer, no research has ever
succeeded in finding. There is no doubt that the first
Captain Donatary was Jobst van Heurter, in Portuguese
named Joz de Utra, Lord of Moerkerke in Flanders,
father-in-law of the celebrated Martin Behaim, from a
legend on whose famous globe, made in 1492, and still
preserved in the ancestral house in Nuremburg, we gather
the following statement respecting the bestower of the
Captaincy:—

" The islands of the Azores were colonized in 1466, when
they were given by the King of Portugal, after much
solicitation, to his sister Isabel, Duchess of Burgundy. A
great war was at that time being carried on in Flanders,
accompanied by severe famine, and the Duchess sent out to
these islands a great number of men and women of all
classes, with priests and everything requisite for the maintenance of religious worship. She also sent out several
vessels laden with materials for the cultivation of the soil
and for building houses, and during ten years she continued

to send out means of subsistence. In 1490 there were some thousands of souls there who had come out with the noble knight, Job de Huerter, Lord of Moerkirchen in Flanders, my dear father-in-law, *to whom and his descendants these islands were given by the Duchess of Burgundy.* In them grows the Portuguese sugar. There are two crops in the year, for there is no winter. All food is cheap, and there would be abundance of subsistence for a large population. In the year 1431, when Prince Pedro was Regent, two vessels were equipped with necessaries for two years by Prince Henry to go to the countries beyond Cape Finisterre, and sailing due west for some five hundred leagues discovered these ten islands, which they found uninhabited, and as there were neither quadrupeds nor men, the birds were so tame that they were not frightened, whence they called these islands the Azores.* In the following year by the King's orders sixteen vessels were sent out with various kinds of domestic animals, that they might breed on each of the islands."

This account does not exactly tally with other documentary evidence. Father Cordeyro, in his " Historia Insulana," writing on the spot with documents before him, and at a time when such documents would be by no means scarce, says nothing of any cession of these islands to the Duchess of Burgundy. His words on the subject run thus :—" Fayal being now in some degree colonized by Portuguese from the island of Terceira, St. Jorge, and Graciosa, the *Royal personages* thought of appointing some Captain Donatary of the island, in order to add to the wealth and dignity of the colony, and as there was then at Lisbon, in the service of the Royal personages, a Fleming of high birth named Joz d'Utra, *the King of Portugal made him Captain Donatary* of the island of Fayal, and gave him in marriage a lady of the court named Brites, of the ancient family of the

---

* According to the Portuguese historians the Formigas only were discovered in this year. It was in 1432 that Gonsalo Velho Cabral landed at Santa Maria. The rest were, as we have seen, discovered later. It is equally plain that Behaim blunders about the origin of the name.

Macedos." Barros says of him that " he was a native of
Bruges, a large landholder, and had come out as a young man,
hearing of the fame of the Portuguese discoveries, with the
view of seeing the world and learning languages, as was the
custom of young men of high family to do. *When he had
received his charter of tenure* he returned to Flanders, sold
all his property, and embarked with a number of relatives
and friends for Lisbon, whence he took his wife and established himself at Fayal, where he had several daughters,
one of whom married Martin Behaim."

There is in existence another piece of evidence which
primâ facie would place in yet another light than the preceding the question as to the first bestower of the commandership of the island of Fayal. This document is a judgment
in a lawsuit respecting the succession to that commandership which still exists in the Torre do Tombo (Gaveta 15,
Maço 16, No. 5), under date of 16th of September, 1571.
The claim lay against the crown on the part of Jeronymo
d'Utra Cortereal, whose allegation was that his grandfather
Joz d'Utra, *at the instance of Dom Fernando, Master of the
Order of Christ,* had come to colonize these islands, which
belonged to that order, and *that the commandership had been
given to him and his descendants,* and afterwards confirmed by
Don Manuel, and that, by the death of the first commander,
the commandership devolved on his son Manuel d'Utra
Cortereal.

From these various statements, no doubt is left of the
appointment, but in each it has been made to emanate from
a different source; in one from the Duchess of Burgundy,
in another from the crown, and in the third from the Grand
Master of the Order of Christ. The sum of the evidence
seems to be that the grant was made by Prince Ferdinand,
Grand Master of the Order of Christ, at the request of his
aunt, the Duchess of Burgundy, and confirmed by the
crown.

In this lawsuit a confusion is made between Jobst de
Huerter, the first captain of the island, who married a

Macedo, and his son, who bore the same name as his father and married a Cortereal. It was the grandson of the latter who instituted the proceedings. All the details tend to confirm the statement of Behaim, on his globe, that the appointment was made in 1466.

The grant of the newly-discovered islands made by King Duarte to Prince Henry in 1433, was transmitted by bequest of the latter to his nephew and adopted son Dom Fernando, and confirmed by Affonso V., by a charter dated at Evora, December 3rd, 1460.

At the same time it is reasonable to infer that the extraordinary expenses incurred by the Duchess of Burgundy, De Huerter, and other Flemings in colonizing these islands, would secure to them privileges and powers that would give some show of plausibility to Behaim's statement, that the islands had been given by the King of Portugal to the Duchess of Burgundy. As to the discrepancy between Behaim's account and that of Cordeyro with respect to the islands being inhabited or not at the time of Huerter's appointment, if credit is to be given to each for a wish to be truthful to the best of his ability, the palm must rest with the former, as his father-in-law was in every sense the highest possible authority on the subject.

Some years after Jobst de Huerter had undertaken the colonization of Fayal, he obtained the commission of the Captaincy of Pico, which island, though lying at only a league's distance from Fayal, is supposed not to have been discovered for many years later than it. This is quite possible, for we have seen that it was some time before the dark spot observed from the island of Porto Santo was made out to be the important island of Madeira, though only at one league distant.

Equal obscurity rests on the date of discovery of the islands of Flores and Corvo, as well as on the discoverer. It is only known that they were first conceded to a lady of Lisbon, named Maria de Vilhena. When the Fleming Willem van der Haagen, *alias* Da Silveira, went from San

Jorge to Fayal, it was by invitation from his compatriot Jobst van Huerter, who had been now four years established there, and promised to give him part of the island. It happened, however, that Silveira became so popular by his virtues and distinguished personal qualities, that Van Huerter, under the influence of jealousy, broke his promise under the pretence that the lands he had referred to had been already given away. Silveira thence passed to Terceira, where he grew great quantities of corn and woad for dying blue, which he exported to Flanders. Returning from a visit to his native country by way of Lisbon, he became the guest of Dona Maria de Vilhena, who proposed to him that he should go out and colonize her two islands of Flores and Corvo, and rule over them in her name. This offer he accepted, but after a trial of seven years, found himself a loser both in property and position ; he therefore once again betook himself to his original locality at the Topo in the island of San Jorge, where he realized great wealth from his corn plantations, and became the ancestor of some of the most noble families in the Atlantic islands.

A tradition which we look for in vain in any Portuguese or Spanish historical document of the fifteenth or sixteenth centuries, has been widely disseminated in almost every work which speaks of the discovery of America, to the effect that an equestrian statue pointing with its right hand to the west, was discovered by the Portuguese in the island of Corvo. A circumstantial account of it is given in the *Epitome de las Historias Portuguezas*, by Manoel de Faria y Souza, published in Madrid, 1628. fol. He says : " On the summit of a mountain called Cuervo was found the statue of a man on horseback without saddle, bare-headed, the left hand on the horse's mane, the right pointing to the west. It stood on a slab of the same stone as itself ; beneath it, on a rock, were engraved some letters in an unknown language." M. Boid, who resided a long time in the Azores, speaking of Corvo in his work entitled " Description of the Azores," London, 1835, 8°, explains how a natural phenonemon has given rise to this fable. He says, "Among a great number of absurdities dealt in by the poor and superstitious inhabitants, they gravely assert that the discovery of the New World is due to their island, because a promontory which stretches far into the sea towards the north-west, *presents the form* of a person with his hand stretched out towards the west." They say that " it was the will of Providence, that this promontory should have this extraordinary form in order to indicate to European navigators the existence of another world, and that Columbus understood and interpreted this sign, and threw himself into the career of Western discovery." We can thus understand how the grotesque configuration of a volcanic rock should have given rise to a story of an equestrian statue, which learned men have not hesitated to attribute to Carthaginians and Phœnicians, who, we know but too well, were very little inclined to point out the road of discovery to rival nations.

CHAPTER XV.

CADAMOSTO.

1455-1456.

WE now reach the period of a very important voyage, made by a Venetian gentleman named Alvise Cadamosto, under the auspices of Prince Henry, and of which a detailed account by himself was published at Vicenza, in 1507, 4°, under the title of " La Prima Navigazione per L'Oceano alle terre de'Negri della Bassa Ethiopia di Luigi Cadamosto." Cardinal Zurla tells us that the name of Cadamosto is synonymous with Casa or famiglia da Mosto, while Alvise is the Venetian form of Luigi or Louis, so that in Alvise or Luigi Cadamosto, we have what in its simpler form might be called Luigi da Mosto. His narrative has always been highly commended for the carefulness of its detail, but I shall have occasion hereafter to show that an account of a second voyage by him to the west coast of Africa is very far from deserving that credit. Messer Alvise Cadamosto, though only twenty-two years of age, had already made one voyage to Flanders on a trading expedition, and his object being, as he expressly declares, to acquire wealth, a knowledge of the world, and, if. possible, fame, he determined to repeat his venture. On the 8th of August, 1454, he set sail in one of the galleys belonging to the Republic, under the command of a Venetian cavalier, named Marco Zeno.*

* The period of this voyage has been variously stated by various authors, but not only is the above date that which is stated in Ramusio, but its correctness is confirmed by a decree of the Venetian senate of the same date to the following effect :—" As on the last voyage, all three of the galleys went to Sluys, for which

Contrary winds detained the vessel off Cape St. Vincent, near which Prince Henry happened to be at the time, at a village named Reposeira, which being a retired and quiet spot, well suited for his studies, was a favourite residence of his. When the Prince heard of their arrival, he sent his own secretary, Antonio Gonsalves, and the Venetian consul, Patricio de' Conti, with samples of Madeira sugar, dragon's blood, and other products of the newly-discovered countries which he had colonized, and commissioned them to assure the Venetians that great things were to be done by those who would make the voyage. All this awakened in Cadamosto a strong desire to go, and he inquired what conditions the Prince made with those who undertook the adventure. He was told that either the adventurers were to equip and freight a caravel at their own expense, and on their return pay the Prince a fourth part of the produce, and retain the remainder themselves; or the Prince would supply the caravel and furnish it with every necessary, in which case the adventurers were to retain only the half of the produce, the Prince, in the event of failure, being at the expense of the entire outlay. Cadamosto was, however, assured that the voyage could scarcely fail of realizing great profits. He then had an interview with the Prince, who received him with great kindness, confirmed all that had been told him, and easily persuaded him to undertake the voyage.

Having made inquiry as to the nature and quantity of the merchandise and provisions he would require, he made the arrangements necessary for his new undertaking, and the Venetian galleys went on their way to Flanders. The Prince kept Cadamosto with him at Reposeira, till he had fitted out for him a new caravel of ninety tons burthen. The sailing captain was Vicente Dias, of whom we have heard already.

port all the merchandise was loaded to evade the duty of two per cent. on goods passing between Venice and England,—the captain of the Flanders galleys, "Ser" Marco Zeno, knight, is ordered to make inquiries as to goods of Venetian subjects unloaded in England, and to exact the two per cent."

See Calendar of State Papers and MS. Brown. Lond. 1864, vol. i. 1202-1509, page 79.

They set sail on the 22nd of March, 1455, and at midday of
the 25th reached the island of Porto Santo.

Cadamosto states that this island had been discovered
twenty-seven years before by the caravels of Prince Henry.
He should have said thirty-seven, but it is probable that
the former figure has been incorrectly transcribed. This
difference of ten years may have had to do with the widely-
accepted mistake that Cadamosto's first voyage took place
ten years earlier than it really did. He found the island,
which thirty-seven years before was uninhabited, tolerably
well peopled, producing sufficient wheat and oats for the use
of the inhabitants, and abounding in cattle, wild boars, and
rabbits, which last were innumerable. The island produced
dragon's blood,* and excellent honey and wax; the coast
abounded in fish.

They left Porto Santo on the 28th of March, and the same
day arrived at Monchrico (doubtless Machico), one of the
ports of Madeira, where they landed. Cadamosto relates
how Prince Henry had, twenty-four years before, colonized
this island, which till then had been uninhabited. He found
four settlements on the island, one named Monchrico
(Machico), another Santa Cruz, a third Funchal, and the
fourth the "Camara dos Lobos." There were inhabitants
elsewhere, but these were the principal localities. The
island could furnish about eight hundred armed men, and of
that number one hundred mounted. Cadamosto describes

---

* Dragon's blood is first mentioned in the account of the voyage of Jean de
Bethencourt to the Canaries in 1402. The produce of the *Dracæna Draco* of the
Canaries has ceased to be the dragon's blood of commerce. That which is now
used as such is the produce of the *Calamus Draco*, and is imported from the
islands of the Indian Archipelago. The famous dragon tree at Orotava at the
foot of the Peak of Teneriffe, the trunk of which ten men can scarcely embrace,
is said to have been almost as large when first found in 1402 as it is now. This
tradition is rendered probable by the slow growth of the tree. Next to the
baobab trees it is regarded as one of the oldest inhabitants of the earth. The
learned botanist Lécluse (Clusius), "Rariorum Plantarum Historia," lib. i. cap. 1,
saw one of these trees in 1564 at Lisbon growing amongst some olives on a hill
behind the monastery of Nossa Senhora de Graça, a tree which had doubtless
been brought over from Porto Santo or the Canaries by a navigator who valued
it as a curiosity more than the monks, by whom it was ignored and neglected.

the fertility as so great that the island produced an average
of thirty thousand Venetian stara or nearly seventy thousand bushels of wheat yearly.   The soil had at first yielded
sixty fold, but at the time of his visit only thirty or forty,
because the land had become impoverished, although well
watered.

On eight or more small rivers which intersected the island
they had set up saw-mills, which were kept constantly at
work in cutting wood for making furniture of various kinds,
which was sent to Portugal and elsewhere.   Two kinds of
wood used for this purpose were held in great esteem: the
one a fragrant cedar like cypress, of which they made tables
of great length and breadth, boxes and other articles; the
other a yew, which was also very exquisite, and of a red
colour.

The sugar canes, which the Prince had caused to be
imported from Sicily, and planted in the island, were producing so abundantly that four hundred cantaros * of sugar
were made at one boiling, and the climate was so favourable
that the quantity was likely to increase.   White sweetmeats
were made in great perfection.   Honey and wax were produced, but in small quantities.   The wines were extremely
good, considering the infancy of the colony.   Among the
vines which the Prince had imported thither were those of
Malvoisie from Candia, which flourished so luxuriantly, in
consequence of the richness of the soil, that they bore as
many grapes as leaves, in bunches two or three or even four
palms in length, which Cadamosto declared was the most
beautiful sight in the world.   There were wild peacocks,
some of which were white; no partridges, or other game,
except quails, and wild boars on the mountains in
great abundance.   There had been immense numbers of
pigeons, and still a great many were to be found, which they
caught by the neck with a kind of lasso, with a weight at
the end, and though they pulled them down from the trees,

* In Portugal the cantaro is the same as the alqueira, which contains about
three gallons.

the birds, having never been hunted, were not frightened. There were plenty of cattle on the island. Many of the inhabitants were wealthy, for the whole country was like a garden. There were Friars Minors of the Observantine order, men of good and holy life.

From Madeira they sailed southward and came to the Canary Islands. Four of them, Lançarote, Fuerteventura, Gomera, and Ferro, were inhabited by Christians; the other three, the Grand Canary, Teneriffe, and Palma, by pagans. The governor of the former was a knight named Herrera, a native of Seville, and a subject of the King of Spain. They had barley-bread, goats' flesh and milk in plenty, for goats were very numerous, but they had no wine nor corn, except what was imported, and the islands produced but little fruit. There were great numbers of wild asses, especially in the island of Ferro. Great quantities of orchil for dyeing were sent from these islands to Cadiz and Seville, and thence to other parts both east and west. The chief products were goats' leather, very good and strong, tallow, and excellent cheeses. The inhabitants of the four Christian islands spoke different languages, so that they could with difficulty understand each other. There were no fortified places in them, only villages; but the inhabitants had retreats in the mountains, to which the passes were so difficult that they could not be taken except by a siege.

Of the three islands inhabited by Pagans two were the largest and most populous of the group, viz., the Grand Canary, in which were about eight or nine thousand inhabitants, and Teneriffe, the largest of all, which contained from fourteen to fifteen thousand. Palma was not so well peopled, being smaller, but a very beautiful island. The Christians had never been able to subdue these three islands, as there were plenty of men of arms to defend them, and the mountain heights were difficult of access.

Teneriffe, of whose Peak Cadamosto speaks as being visible according to some sailors' accounts at a distance of

two hundred and fifty Italian miles, and sixty miles high *
from the foot to the summit, was governed by nine chiefs,
bearing the title of Dukes, who did not obtain the succession
by inheritance, but by force. Their weapons were stones,
and javelins pointed with sharpened horn instead of iron,
and sometimes the wood itself hardened by fire till it was
as hard as iron itself.

The inhabitants went naked, except some few who wore
goats' skins. They anointed their bodies with goats' fat,
mixed with the juice of certain herbs to harden their skins
and defend them from cold, although the climate is mild.
They also painted their bodies with the juice of herbs, green,
red, and yellow, producing beautiful devices, and in this
manner showed their individual character, much as civilized
people do by their style of dress. They were wonderfully
strong and active, could take enormous leaps, and throw
with great strength and skill. They dwelt in caverns in
the mountains. Their food was barley, goats' flesh, and
milk, which was plentiful. They had some fruits, chiefly
figs, and the climate was so warm that they gathered in
their harvest in March or April. They had no fixed religion,
but some worshipped the sun, some the moon, and others
the planets, with various forms of idolatry.

The women were not taken in common among them; but
each man might have as many wives as he liked. No
maiden, however, was taken till she had passed a night with
the chief, which was held as a very great honour. These
accounts were had from Christians of the four islands, who
would occasionally go to Teneriffe by night, and carry off
men and women, whom they sent to Spain to be sold as
slaves. It sometimes happened that the Christians were
captured in these expeditions, but the natives, instead of
killing them, thought it sufficient punishment to make
them butcher their goats, and skin them, and cut them

---

* The perpendicular height is twelve thousand one hundred and eighty feet,
but the distance in ascending from the foot to the summit may fairly be com-
puted at sixty miles.

up, an occupation which they looked upon as the most
degrading that a man could be put to; and at this work
they kept them till they might be able to obtain their
ransom.

Another of their customs was, that when one of their
chiefs came into possession of his estate, some one among
them would offer himself to die in honour of the festival. On
the day appointed, they assembled in a deep valley, when,
after certain ceremonies had been performed, the self-devoted
victim of this hideous custom threw himself from a great
height into the valley, and was dashed to pieces. The chief
was held bound in gratitude to do the victim great honour,
and to reward his 'family with ample gifts. Cadamosto was
told of this inhuman custom, not only by the natives, but
also by Christians who had been kept prisoners in the island.

Cadamosto visited the islands Gomera and Ferro, and
also touched at Palma, but did not land, because he was
anxious to continue his voyage.

Sailing southwards, in a few days they reached Cape
Branco, eight hundred and seventy (say rather five hundred
and seventy) miles from the Canaries, during one-third of
which passage they were out of sight of land. They then
steered for the coast, lest they might pass the Cape without
seeing it.

Cadamosto here observes that from the Straits of Gibraltar
to Cape Cantin, no habitations were found, and between
Cape Cantin and Cape Branco, the desert of Sahara begins,
which on the north is bounded by the mountains, and on
the south by the country of the blacks.

This desert, he states, is fifty or sixty days' journey on
horseback, in some places more, in others less. It reaches
to the sea-coast, where it is sandy, white, and arid, and is a
perfect level as far as Cape Branco. This Cape was so
called by the Portuguese, who first discovered it, on account
of the whiteness of the sand, on which there was no sign of
grass or of any vegetation whatever. The Cape itself is a
very beautiful object when seen from the front, and forms a

triangle, whose three points are about a mile distant from
each other.

On all this coast they found abundance of fish. The Gulf
of Arguin is very shallow throughout, and there are many
shoals in it both of sand and rock. The currents are so very
strong that they did not venture to sail except in the day-
time, and then constantly heaving the lead, and going with
the current. Two ships had already been wrecked on these
shoals. Cape Branco is situated S.E. of Cape Cantin.

Beyond Cape Branco was a place named Hoden, about
six days' camel-journey inland. It was not enclosed with
walls, but was a place of resort for the Arabs and caravans
trading between Timbuctoo and other places belonging to
the negroes, and the western parts of Barbary. The inhabitants of this place lived on dates and barley, which they had
in abundance. Their drink was camels' milk. They kept
cows and goats, but no great number, as the soil was so
barren, and their cattle were small compared with those of
Venice. The people were Mahometans, and great enemies
of Christianity. They had no settled habitations, but
wandered continually over the deserts, travelling between
the country of the negroes and the western parts of Barbary.
They travelled in great numbers, with long trains of camels,
conveying brass and silver and other things from Barbary to
Timbuctoo, and the country of the Blacks, and bringing
back in exchange gold and malaguette pepper. These
people were tawny, and both sexes wore white dresses with
red borders, without any linen under; the men wore turbans
like the Moors, and always went barefoot. Lions, leopards,
and ostriches abounded in these deserts, the eggs of the
latter Cadamosto found very good food.

Prince Henry farmed out the trade of the island of
Arguin for ten years in the following manner. None were
to enter the gulf to trade with the Arabs who came to the
coast excepting those who held a grant from him, and who
were to be residents in the island, and have agents for that
purpose. Their merchandise consisted of linen and woollen

cloths, silver, *alkhizeli* or cloaks, carpets, &c., but especially wheat, which was eagerly sought after. In return the Arabs gave slaves brought from the lands of the negroes, and gold.

The Prince consequently had a fort built in the island that this trade might be permanently established, and with this object the caravels of Portugal made a yearly voyage to this island. The Arabs had a great number of Barbary horses, which they took to the land of the negroes to barter for slaves, a good horse being often valued at twelve or fifteen slaves. They brought also Moorish fabrics of silk made in Granada and in Tunis, with silver and a variety of other things, for which they received in exchange a great number of slaves and a small quantity of gold. These they took to Hoden to divide. Part went to Barca, and thence to Sicily, and part to Tunis and the whole coast of Barbary. The rest were taken to Arguin and there sold to the licensed Portuguese traders, who purchased every year seven or eight hundred slaves to send to Portugal. Before the establishment of this traffic the Portuguese sent out every year four or more caravels to the Bay of Arguin, the crews of which attacked the fishing villages, and carried off both men and women to sell in Portugal. They did the same all along the coast from Cape Branco to the Senegal.

The Azanegues are tawny, or rather dark brown. They inhabit the part of the coast beyond Cape Branco, and their district is bordered by that of the above-named Arabs of Hoden. Their food was dates, barley, and camels' milk, they also procured millet and beans from their neighbours the negroes, and thus supported life, for they required but very little. The Portuguese, as just stated, used to seize and sell them, as the best kind of slaves, but Cadamosto bears witness that for some time this had been prevented by Prince Henry, and the traffic confined to merchandise, because the Prince hoped that by kindness these people might be converted to Christianity. These Azanegues have a curious custom of covering their mouths with a piece of

linen, which is first twisted round their heads, and the end
left to hang over their mouths; they do this because
they say that the mouth is an unseemly thing, from which
bad odours are emitted, and therefore ought to be concealed.
They have no chiefs among them, but any that are better
off than the rest are treated with deference and obedience.
They are a poor race, and the most lying, thievish, and
treacherous people in the world. They are of middle height
and thin; they wear their black hair flowing down over
their shoulders, and anoint it daily with fish oil, which
causes a most· offensive smell, but is looked upon as a great
embellishment. They had never seen any Christians but
the Portuguese. They thought the ships were great birds
with white wings floating on the sea, then seeing them with
the sails furled, they took them for fish; some thought they
were phantoms wandering through the night, which caused
them great fear, the more so that they could not understand
being attacked at different places at a great distance, within
so short a time.

About six days' journey from Hoden, there is a place
named Tegazza (which signifies a chest of gold), whence
rock-salt is obtained in great quantities, and carried
by the Arabs and Azanegues on the backs of camels to
Timbuctoo, and thence to Melli, in the empire of the
negroes, where it is sold at two or three hundred mitigals*
the load, in exchange for gold. The Melli country is very hot,
and affords but poor sustenance for cattle; the climate is bad
even for the natives, and many who go with the caravans
never return. From Tegazza to Timbuctoo is forty days'
journey on horseback, and from Timbuctoo to Melli is thirty
days'. In reply to inquiries about the consumption of salt,
the natives said, that it was used in great quantities by the
people who lived so near the equinox, as a purifier of
blood, because the excessive heat caused disease for which
the salt was a remedy.

A curious account is given of the transport of the salt

* The mitigal or miscal is equal to about a drachm and a half.

when the heat becomes too great to be endured by camels. It is then carried by negroes, who go in a long procession, each with a large block on his head, and carrying in his hand a fork on which he rests the block when he is tired. In this way they reach a piece of water which Cadamosto supposed to be river, and here a singular traffic commences with another tribe of negroes. When the first party reach the water, they pile the salt in mounds, each marking his own pile. They then retire half a day's journey to give place to the purchasers, who will not be seen or spoken to; these come in large boats as if from an island; they examine the salt, and put a quantity of gold by the side of it, and then retire leaving the gold and the salt together. When they have left, the others return and take the gold, if they find it enough; if not, they again withdraw. The purchasers come back, take the salt for which the gold has been accepted, and leave more gold with the remainder, if they think it worth more. And so the traffic goes on till they are mutually satisfied, without either party seeing the other. This was an ancient custom of which Cadamosto was informed by Arab and Azanegue merchants, on whose word he could rely. Cadamosto inquired of the same merchants why the Emperor of Melli, being a great and powerful lord, had not tried to discover who these people were. They replied, that not long ago the attempt had been made, and one of the blacks was captured in order to bring him before their own prince, but the man would not utter a word, either not understanding them, or resolving not to speak, nor would he touch any food, so that after four days he died. The chieftain was greatly vexed at the result, but those engaged in the capture were able to give him some account of these people, and told him that they were very black and well made, taller than themselves by a hand's breadth, and had the upper lip small like their own, but the under lip was large and red, showing the gums, so that it seemed to have blood oozing out of it. Their teeth were large, and they had two on each side of extraordinary size. Their eyes were black, and very open, which gave them a very fierce and savage look. After the capture and death of this negro, the others were so much offended, that for three years they bought no more salt, and when they returned, the blacks of Melli concluded that they found they could not exist without the salt, which kept their lips from corrupting. This was all that Cadamosto had been able to learn on this subject, but he believed it from the number and credibility of the witnesses.

The gold taken to Melli was divided into three parts. The first was sent by caravan to a place called Cochia,* which is on the road to Syria and Cairo; the two others to Timbuctoo, whence the one was sent to Toet, and so to Tunis; the other part to Hoden, and thence to Oran and Hona, in Barbary within the straits, and to Fez, Marocco, Arzilla, Saffi, and Messa without the straits. It was taken hence by Italian merchants, in exchange for a variety of merchandise. The greatest advantage which the Portuguese obtained from the country of the Azanegues was the gold which was yearly sent from Hoden to the island of Arguin, and which they got by barter with the negroes.

The Azanegues used no coin, but in some of the inland towns the Arabs used cowries for small purchases; these were brought from the Levant to Venice, and sent thence to Africa. The gold was sold by the mitigal, as in Barbary. The women were brown, and they had little petticoats or *alkhizeli*, which were brought from the country of the negroes, and some wore these *alkhizeli* without any other dress. Those who had the longest breasts were considered the most beautiful, and so anxious were they for this dis-

---

* Kukia, or Kúgha, the ancient capital of the Songhay empire. It must not be confounded with Gógó, the present capital, for El Bekri, besides Gógó, gives an account of Kúgha, but unfortunately says nothing of its situation, except the distance of fifteen days froм Ghánata, nor does he show its position with regard to Gógó. Kukia, lying at the very outset of the Egyptian caravan road, was inhabited exclusively by Mahometans, while all around were idolaters. It was the greatest market for gold in all Negroland, although the quality of the gold brought to Audagost was better than that exported from Kukia. See Barth, "Travels in North and Central Africa," vol. iv. p. 583.

tinction, that girls of seventeen or eighteen submitted to
have a cord drawn tightly round each breast, so as to break
them, and make them hang down; and by frequently pulling
these cords, they made them grow so long that they sometimes reached the navel. These people were good horsemen,
like the Moors, but they could not keep many horses, on
account of the barrenness of the land and the great heat,
and those they had did not live long. There were no rains,
except in August, September, and October. The locusts,
which were of a finger's length, and of a red and yellow
colour, sometimes rose in the air in such numbers that for ten
or twelve miles nothing else could be seen on the earth or in
the air, and nothing remained undestroyed wherever they
passed. These creatures came only once in three or four
years, or the country would have become unfit for habitation.
When Cadamosto was there, he saw numbers on the ships.

He now came to the Senegal,* which he describes as more
than a mile wide at the mouth, and deep. A little further
on it has another entrance, and between the two there is an
island which forms a cape, running into the sea; there are
sand-banks at each mouth, that extend about a mile from
the shore. The flux and reflux of the tide extends more
than sixty miles up the river, as Cadamosto learned from
Portuguese who had ascended it in their caravels. In entering the river it is necessary to go with the tide to avoid
the sand-banks at the mouth. It is three hundred and
eighty miles from Cape Blanco; the coast is sandy for twenty
miles up the river, and was called the Anterote coast, and
belonged to the Azanegues.

Cadamosto was surprised to find so great a difference
between the inhabitants on the two sides of the river. On
the south side the people were very black, stout, and well
made, and the country verdant, woody, and fertile; while,

* Cadamosto says that five years before his voyage, this river was discovered
by three of Prince Henry's caravels, and that a commercial treaty was made
with the blacks, so that in his time many ships went there. He is inaccurate
in this statement. According to Azurara the Senegal was discovered in 1445.

on the north side, the men were thin, tawny, and short, and
the country dry and sterile. It was believed by some that
this river was a branch of the Gihon, which rises in the
terrestrial Paradise. The ancients named this branch Niger,
and say that after watering Ethiopia it runs westward, and
dividing into several branches, falls into the ocean; and
that the Nile is another branch, which waters Egypt, and
falls into the Mediterranean.

The first kingdom of the negroes bears the same name as
the river, the Senegal, and the people are called Jaloffs.
The country is quite flat as far as Cape Verde, which is the
highest land on the whole coast, and is four hundred miles
from Cape Branco. The kingdom of Senegal is bounded on
the east by the country of Tukhusor, on the south by the
kingdom of Gambra, on the west by the ocean, and on the
north by the river. When Cadamosto was there, the King
of Senegal was named Zucholin; he was about twenty-two
years of age. The succession was not hereditary, but the
nobles chose a king from among their number, who remained
on the throne as long as he pleased them. If they were dissatisfied, they dethroned him by force, unless he had made
himself powerful enough to resist them.

The people were poor and ferocious; they had no walled
towns, only miserable villages, with houses covered with
thatch. They did not understand masonry, or brick-making.
The kingdom was very small, being only about two hundred
miles square. The king had no fixed revenue, but the nobles
made him presents of horses and cattle, and different kinds
of vegetables and grains. The principal part of his wealth,
however, was got by pillage. He carried off the neighbouring
people for slaves; some to cultivate the land, and some for
sale to the Azanegues and Arab merchants, in exchange for
horses and other merchandise, besides the traffic with the
Christians, since the trade was opened with them. Each
negro was allowed as many wives as he pleased. The king
had never less than thirty, who were honoured according to
the rank of their fathers. These wives were distributed by

tens and twelves in different villages, where each had a
house to herself, with women to wait upon her, and slaves
to cultivate the land assigned her by her lord.   They also
had cows and goats, with slaves to keep them.   When the
king visited them, he took no provision with him for himself
or his retinue.   At sun-rising, each wife at the place where
he arrived prepared for him food and delicacies, and after
the king had stayed his appetite, the remainder was distributed among his followers; but they were so numerous,
that there were always some left unsatisfied.   The king
travelled in this way from place to place, to visit all his
wives in succession, and in consequence his children were
very numerous.   As soon as he knew one of his wives to be
pregnant, he left her; which custom was observed by all his
nobles.

These negroes professed Mahometanism, but were not so
strict as the white Moors.   The nobles having most intercourse with the Azanegues or Arabs, paid more attention
to religion than the people, but since they had become
acquainted with Christians, they had less respect for Mahometanism.

The common people wore nothing but goats' skins made
in the shape of breeches.   The nobles wore shirts of cotton,
spun by the women.   The width of the cloth was only a hands-
breadth; they did not know how to make it wider, and were
obliged to sew several pieces together to make it the required
width.   These shirts reached half way down the thigh, and
had wide sleeves which covered half the arm.   Beside this
they had hose of the same cloth, which reached from the
waist to the instep, and were exceedingly broad, some of them
containing thirty or even forty hands-breadths of cloth, which
hung in many folds, like a sack in front, and dragged on
the ground behind.   The women wore nothing above the
waist.   Whether married or not, they had only a short petti-
coat reaching from the waist to the middle of the leg.   Both
sexes went bare-foot, and wore nothing on their heads.   Their
hair was well-dressed, and fastened up tastefully, though it

was very short. The men worked like the women, at spinning, washing, &c.

The climate is very hot, their January being warmer than April in Italy, and later in the season the heat becomes insupportable. It was the custom to wash three or four times a day, so that the people were extremely clean in their persons, but the reverse in their food. Though they were very stupid and awkward in matters that they were not accustomed to, they showed considerable skill in those they had been used to. They were great talkers and great liars, but so hospitable that the poorest would give food and lodging to strangers, looking for no reward. They were often at war among themselves or with their neighbours. They fought on foot, the heat preventing them from keeping war horses. The same cause prevented their wearing armour. They used round shields, covered with the skin of an animal called the danta, which was very difficult to pierce. Their offensive arms were azagays, or light darts, having barbed iron points, which they threw with admirable skill, and inflicted very dangerous wounds; and a kind of scimitar, which they got from the negroes of Gambra; they had iron in their country, but did not know how to work it. They had also a kind of javelin. Their wars were very deadly, because, their bodies being unprotected, all their blows took effect. They were a bold and savage people, with no fear of death, which they infinitely preferred to flight. They knew nothing of navigation, and never saw a ship till the coming of the Portuguese.* Those who lived on the banks of the river, or on the sea-shore, had canoes made of one piece of wood, the largest of which could contain only three or four men, and which they used for fishing and other purposes. They were the finest swimmers in the world.

After having passed the river Senegal, Cadamosto reached the country of Budomel, which is about fifty miles further. Budomel was the title of the Prince, but it gave the name

---

* In this short sentence we have a summary disproval of the claims to prior exploration by the Genoese, Catalans and Frenchmen.

to the country, as in Europe we should say the territory of
such a Count or such a Lord. The country is flat all along
the coast. Cadamosto, having heard that the Prince was a
courteous and honourable man, stopped here. He had on
board the caravel some Spanish horses, much valued by the
negroes, linen cloths, Moorish silks, and other merchandise.
Having anchored in the bay called the Palma de Budomel,
he sent his interpreter on shore, to give notice of his arrival
and make proposals of commerce.

The following day the negro Prince appeared with a
retinue of fifteen horse and a hundred and fifty foot soldiers.
He invited the Portuguese to land, and Cadamosto came on
shore in a sloop, and was very well received by the Prince,
to whom he offered seven horses in harness, and other
merchandise to the value of about three hundred ducats.
The payment was to be made at the house of the Prince,
which was twenty-five miles inland, and Cadamosto was
invited to receive it himself, and to be the Prince's guest
for some days. Before setting out Budomel presented him
with a young girl twelve or thirteen years old, to serve, he
said, in his cabin ; she was very black, and on that account
was considered very beautiful. The Prince also furnished
him with horses and all things necessary for the journey.
When they were within four miles of his house he consigned
him to the care of one of his nephews named Bisboror, the
lord of a neighbouring village, who received him into his
house and entertained him honourably.

It was now November. Cadamosto remained there
twenty-eight days, and made frequent visits to the
Prince, which gave him excellent opportunities of observing the customs of the country. He had still more
opportunity of doing this when he was obliged to return
to Senegal by land ; the weather was so stormy that he
could not return to the ships without danger, and in
consequence he sent them to the entrance of the river, and
made the journey himself on horseback. In order to communicate with his caravel, and send orders to his men to

meet him at Senegal, he had to put the swimming powers
of the natives to the test. He says the vessel was three
miles out at sea, and it appeared impossible to execute his
commission on account of the violence of the waves breaking on the sand-banks, but in spite of this the negroes were
eager in offering their services to carry his letter on board.
He asked two of them what he should give them for the
enterprise, and they only asked two mavulgies of tin apiece,
the mavulgi being worth something less than a penny. " I
cannot describe," says the author, " the difficulty they had
to pass the sand-banks in so furious a sea. Sometimes I
lost sight of them, and thought they were swallowed up by
the waves. At last one of the two could no longer resist
the force of the water, turned his back on the danger, and
returned to the shore. The other, more vigorous, after
battling for more than an hour with the wind and the
waves, passed the bank; carried my letter to the ship, and
brought me the answer. I dared hardly touch it, looking
upon it as a wonderful and sacred thing. And thus I
learned that the negroes of Budomel are the best swimmers
in the world."

The negro kings and nobles had neither cities nor forts,
their richest habitations were but miserable villages. The
Prince Budomel's authority depended chiefly on the respect
the negroes had for his riches, so little was the subordina-
tion to rank understood. Personal merit, strength, sense,
justice, courage, and good looks also produced an effect, and
Budomel possessed these advantages. He had assigned to
him, for himself and his wives, a certain number of villages,
which he visited in succession. The one in which Cadamosto stayed contained between forty and fifty houses
covered with thatch, built close to each other in a round,
encompassed by a ditch and screens of large trees, with
two or three passages for entrance; each house had an
enclosed court. Budomel had nine wives in this place, and
more or less in his other villages. Each wife had five or six
young girls for her service, with whom their lord was

permitted to live as with his wives, who did not consider this an injury, as it was the custom. Jealousy was a common vice among them, and it was an insult to a negro to enter the house of his wife; even his sons were excluded.

Budomel had always about two hundred negroes in attendance upon his person, when one left, another supplying his place; beside which there were always a number of people who came to attend his court. Between the entrance of his house and his own private apartment there were seven courts, and in the midst of each was a large tree, to shelter those who waited for an audience. In these courts his retinue were distributed, according to their rank and employments; those in the courts nearest. the Prince being the most distinguished. Few, however, dared approach the person of the Prince; the Azanegues and the Christians had almost the exclusive privilege of entering his apartment, and speaking to him. He maintained great state towards his subjects, and showed himself only for one hour in the morning, and again for a short time in the evening near the door of the outermost court.

He required great ceremony when giving audiences to his subjects. However high the rank of a suitor, he had to take off his garments, with the exception of a covering round the middle, and when he entered the last court he threw himself on his knees, with his forehead on the earth, casting sand over his head and shoulders. Even the Prince's relatives were not exempt from these humiliations. The suppliant remained a long time in this posture, sprinkling himself with sand. He approached his lord on his knees, still throwing sand on his bowed head; when about two paces from him, he stopped and offered his petition. The reply was given as shortly as possible, and with scarcely a glance towards him. Cadamosto witnessed this scene several times, and accounts for the excess of submission by the excess of fear; the negroes knew that their tyrants could carry off their wives and children, and sell them for slaves at their pleasure, and they trembled before them, and feared them more than God himself, with whose name indeed they were scarcely acquainted.

Budomel was so gracious to Cadamosto, that he allowed him to enter his mosque at the hour of prayer. The Azanegues and Arabs, who were his priests, were summoned to attend, and Budomel performed his orisons in the following manner. Standing up, he raised his eyes towards heaven, then walked forward two steps, uttered a few words in a low tone, and prostrated himself on the ground, which he kissed respectfully. In all this he was followed by the Azanegues and the rest of his retinue. He continued in prayer about half an hour, repeating the same ceremonies ten or twelve times. Having finished, he turned to Cadamosto, asked him what he thought of it, and desired him to give him some idea of the Christian religion. Cadamosto had the courage to tell him, in the presence of the priests, that the Mahometan religion was false, and that the Catholic was the only true faith. This enraged the priests, but Budomel only laughed, and said that the Christian faith must be good, because God alone could have bestowed such riches and knowledge as we possess. He added that he thought the Mahometan religion was good also, and that the negroes must have a better chance of salvation than the Christians, because God being a just master, and having given the Christians so many advantages in this world that they had a paradise here, it followed that great compensation awaited the negroes in the next world, and they might expect their paradise there. Budomel showed much good sense and reflection in his remarks, and took pleasure in conversing about religion. Cadamosto thought he would easily have been induced to embrace Christianity, had he not been afraid of offending the people. His nephew told Cadamosto this, and took great delight himself in conversing on the subject.

The table of Budomel was supplied in the same manner as that of the King of Senegal. The negro nobles ate lying upon the ground, without ceremony, and no one might eat with them but the Moors, whom they looked upon as instructors. The common people ate in companies of ten or twelve, round a copper full of meat, in which they all put their hands. They ate little at a time, but had frequent meals.

The climate is so hot that they could not grow wheat, rye, barley, oats, or vines, for there is no rain for nine months, that is, from October to June. However, they had millet, large and small, and two kinds of beans. The beans were very fine, both red and white. They sow in July to reap in September, as this is the rainy season when the rivers overflow and fertilise the land; and thus all the work of agriculture was done in the three months. They only planted as much as they thought necessary for the year, not caring to raise provisions for sale, as they were bad economists and very idle. Their method of cultivating the ground was for five or six to work in a field, and they used their swords for tools; they scarcely dug four inches deep, but the rains made the land so fertile that it gave an abundant harvest.

Their drinks were water, milk, and palm-wine, which was distilled from a tree found in abundance in the country, but not the same that produces the date, though it is like it. This wine, which was called *mignol*, was distilled into calabashes from two or three openings in the trunk of the tree : from morning till night a tree would not fill more than two calabashes. This liquor had a good flavour, and without any mixture was as intoxicating as wine. Cadamosto says that the first day it was as good as the best European wine, but daily it lost its flavour, till it became sour. It was, however, more wholesome on the third or fourth day than the first; for in losing some of its sweetness, it became purgative. It was not so abundant that all could have as much as they liked, but as the trees which produced it were spread through the fields and forests, every one procured what he could by his labour, and the nobles employed people to collect it for them.

The land produced many kinds of fruit, differing more or

less from those of Europe; they were excellent growing wild, but might have been much improved by cultivation.

The country was very fertile and woody, and abounding in small but very deep lakes, full of fish and water snakes, called *calcatrici*. There was a kind of oil used by the natives to flavour their food, the ingredients of which Cadamosto was unable to discover. It had the scent of violets, the taste of olives, and the colour of saffron. There was also a tree which produced little red beans, with black specks, in great abundance.

The country abounded with animals, and there were a prodigious number of serpents, some venomous, and some so large that they would swallow a goat whole. Cadamosto was told by the negroes that these creatures went in great numbers to the mountainous parts of the country, which also abounded in white ants. The ants by a wonderful instinct build houses for these terrible neighbours with earth, which they carry in their mouths. The negroes being great magicians, had recourse to charms on all occasions, especially to defend themselves against serpents. Cadamosto relates an anecdote which he had from a Genoese, a trustworthy man, who told him that the year before he had been in the country of Budomel and was staying in the house of his nephew Bisboror. Once at midnight he was awakened by hisses all round the house, and saw his host rise and give orders to two negroes to bring his camel. The Genoese asked where he was going so late, and was only told that he was going on business, and would soon return. He came back early in the morning, and the Genoese, curious to know the end of the adventure, renewed his questions. " Did you not hear at midnight," said Bisboror, " hissings all round the house? It was surrounded by serpents, and if I had not employed charms to make them return to their own place, they would have destroyed much cattle." The Genoese was greatly surprised, but Bisboror told him that his uncle could do much more wonderful things. When he wanted to obtain venom to poison his arrows, he had a large circle formed, into which

he charmed all the serpents in the neighbourhood; and when
he had selected the most venomous, he killed them with his
own hands, and let the others go. He then mixed their
venom with the seed of a certain plant, which made a poison
so powerful that a wound from a weapon dipped in it was
fatal in a quarter of an hour. The Genoese added that
Bisboror offered to show him some charms, but he had no
taste for such things, and declined having anything to do
with them. Cadamosto says that he believed in the negroes'
power of charming serpents the more readily, because in
Italy there were Christians who practised enchantments.

There were no domestic animals nor sheep in Senegal, but
there were oxen, cows, and goats. The cattle were thinner
than in Europe, and red was a rare colour among them, the
usual colours being black, white, or a mixture of both.
There were great numbers of beasts of prey; lions, panthers,
leopards, wolves, and other wild animals. The wild elephants
went in herds; they were of great size, as the tusks imported into Europe proved.

Cadamosto saw no other beasts than those here mentioned,
but there were a great number of birds, especially paroquets,
which the negroes hated because they destroyed their millet
and vegetables. They said that there were several kinds,
but Cadamosto only saw two: one like those of Alexandria,
but a little smaller; the other much larger, with brown
heads, and the rest of the body mingled with green and
yellow. He took many of both kinds, but lost a number of
them on the voyage home. The caravel which accompanied
him took a hundred and fifty more, which sold for half a
ducat each in Portugal. These birds build their nests very
cleverly: they collect a quantity of reeds and twigs, and
make them into the shape of a ball, with a hole for the
entrance; these they hang from the slenderest branches
they can find, as a protection from serpents, the weight of
these creatures preventing them from attacking the nests in
such a position. There were great numbers of the birds
called in Europe Pharaoh's hens, which came from the East.

They were dark birds, and marked with black and white spots. They had likewise other birds, very different from those in Italy.

During Cadamosto's sojourn with Bisboror, he went to a market or fair, which was held on Thursday and Friday in a meadow near, and which was attended by numbers of both sexes from five or six miles round. Those who lived at a greater distance had other similar markets. The poverty of the people was shown by their merchandise, consisting of cotton in small quantities, nets and cotton cloths, vegetables, oil, millet, wooden bowls, and palm mats. Sometimes they brought a little gold, but in very small quantities. They had no money, and all the traffic was by barter. The people who came from the interior were very much astonished at the whiteness of Cadamosto's skin, and the fashion of his dress. He wore a Spanish dress of black damask; his mantle, being of wool, greatly surprised them, as they have no wool in their country; and some of them rubbed his hands with saliva to find out if they were painted white. His object in going to these markets was to see if any quantity of gold was brought there.

Horses were valued by the negroes in proportion to their rarity. The Arabs and Azanegues imported them from Barbary and the countries bordering upon Europe, but the extreme heat soon killed them; besides, the beans and millet, which were their only food, made them so fat that it became a disease. A horse with its harness was worth from nine to fourteen slaves, according to its beauty. When a noble purchased a horse, he went to the sorcerers, who lighted a fire of dried herbs, over the smoke of which they held the horse's head by the bridle, and repeated their charms. They anointed him with the best oil, shut him up for eighteen or twenty days, so that no one might see him, and tied round his neck certain charms folded square, and covered with red leather, and having done this the master believed him to be secured from danger.

The negro women were very gay, especially the young ones, and very fond of singing and dancing. Their time for

dancing was at night, by moonlight, and their dances were
quite different from the Italian.

Nothing caused so much astonishment to the natives as
the discharges of artillery from the caravel. Cadamosto
caused a cannon to be fired when some of the negroes were
on board, the noise of which terrified them extremely, but
they were still more frightened when they were told that one
discharge of this dreadful machine would kill a hundred of
them. After they had recovered from their fright, they
declared that so destructive an engine could only be the work
of the devil. They were greatly pleased with the sounds of
the bagpipe, and thought it was an animal which sung the
different tunes. Cadamosto, amused with their simplicity,
placed the instrument in their hands, and when they saw
that it really was a work of art, they thought it must be
made by divine skill, for they had never heard such sweet
sounds. The most simple instruments about the vessel
excited their admiration, and they thought the eyes painted
on the prow of the vessel were real eyes by which it saw its
way through the water. They repeated incessantly that the
Europeans must have much more skilful sorcerers than
theirs, and little inferior to the devil himself, for travellers
by land found it difficult enough to keep the right road
from one place to another, while they, in their vessels, could
find their way on the sea, however distant they might be
from the land.

Though the country abounded in honey, the negroes had
no idea of making any use of the wax, and Cadamosto
greatly surprised and delighted them by making some
honeycomb, drained of the honey, into candles before their
eyes. " The white people," they exclaimed, "know everything." They had two kinds of musical instruments—the
one was a sort of Moorish drum, and the other a kind of
violin with two strings, played with the fingers, but there
was little music to be got out of them.

After this long sojourn in Budomel's country, Cadamosto
resolved, having bought some slaves, to proceed on his way

to double Cape Verde, and make further discoveries. He remembered to have heard from Prince Henry that beyond Senegal there was another river called the Gambia, from which a quantity of gold had already been brought, and that no one could go there without amassing great riches. With this inviting prospect he took leave of Budomel, and again set sail. One morning he came in sight of two vessels, which proved to belong, the one to Antonio Uso di Mare, a Genoese gentleman, and the other to some Portuguese in the service of Prince Henry. They were going together towards the coast of Africa, with the intention of passing Cape Verde and making new discoveries. Cadamosto, whose objects were the same, joined company; they sailed together towards the south, keeping sight of land, and the day following they came to the Cape. The name of Cape Verde had been given the year before,* when it was discovered by the Portuguese, because they found it covered with trees which never lost their verdure. It projects far into the sea, and has two small mountains at the point. In passing they could see many villages of the Senegal negroes, consisting of thatched cottages. Above the Cape there are sand-banks, extending for half-a-mile into the sea. After doubling the Cape, the ships came upon three islands, filled with large trees; they anchored at the largest, hoping to take in water, but they could find no spring. However, as there were quantities of birds'-nests and eggs, of an unknown species, they stayed there one day, which they spent in hunting and fishing. They took an incredible number of fish, among which were dentali and orate vecchie, which weighed twelve or fifteen pounds.

This was in the month of June. The following day they continued their course, always in sight of land. Beyond the Cape was a gulf; the coast was low, and covered with fine large trees, which were always green, the fresh leaves supplying the place of those that fell, without the trees ever becoming bare,

* This error will be readily noticed by the reader, who has seen that the discovery was made by Diniz Dias in 1445.

as in Europe. They grow so close to the sea, that they seemed
to be watered by it. The prospect was so beautiful that
Cadamosto declared that he had never seen anything to
compare to it. The land was watered by several small rivers,
but as it was impossible for the vessels to enter, they could
not take in water.

Beyond the little gulf, the coast was peopled by two
nations of negroes, the Barbacini and the Serreri, both
independent of the King of Senegal. They had no dis-
tinctions of rank among them, but only of riches and
personal qualities. They were idolaters, lawless, and very
cruel. They fought with poisoned arrows, the least scratch
of which that fetched blood caused instant death. They
were very black, and very well made. The country was full
of wood, lakes, and rivers, and could only be approached
through very narrow defiles, which had helped them to preserve their independence. The Kings of Senegal had often
tried to conquer them, but had always been foiled by the
difficulties of the country.

Advancing along the coast with a favourable wind, they
discovered the mouth of a river, about a bow shot in width and
very shallow (the Joal?). They gave it the name of Barbacins;
which name it bears in Cadamosto's chart. This river is
sixty miles from Cape Verde. They continued to follow the
coast all day, and at evening cast anchor four or five miles
from shore; in the morning they continued their course,
taking care to keep a man at the mast-head, and two in the
fore part of the vessel, to keep watch for rocks and sand-
banks. They arrived at the mouth of another river as large
as the Senegal (the Joombas), which was so beautiful, with
the trees growing down to the water's edge, that they determined to send one of their negro interpreters on shore.
Each ship had some on board whom they had brought from
Portugal; slaves that had been taken in the first voyages,
who having learned the language, had come out as interpreters with the promise that they should be made free.
They drew lots to find which of the three ships should send

to the shore, and it fell to that of the Genoese. He despatched an armed barque, with orders to his people not to land till the interpreter had obtained information respecting the government and riches of the country. They set him on shore, and when they had put off to a little distance, saw several negroes advance to meet him, who had been waiting in ambush. After some questioning, which the men in the boat could not hear, they attacked and killed him, before the others could come to his rescue. The boat returned to the ships with the news, and the commanders thinking that a people who had shown themselves so cruel to one of their own countrymen would be still more barbarous to strangers, continued their course along the coast, which increased in beauty and verdure the further they went, but was very flat and low.

At length they came to the mouth of a very large river, at the narrowest part not less than three or four miles wide, and the ships could enter it with safety. The next day they learned that this was the much desired Gambia. They sent on the smallest caravel, well equipped with men and arms, to sound the river, and find out whether the larger vessels could follow. Finding that at the shallowest it was four feet deep, they resolved to send sloops well armed with the caravel, with instructions that if the negroes came to attack them, they were to return without fighting, because, their object being to establish commerce, they could only do this by gaining the confidence of the people. Two miles up the river the sloops found twelve and sixteen feet of water. The banks of the river were extremely beautiful and covered with magnificent trees, but, as they proceeded, it became so winding that they did not care to go further. On their way back, they saw, at the entrance of a small river which ran into the large one, three canoes made each of a single piece of wood. Though the men in the sloops were strong enough to defend themselves, they rowed back with great speed, in accordance with their orders, and when they reached the caravel and had got on board, they saw the blacks about a bow-shot behind. The negroes were about twenty-five or thirty in number, and seemed much surprised at the sight of the caravel. They stayed some time to satisfy their curiosity ; but made no reply to the signs made to them, and at last departed as they had come.

The following day, at three in the morning, the two caravels which had remained at the mouth of the river, took advantage of wind and tide being in their favour, to enter the river and rejoin their companion. They had scarcely gone three or four miles, one after the other, when they perceived that they were followed by a great number of canoes, but could not understand where they came from. Seeing this, they turned upon the negroes, and a battle appearing inevitable, covered themselves as well as they could as a protection against the poisoned arrows. The canoes surrounded the prow of Cadamosto's ship, which was in advance of the rest. There were fifteen of them, containing about a hundred and fifty negroes, all tall, well-made men. They had on shirts of white cotton, and white hats with a plume, which gave them a war-like air. At the prow of each canoe was a negro on the look-out, with a round shield that seemed made of leather. When close to the caravel, they remained with their oars raised, looking at it with admiration, till the other caravels came hastening up at the sight of danger. As soon as they came quite near, the negroes laid down their oars, and took to their bows, from which they discharged a heavy shower of arrows. The three caravels remained stationary, but fired off four cannon, which astonished the negroes so much that they threw down their bows, and looked on all sides in the greatest terror for the cause of so frightful a sound. When the noise ceased they again took courage, and resumed their bows, coming within a stone's throw of the ships, and bearing the fire of the crossbow-men very bravely. One of them was killed by a shot from the son of the Genoese gentleman, but they continued their attack till a great number had been slain, without the loss of a single man on board the caravels. When the negroes became aware of their loss, and found their canoes likely to sink, they threw themselves on the smallest caravel, which was badly armed, and attacked it violently. Cadamosto seeing their intention, placed the smaller vessel between the other two, and gave orders for a general discharge of artillery. Though they took care not to fire on the canoes, the noise and the agitation of the water so terrified the negroes, that they fled in disorder. The three caravels were then made fast to each other, and by means of a single anchor remained as firm as a vessel in the greatest calm.

During several days following, Cadamosto sought occasion to convince the natives that he had no intention of hurting them. The interpreters went on shore in a canoe, spoke to the people, and asked them why they attacked strangers who only wished to make conditions of peace and commerce with them, as they had already done with the people of Senegal, and who had come from a far distant land with presents from the King of Portugal to them. They asked the name of their country and river, and invited them to come to the vessels and make exchanges of merchandise, according to their own will and pleasure. To all this the negroes replied, that they had heard of the arrival of the white people at Senegal, and that they despised their neighbours of Senegal for entering into any treaty with the Christians, who, they believed, lived on human flesh, and only bought negroes to devour them. They declared their intention to kill the Christians if possible, and take their spoil to their sovereign, who was three days' journey inland. Their country was called Gambra, and the large river had a name which Cadamosto could not remember. The wind having risen, during this conference, the caravels took advantage of it to bear down upon the negroes, who escaped to the shore, and thus ended the encounter.

The commanders then consulted whether they should sail further up the river, in the hope of finding some more hospitable people, but the sailors were so anxious to return home, that they declared they would not go on. The commanders were obliged to submit, fearing a mutiny. On the following day they set out on their homeward voyage and sailed towards Cape Verde on their way to Portugal.

All the time they remained at the mouth of the river, they saw the north star only once, when it seemed to be low down, about a span's length above the sea. They observed also six other stars about the same height—large, clear, and brilliant—which were placed thus * : * * *, and which they took for the southern chariot, but they did not see the principal star, not being far enough to the south to lose sight of the north star. In the same place they found the night to be eleven hours and a half on the first and second days of July. The heat was excessive, though it was a little more temperate during the short time which the natives called winter, from the beginning of July to the end of October. During this time it rained every day; the clouds rose in the north-east quarter east, or east-south-east, and the rain was accompanied by violent thunder. This was the time for planting and sowing as among the negroes of Senegal. Their food was milk, honey, and vegetables. Cadamosto was told that in the interior the heat was so great, that even the rain was very hot. There was no twilight as in Europe, but the darkness did not disappear at once when the sun rose; for about half an hour the sky was obscured as by a thick smoke. Cadamosto believed that the flatness of the country was the cause of the sun appearing so suddenly.

CHAPTER XVI.

THE CAPE VERDE ISLANDS.

1460.

In the present chapter will be given for the first time in the English language the correct statement of the discovery of the Cape Verde Islands, and that by their first discoverer. Hitherto we have had to content ourselves with the narrative transmitted to us by Cadamosto in his description of his own second voyage to the west coast of Africa, to which has been erroneously assigned the date of 1456. But it will be my duty first to show that that narrative is so full of errors, contradictions, and incoherences that we must look elsewhere for the truth; and secondly, to produce from a hitherto untranslated manuscript, recently discovered in Munich, the truth as related to us by the original discoverer. It is a happy circumstance that the details supplied by this latter document coincide with and confirm the careful corrective criticisms on Cadamosto's narrative, which had been made, in entire ignorance of the MS. in question, by a learned Portuguese writer in 1844.*

But while one ancient document is thus introduced to correct another, it would be unjust alike to the authors of those documents and to the reader not to give their own statements as far as is consistent with the avoidance of prolixity. A verbatim rendering of such documents would

* See "Ensaios sobre a statistica das possessoës Portuguesas," &c., por José Joaquim Lopes de Lima, Lisbon, 1844. Liv. i., part 2, cap. 1, page 4.

weary most readers, but it is at the same time necessary
that no substantive statement be omitted. The following,
with such limited modification, is the narrative of Cadamosto's second voyage. Of its date we shall speak hereafter.

CADAMOSTO'S SECOND VOYAGE.

The next year Cadamosto, together with the Genoese
gentleman, Uso di Mare, undertook a second voyage, with
the view of following up his discoveries in the country of
Gambra, which had been frustrated before by the barbarity
of the natives and the opposition of the Portuguese sailors.
Prince Henry warmly approved of the expedition, and
fitted out a caravel in his own name to accompany them.

The three ships set out from Lagos in the beginning
of May, and the wind being favourable, they reached the
Canaries in a few days, and without stopping went on to
Cape Branco. When they had doubled the Cape they put
out into the open sea, but the night following were surprised
by a storm from the south-west, which carried them west-
north-west during three days and two nights.

The third day they discovered land, to their great joy, and
two men being sent to the mast-head saw two large islands,
at which there was great rejoicing, for they believed them to
be hitherto unknown, and hoped to find them inhabited.
They sailed towards them, and having found good anchorage,
sent on shore a skiff with men well armed to explore, but
they found no sign of habitation. The next day, to make
quite sure, Cadamosto sent ten men, armed with crossbows,
with orders to ascend a mountain and see if there were any
sign of habitation, or if there were any other islands within
sight. They could see no dwellings of any kind, but found
an immense number of pigeons, so tame that they could be
taken by the hand. From the mountain they could see
three other islands, one toward the north, and two in a
southward direction. They thought they could see islands in
the west, but so far off that they could not distinguish them,

and Cadamosto did not care to spend the time required to go to them, as he thought they would be all alike wild and uninhabited. Afterwards others, attracted by the discovery of the four islands, went further and found ten islands of different sizes, inhabited only by pigeons and other birds.

The three caravels then weighed anchor and went to one of the other islands, which appeared covered with trees, and finding the mouth of a river, they anchored there in order to get water for the ships. Some of the sailors went up the river a good distance in the sloop, and found some small lakes of salt, fine and white, which they brought into the vessel in great quantity, and took some of the water, which seemed very good. They found many turtles, and put some of them in the caravels, the upper shells of which were larger than a shield. The sailors killed a great number, and dressed them in different ways, observing that they had formerly eaten them in the Gulf of Arguin, where there are some of the same sort, but not so large. Cadamosto tasted them, and found them very savoury. They salted a good number, which proved very useful in their voyage. At the mouth of the river and further up, they found fish in incredible numbers and great variety. The river is very broad, so that a ship of one hundred and fifty tons can get into it easily, being a bow-shot in breadth.

They remained two days to refresh themselves, agreeing to name the first island they had found Boavista, because it was the first they had discovered. The larger one they named Santiago, because they came to anchor there on the feast of St. James and St. Philip.

They again set sail and came in sight of land at a place called Spedegar, and followed the coast till they came to The two Palms, a place situated between Cape Verde and the River Senegal. Without any further difficulty they proceeded on their way to the River Gambra, which they entered without interruption. They sailed on, sounding the river for about ten miles, the few negroes that they saw not daring to approach. They then anchored one Sunday near an island, where they buried one of the sailors who had died of a fever, and as he was much beloved and lamented, they named the island after him, S. André. They now continued their course up the river, followed by some canoes at a distance. Cadamosto sent interpreters to them to tell them they might come on board in safety. The interpreters showed them stuffs and toys which had been brought for the purpose, and offered to give them some if they would come on board.

At length overcoming their fears, they came to the caravel, and one of the negroes was able to talk with Cadamosto's interpreter. They were very much astonished at everything they saw on board the caravel, and especially with the sails, for they had only been accustomed to use oars. The dress and colour of the Europeans amazed them, their own dress being only a white cotton shirt. Cadamosto received them with great kindness, and asked them the name of their country and of their prince, to which they replied that their country was called Gambra, and that their prince was named Forosangoli; that he lived about ten days' journey from the river, between the south and south-west; that he was a vassal of the Emperor of Melli, who was chief of all the negroes; that there were many other princes who lived nearer, and that if Cadamosto wished it, they would take him to one named Batti Mansa (i.e. King Batti, Mansa being the Mandingo for king). This offer was thankfully accepted, and the caravel proceeded up the river according to the direction of the negroes, till they reached the residence of this prince, which Cadamosto believed to be about sixty miles from the river's mouth. It must be remarked, that they sailed up the river in an easterly direction, and saw many tributaries which flowed into it; the place where they anchored was much narrower than the mouth, being only about a mile in breadth.

When they had cast anchor, Cadamosto sent one of the interpreters with the negroes to Batti Mansa, bearing a very handsome Moorish dress as a present, and charged with a message to the effect that they had come from the Christian

King of Portugal, to make a treaty with him. The messengers were favourably received by Batti Mansa, who sent some of his people to the caravel. A treaty was made, and European goods were exchanged for slaves and gold, but the quantity of gold was not at all equal to the expectations raised by the accounts given by the people of Senegal, who being very poor themselves, thought their neighbours richer than they were. The negroes valued their gold as highly as the Portuguese did, but showed how much they admired the European trifles by their willingness to give a large price for them. The Portuguese remained there eleven days, during which many negroes came on board, some only from curiosity, others to sell their merchandise, cotton cloths, white and striped nets, gold rings, &c. They also brought baboons and marmots, civet and skins of the civet cat, all which they sold very cheap. An ounce of civet they would give for forty or fifty marchetti. Others brought fruits, especially dates, which the sailors found very good, but which Cadamosto would not touch, fearing they were not wholesome.

Every day the caravels were visited by negroes differing in race and language, both men and women, who came and went in their canoes with the utmost confidence. They only used oars and rowed standing, having always a second in the boat to steer with his oar. The oars were in the form of a half lance, between seven and eight feet long, with a round board like a trencher at the end; with these they managed their canoes very skilfully, keeping close to the coast, not venturing far for fear of being taken by the neighbouring people and sold for slaves.

At the end of eleven days, they resolved to return to the river's mouth, as fever began to show itself among them. Cadamosto had not failed to make his observations on the religion and customs of the people. They were generally idolaters, and superstitious with regard to charms and enchantments; but they believed in a God, and there were some Mahometans among them, who travelled about, and

traded with other countries.   There was but little difference
between the food of these people and that of the natives of
Senegal, except that they ate dogs' flesh, which Cadamosto
had never seen done elsewhere. They dressed in cotton, which
they had in abundance, while the natives of Senegal, whose
cotton was scarce, often wore nothing at all.   The women
dressed like the men, but for ornament tattooed their skins
when they were young with a hot needle.   The heat of the
climate was extreme, and increased towards the south; and
was greater on the river than on the sea, on account of the
great quantity of trees which grew on the banks and kept
the air confined.   As an instance of the size of these trees,
Cadamosto mentions one which measured seventeen fathoms
round ; the trunk was pierced and hollowed out in many
places, but the foliage was green, and the branches spread
out so as to afford an immense shade.   There were others
still larger, showing that the country was wonderfully fertile and well watered.

    There were great numbers of elephants, which the natives
did not know how to tame.   While the caravels were at
anchor, these elephants came out of the neighbouring
wood, down to the banks of the river.   Some of the sailors
got into the skiff, but before they could reach the bank,
the elephants saw them and went back to the wood.   These
were the only living elephants that Cadamosto saw.   A
negro chief named Guumi Mansa, who lived near the
mouth of the river, showed him a small one that he had
killed after a hunt of two days.   The negroes hunted on
foot with bows and poisoned darts or javelins.   They hid
behind the trees, and sometimes climbed up into them, and
from their hiding-places threw their poisoned weapons at
the animals, leaping from tree to tree in pursuit, and the
elephants being large and unwieldy were struck many
times before they could escape or defend themselves.   They
never dared attack an elephant in the open country,
for however active a man might be, he could not hope
to outrun him.   But the elephant never attacks a man

unless in self-defence, for he is naturally gentle and docile.

The tusks of the little elephant which Cadamosto saw dead were not more than three palms long, one-third of the length being buried in the jaw; this showed it was quite a young one, for the full-grown animals have tusks from ten to twelve palms long. Young as this one was, it had as much flesh as five or six oxen. The negro chief presented Cadamosto with the best part, and gave the rest to the hunters. Cadamosto's portion was taken to the caravel to be cooked, as he was curious to taste the flesh of an animal so new to him, but he found it hard and disagreeable; however, he had some salted to take home to Prince Henry on his return. He sent on board the caravel one of the feet and a part of the trunk, with some of the skin, which was black and very coarse; all of which with the salted flesh he presented to the Prince, who received them as great curiosities.

This chief also gave him another elephant's foot, which measured three palms and a finger each way, and a tooth twelve palms in length, which were afterwards presented by the Prince to his sister the Duchess of Burgundy.

In the river Gambra and other rivers of the country, besides the serpents called "calcatrici," and other animals, Cadamosto saw the horse-fish (the hippopotamus), which he thus describes: " It is amphibious, and its body is as large as that of a cow, with very short legs and its feet cloven; the head is large, and like that of a horse. It has two tusks like those of the wild boar, some two palms and a half long. It comes out of the water, and walks like other quadrupeds." Cadamosto says that it had not been seen by Christians before, except on the banks of the Nile. He saw also bats three palms long or more, a number of birds, different from those of Europe, and a multitude of fishes, also differing from the European, but almost all very good to eat.

The sickness of the men now compelled them to leave the country of Batti Mansa. They descended the river, and being well-furnished with provisions, determined to go further along the coast. The current of the Gàmbra carried them far beyond the mouth of the river, and the land stretched south-south-west, in the form of a cape (Cape St. Mary). They therefore stood out well to sea, but found the point of land was no cape to speak of, for the shore was quite straight beyond it. But as they saw breakers at four miles distance, they kept out to sea, to escape the sand-banks and rocks, and two men continued on the look-out, one at the prow, and one at the mast-head. Besides which they only sailed by day and cast anchor at night. To avoid disputes, they every day cast lots which caravel should go first, and in this way they coasted along two days. On the third, they discovered the mouth of a river, half a mile wide, and towards evening they saw a little gulf, which had been taken for another river, but as it was late they cast anchor.

The next morning they sailed on, and came to the mouth of a very large river, but somewhat smaller than the river Gambra, the banks of which were covered with trees of extraordinary size and beauty. They sent on shore two armed sloops, with interpreters, to reconnoitre, and they returned with the information that the river was called Casa Mansa, from the name of a negro chief, who resided thirty miles up the river and was then at war with a neigh-bouring chief. They departed the following day. The distance from the river Gambra was about a hundred miles.

They continued their course, following the coast, till they came to a cape about twenty miles further on, to which they gave the name of Capo Roxo, from the red colour of the earth. They next came to the mouth of a river, about a bow-shot in width, which they did not enter, but gave it the name of Rio de Santa Anna (the Cacheo). Further on they found another river of the same size, which they named

S. Domingo (the Rio de Jatte), which was about fifty-five or sixty miles from Cape Roxo.

A day's journey beyond, they came to the mouth of a river, so wide that they thought it was a gulf; they were some time crossing, for it was twenty miles over. The south bank was covered with beautiful trees, and when they arrived there, they discovered some islands out at sea. They then cast anchor, resolving to gain more information before they went on. The following day two canoes approached the caravels, the largest containing about thirty negroes, and the other sixteen. All on board the caravels took to their arms, expecting an attack, but the negroes raised a piece of white linen fastened to an oar, in sign of peace. The Portuguese replied in the same manner, and the negroes came on board Cadamosto's caravel, where they showed great surprise at everything they saw, for all was new to them, the whiteness of the men, the form of the vessel, the masts, the sails, and cordage, &c. It was a great disappointment to find that the interpreters were no longer of use, for they did not understand the language of the natives. This induced Cadamosto to think of turning back, as they could not get any further information, and to this the other commanders agreed. During their stay, which lasted two days, one of the negroes gave bracelets of gold in exchange for other things, without speaking a word, only making use of signs.

They named the river, Rio Grande.* The north star appeared to them very low. They also found that the tides were different here from anything they had observed in other countries. Instead of the flux and reflux being six hours each, as at Venice and other countries of Europe, the tide here rose in four hours, and took eight to subside; and so great was the impetuosity of the tide, that three anchors scarcely sufficed to keep each caravel steady, and they set sail with great danger, for the force of the sea was greater than that of the wind, though all the sails were set.

* From the extreme breadth of its mouth this would seem to be the river Jeba.

In returning to Portugal, Cadamosto visited two large islands and some small ones, which they saw about thirty miles from the main land. The large islands were inhabited by negroes, the land was low and covered with fine trees, but the language of the people was unknown to the interpreters, so they made a very short stay, and steered homewards, arriving safely in Portugal after a good voyage.

Now it will not be difficult to prove from the very words of the foregoing document, the utter impossibility of its own correctness as to the details of the asserted discovery of the Cape Verde Islands, while the incorrectness of the date can be demonstrated from other data. Cadamosto says, that he sailed from Lagos in the beginning of the month of May; then went to the Canary Islands, thence proceeded to Cape Branco, off which he was assailed at night by a furious storm from the south-west. It went on increasing, and for two nights the ship was driven in a west-north-west direction, when on the third day they sighted two large islands, on one of which they landed and named it Boavista. From the top of a mountain in this island the explorers gained sight of three more islands, one to the north and two to the south, and they thought they saw others to the west. On the following day, Cadamosto says, that he came in sight of the two to the south, and went to one of them, which he named Santiago, because he cast anchor there on St. Philip and St. James's day. He found in that island a river of fresh water, so large that a ship of seventy-five tons could enter it with ease; and along this river his men found some small lagoons of remarkably white and beautiful salt, a great quantity of which they brought away. They also found many turtles, which they cooked in various ways, and of which he tasted.

Now all this is simply impossible. 1st. A man who sailed from Lagos in the beginning of May, could not, after a voyage of some days, anchor off one of the Cape Verde Islands on St. Philip and St. James's day, which throughout

Christendom is kept on the 1st of May. 2dly. Three days driving before a furious tempest in a west-north-west direction, from Cape Branco, could not bring a vessel to the island of Boavista, which lies a hundred leagues to the south-west. 3rdly. From Boavista may be seen in clear weather the island of Sal, which is eight leagues off; but, from Cadamosto's time till now, no human being has been able from that island to sight Santiago, which is more than twenty-five leagues to the south-west. To the west lies St. Nicolas, nearly as far off as Santiago, so that the explorers could not possibly see from Boavista more than one island to the north. 4thly. Neither in Santiago, nor in any one of the Cape Verde Islands, is there a single river of fresh water, nor any stream big enough to float a canoe; and the inhabitants of Santiago would be only too glad to realize the pleasant dream of Cadamosto, especially if their fresh water river, which would easily admit a vessel of five-and-seventy tons, were fringed with lagoons of remarkably white and beautiful salt, a commodity of which the island is as lacking as it is of the turtles with which the Venetian had also blessed them.

We know from a statement of Cadamosto's elsewhere that he remained in Lisbon till 1463, and it is tolerably clear that he made capital of discoveries that had taken place in the interval, and appropriated them by an alteration of the date. His own first voyage, on which he started on March 22nd, 1455, would appear to have lasted till 1456, for in it the mention of a month of November is followed by a month of June; and his second voyage being in the "anno sequente," would be 1457. But there is evidence to show that the discovery of Santiago was not made till 1460, and it would also seem from the recently discovered manuscript of which I am about to give the translation, that the honour of that discovery belongs, not as has been hitherto supposed by the severest critics, to the Genoese, Antonio de Nolle, but more properly to the Portuguese Diogo Gomez, who claims it as his own, and shows

how the Genoese took advantage of his first arrival in
Portugal to claim the honour and emoluments of the discovery at the hands of the King.

The following is Gomez' narrative :—

THE NARRATIVE OF DIOGO GOMEZ.

"Not a long time after (the disaster of Adalbert, or Vallarte,
as related by Azurara, see page 224) * the Prince equipped
at Lagos a caravel, named *Picanso*,† and appointed (the
writer) Diogo Gomez captain, together with two other caravels, of which he appointed Diogo Gomez captain-in-chief.
The captain of one of these was João Gonsalvez Ribeiro, of
the Prince's household, and of the other Nuño Fernandez de
Baya, the Prince's esquire-at-arms. The Prince gave them
orders to proceed as far as they could. After passing the
River of S. Dominick (the S. Domingo), and another great
river called Fancaso (the present Rio Grande) beyond the
Rio Grande (the Jeba), we encountered strong currents in
the sea, so that no anchor could hold, and which were called
Macareo. The other captains, therefore, and their men
were greatly alarmed, thinking that they were at the extremity of the ocean, and they begged me to return. In the
middle of the current the sea was very clear, and the natives
came from the shore and brought us their merchandise, viz.,
cotton cloth, elephants' teeth, and a quart measure of malaguette, in grain and in its pods as it grows, with which I was
much delighted. The current prevented our proceeding
further, and in fact increased so much that it obliged us to
put back.

We came to a land where, near the shore, were many
palm trees, with their branches broken, and so tall that

---

* The earlier portion of Diogo Gomez' narrative is omitted, because it contains none of his own adventures, but a hearsay account of Prince Henry's previous expeditions, which he related to Martin Behaim in a far less accurate and
authentic form than has already been laid before the reader from Azurara's
chronicle.

† This is doubtless the same vessel, the "Picanço" or "Wren," already
mentioned by Azurara. (See page 203.)

from a distance we thought that they were masts or spars of negro vessels.* Thither we went, and found an extensive plain full of hay, and more than five thousand animals called in the negro language *myongas.* These are beasts a little larger than stags, which showed no fear at sight of us. We also observed five elephants come out of a small river sheltered with trees. Three of them were large, with two young ones, and they fled from the *myongas.* On the sea-shore we saw many crocodiles' holes. We returned to the ships, and on the next day made our way from Cape Verde, and we saw the broad mouth of a river, three leagues in width, which we entered, and from its size correctly con-cluded that it was the River Gambia. We entered it with the wind and tide in our favour, and came to a small island in the middle of the river, and there remained that night. In the morning, however, we went further in, and saw many canoes full of men, who fled at sight of us, for it seems they were the same who had slain Nuño Tristam and his men. The next day, however, we saw beyond the point ? [Caput] of the river some people on the right hand side of it, to whom we went, and were received in a friendly manner. Their chief was called Frangazick, and was the nephew of Farisangul,† the great Prince of the negroes. There I received from the negroes one hundred and eighty pounds weight of gold, in exchange for our merchandise; such as cloths, necklaces, &c. They told us that the negroes on the left shore would not hold intercourse with us because they had slain the Christians. The lord of that country had a certain negro, named Bucker, who was acquainted with the whole country of the negroes, and finding him perfectly truthful, I asked him to go with me to Cantor, and promised to give him a mantle and shirts, and every necessary. I made also a similar promise to his chief, which I kept. We ascended the river, and I sent a captain with his caravel into a certain harbour, named Ulimays [doubtless Ollimansa,

* The Cabo dos Mastos.
† See Cadamosto, page 280.

see *post*]. The other remained in Animays [Nomimansa?], and I went up the river as far as Cantor, which is a large town near that river's side. On account of the thick growth of the trees on both sides of the river, the vessels could proceed no farther, and I sent out the negro whom we had brought with us, to make it known to the people of the country that I had come thither for the purpose of exchanging merchandise, and, in consequence, the negroes came in very great numbers. When the report spread throughout the country round, that the Christians were in Cantor, the natives came together from all quarters, viz., from Tambucatu [Timbuctoo] in the north, from the Sierra Geley * in the south, and there came also people from Quioquun (*sic*), [Kukia], which is a great city, surrounded by a wall of baked tiles, and where I understood there was abundance of gold, and that caravans of camels and dromedaries crossed over thither with merchandise from Carthage or Tunis, from Fez, from Cairo, and from all the land of the Saracens, in exchange for gold. They said that the gold was brought from the mines of Mount Gelu, which is the opposite side of the range called Sierra Lyoa. They said that that range of mountains began at Albafur, and ran southwards, which pleased me much, because all the rivers, large and small, descending from those mountains (which had been as yet observed) ran westward ; but they told me that other very large rivers ran eastward from them, and that near that city was a certain great river, named Emiu, and that there was also a great lake (mare), but not very broad, on which were many canoes, like ships, and that the people on the opposite sides were in constant warfare with each other, those on the eastern side being white men. On my inquiring what sovereigns ruled in those parts, they answered, that the chief of that part, which was inhabited by the negroes, was named Sambegeny, and that the lord of the eastern part

---

* The same as Mount Gelu eight lines on. It is most probably the " Djalout " of Abulfida. See Reinaud's edition, tom. ii. page 86. He speaks of the treasure that it produces.

was called Semanagu, and that a short time before they had a great battle, in which Semanagu was the conqueror. And a certain Saracen of Termezen [Tlemsen ?], named Admedi, told me that he had been through all that land, and had been present at the battle, both by sea and land. When I afterwards related all these things to the Prince, he told me that a merchant in Oran had written to him two months before respecting this engagement, which had taken place between Semanagu and Sambegeny, and, therefore, he believed the account. Such are the things which were told me by the negroes who were with me at Cantor. I questioned the negroes at Cantor as to the road which led to the countries where there was gold, and asked who were the lords of that country. They told me that the king's name was Bormelli, and that the whole land of the negroes on the right side of the river was under his dominion, and that he lived in the city Quioquia [Kukia]. They said further, that he was lord of all the mines, and that he had before the door of his palace a mass of gold just as it was taken from the earth, so large that twenty men could scarcely move it, and that the king always fastened his horse to it, and kept it as a curiosity on account of its being found just as it was, and of so great size and purity. The nobles of his court wore in their nostrils and ears ornaments of gold. They said also, that the parts to the east were full of gold mines, and that the men who went into the pits to get the gold did not live long, on account of the impure air. The gold sand was afterwards given to women to wash the gold from it.*

I inquired the road from Cantor to Quioquia [Kukia], and was told that to Morbomelli [Bormelli †] from Cantor

---

* The mountainous country of Bouré on the Tankisso, an affluent of the Joliba, is doubtless here referred to. It contains many very abundant gold mines. The gold of Bouré circulates throughout the whole interior, and finds its way to the French and English settlements on the coasts; while Jenné, which was formerly considered as the country most plentifully supplied with this precious metal, has none except what is brought from this rich tract. See Caillié's "Travels through Central Africa to Timbuctoo." London, 1830. Vol. i. p. 284.

† The name of the resident, Bormelli, put in lieu of his place of residence, Kukia.

the road is eastward to Somandu, and from Somandu to
Conmuberta and to Cereculle and other places, the names of
which I have forgotten. And in these aforenamed places is
great abundance of gold, as I can well believe, for I saw the
negroes at the time who went by those roads come loaded
with gold. And they said that Forisangul [*sic*] was subject
to Mormelli [Bormelli], who was lord of the right part of
the river Gambia.

While thus holding peaceful intercourse with these negroes
of Cantor, my men became worn out with the heat, and so
we returned in search of the other two caravels, and in the
caravel which had remained in Ollimansa [the same named
previously Ulimays], I found nine men had died, and the
captain, [João] Gonzalo [Alphonso], very ill, and all the
rest of his men sick, except three. I found the other caravel
fifty leagues lower down towards the ocean, and in it five
men had died. We immediately withdrew, and made for
the sea, and I went to the place where I had hired the negro
traveller, and gave him what I had promised him.

They then informed me that on the other, that is, the left
or south side of the river, there was a certain great chief,
named Batimansa, and I desired to make peace with him,
and I sent to him that negro who had been with me at
Cantor. That chieftain desired to speak with me in a great
wood on the bank of the river, and brought with him an
immense throng of people armed with poisoned arrows,
azagays, and swords and shields [dargis]. And I went to
him, carrying him some presents and biscuit, and some of
our wine, for they have no wine except what is made from
the date palm, and he gave me three negroes, one male and
two female, and he was pleased and extremely gracious,
making merry with me and swearing to me by the one only
God that he would never again make war against the
Christians, but that they might travel safely through his
land and interchange their merchandise. Being desirous of
putting this to the proof, I sent a certain Indian named
Jacob, whom the Prince had sent with us, in order that, in

the event of our reaching India, he might be able to hold
speech with the natives, and I ordered him to go to the
place which is called Alcuzet, with the lord of that country,
whither, on a former occasion, a knight had gone with him,
through the land of Geloffa to find the Sierra de Gelu and
Tambucutu. This Jacob, the Indian, related to me that
Alcuzet is a very vicious land [multum viciosa], having a river
of sweet water and abundance of lemons, some of which he
brought with him to me. And the lord of that country sent
me elephants' teeth, one of them very large, and four negroes, who carried the tooth to the ship. And so they came
peacefully up to our ships, and thus I was safe from them.
Afterwards I sent to his abode, which was surrounded by
many negro habitations. Their houses are made of seaweeds,
covered with straw, and I remained with them for three days.
Here were many parrots and many ounces, and he gave me
six skins, and ordered that an elephant should be killed and
its flesh carried on board the caravels.

It was here that I learned the fact that all the mischief that
had been done to the Christians had been done by a certain
king, called Nomymans [Nomimansa], who possesses the land
which lies on that promontory. I took great pains to make
peace with him, and sent him many presents by his own men
in his own canoes, which were going for salt to his own
country. This salt is plentiful there, and of a red colour. He
greatly feared the Christians, on account of the injury which
he had done them. I went by the river towards the ocean,
as far as the harbour, near the mouth of the river, and he
sent to me many times men and women to try me, whether
I would do them any harm, but, on the contrary, I always
gave them a friendly reception. When the King heard this,
he came to the river-side with a great force, and sitting down
on the bank, sent for me to come to him, which I did,
paying him all ceremonious respect in the best fashion I
could. There was a certain Bishop there of his native
church, who put questions to me with respect to the God of
the Christians, and I answered him according to the intelli-

gence which God had given me, and at last I questioned him respecting Mahomet, in whom they believe. What I said pleased the King so much, that he ordered the Bishop within three days to take his departure out of his kingdom, and springing to his feet, he declared that no one, on pain of death, should dare any more to utter the name of Mahomet, for that he only believed in the one only God, and that there was no other God but He, whom his brother, the Prince Henry, said that he believed in. Calling the Infant his brother, he desired that I should baptize him, and so said also all the lords of his household, and his women likewise. The King himself declared that he would have no other name than Henry, but his nobles took our names, such as Jacob, Nuño, &c., as Christian names. I remained that night on shore with the King and his chiefs, but I did not dare to baptize them, because I was a layman.

On the next day, however, I begged the King with his twelve principal chiefs and eight of his wives to come to dine with me on board the caravel, which they all did unarmed, and I gave them fowls and meat cooked after our own fashion, and wine, both white and red, as much as they pleased to drink; and they said to each other that no nation was better than the Christians.

Afterwards, when we were on shore, he desired that I would baptize him; but I answered that I had not received authority from the supreme pontiff. I told him, however, that if he so desired, I would convey his wishes to the Prince, who would send a priest to baptize them. He immediately wrote to the Prince to send him a priest, and some one to inform him respecting the faith, and begged the Prince to send him a falcon for hunting, for he wondered greatly when I told him that the Christians carried a bird on the hand which caught other birds. He wished him also to send two rams, and sheep, and ganders and geese, and a pig, as well as two men who would know how to construct houses and make a survey of his city. All these requirements I promised that the prince would fulfil. At my departure he and all his people lamented, so great was the friendship which had sprung up between him and me.

It so happened that for two years no one went back to Guinea because King Alphonso was gone, with a fleet of three hundred and fifty-two ships, to Africa, and took the powerful city of Alcacer al Quivir [or rather El Seguer], for which reason the Prince, being fully occupied, gave no attention to Guinea.

After leaving the King at Gambia I pursued my way to Portugal, and sent one caravel with those who were in the best health straight home. The other remained with me, because the people on board of her were sick. And I ordered the captain of the first vessel, if he had a favourable wind, to go straight to Portugal, if not, to wait for me at Arguin, and so he departed; but I with the other caravel sailed with a favourable wind to Cape Verde. As we came near the sea-shore we saw two canoes putting out to sea. We sailed between them and the land, and came up to them, and in one of the canoes we counted thirty-eight men, and the interpreter came to me, and said in my ear, that that was Beseghichi, lord of that land, a malicious man, of whom I have already spoken. I made them come into the caravel, and gave them to eat and drink, and a double portion of presents, and pretending that I did not know that he was the chief, said to him by way of trying him, " Is this the land of Beseghichi?" He said "Yes." I replied, " Why is he then so malignant against the Christians? It would be better for him to make peace with them, and that both might interchange merchandise, and that he might have horses, &c., as Burbruck in Budumel, and other lords of the negroes did. Tell him that I have taken you in this sea, and for love of him have set you free to go on shore." He much rejoiced, and I told him to go into their canoes, which they did, and as they all stood in their canoes, I said to the chief, " Beseghichi, Beseghichi, do not think that I did not know thee. It was certainly in my power to do with thee whatever I wished, and since I have acted kindly by thee, do thou do likewise with our Christians," and so they went their way.

A few days after, we came to Cape Tofia, and to Anterot,* and entered Arguin. And not far from the coast we came to the island called the Ilha de Garças. It was not inhabited, and was only one league in circumference. On it we found an innumerable multitude of birds of every kind, and on the ground pelicans' nests, and many dead pelicans. These are not as the painters represent them, but have a broad beak, and a stomach large enough to hold a measure of wheat, such as is commonly called an *alqueiro*.† The number of birds there was so great, that we killed as many as we could carry in our boat, and so we went into Arguin. We thus sailed for Portugal, and came to Algarve, to the great city named Lagos, where the Prince at that time was, and he rejoiced greatly at our arrival.

After the Prince returned from the fleet with King Alphonso, I reminded him of what King Nominaus [Nomimansa] had said, so that he should send to him all those things which had been promised. This the Prince did, and sent thither a certain priest, a relation of the cardinal's, the Abbot of Soto de Cassa, that he should remain with that King and instruct him in the faith. He also sent with him a young man of his household, named John Delgado. This was in the year 1458.‡ .... Two years afterwards King Alphonso equipped a large caravel, in which he sent me out as captain, and I took with me ten horses and went to the land of Barbacins, which is between Serreos and King Nomemans [*sic*]. Those Barbacins had two kings, viz. Barbacin Dun and Barbacin Negor. And the King gave me authority over the shores of that sea, that whatever caravels I might find off the land of Guinea should

* The coast from Arguin to the Senegal was called Anterote.

† See *ante*, page 201.

‡ Here occurs an account of the death and burial of Prince Henry in 1460, which will be found in the following chapter; but that the immediately ensuing words, "Two years afterwards," bear reference not to the Prince's death, but to Gomez' last explorations, will be demonstrated at the close of the present chapter.

be under my command and rule, for he knew that there were caravels there which carried arms and swords to the Moors, and he ordered me to take such prisoners and bring them to him to Portugal. And by the help of God in twelve days I arrived at Barbacins, and found there two caravels: one, in which was Gonzalo Ferreira, of the household of Prince Henry, a native of Oporto, who was conveying horses thither; and in the other caravel was Captain Antonio de Noli, a Genoese, who was also a merchant conveying horses. This was in the port of Zaza. I found there also Borgebil, who had been King of Geloffa, and who had fled thither from fear of the King of Burbuck, who had taken his country from him. The aforesaid merchants with their caravels greatly damaged the traffic in those parts, for whereas the Moors used to give twelve negroes for one horse, they gave them now no more than six. Then I summoned those captains to me, and on behalf of the King gave them seven negroes for one horse, but myself exchanged every horse for fourteen or fifteen negroes. While we were there, there came a caravel from Gambia, which brought us information that a certain man named De Prado, was coming with a very richly laden caravel, whereupon I immediately fitted out the caravel of Gonzalo Ferreira, and ordered him in the King's name, on pain of death, and confiscation of all his goods, to go to Cape Verde, and to look out for that caravel, which he did, and took it, and we found great booty in it. I forthwith despatched the captain, together with Gonzalo Ferreira, to the King, and wrote to the King an account of all these events.

I and Antonio de Noli then left that port of Zaza, and sailed two days and one night towards Portugal, and we saw some islands in the sea, and, as my caravel was a lighter sailer than the other, *I came first to one of those islands*, and saw white sand, and it seemed to me a good harbour, and I cast anchor there, and so also did Antonio. *I told them that I wished to be the first to land, and so I did.* We saw no sign of any man there, and we *called the island Santiago : it is so called*

to this day. There was abundance of fish to be caught there.
On shore we found many strange birds and streamlets of
fresh water. The birds were so tame, that we killed them
with sticks; and there were many geese there. There were
also an abundance of figs, but they do not grow on the trees in
the same manner as in our parts, for our figs grow near the
leaf, but these all over the bark from the foot of the tree to
the top. These trees grow in great numbers, and there was
great quantity of hay there. And I had a quadrant when I
went to these parts, and I wrote on the table of the quadrant
the altitude of the Arctic Pole, and I found it better than the
chart. It is true that the course of sailing is seen on the
chart, but when once you get wrong, you do not recover your
true position. We afterwards saw one of the Canary
Islands, called Palma, and after that we went to the Island
of Madeira. Though I was anxious to go to Portugal, I
was driven by a contrary wind to the Azores, but Antonio
de Noli remained at Madeira, and *availing himself of a more
favourable wind, reached Portugal before me. And he begged of
the King the captaincy of the Island of Santiago, which I had
discovered, and the King gave it him, and he kept it till his
death.* And I with extreme labour made my way to Lisbon,
and after some time the King went to Oporto, where that
De Prado, who had carried arms to the Moors, and whom
Gonzalo Ferreira had taken prisoner, lay in irons, and the
King ordered that they should martyrize him in a cart, and
that they should make a furnace of fire, and throw him into
it with his sword and gold."

In the above narrative of Gomez we have an interesting
supplement to Cadamosto's account of those explorations
along the west coast of Africa which, till now, had been the
latest recorded as occurring during the life of Prince Henry.
But this narrative brings us to 1460, the year of the Prince's
death, since not only does Gomez use the words "two years
afterwards," after the mention of the date 1458, but it can
be shown that he could not mean "two years after" the

Prince's death, of which he had spoken in the interim, because if we revert to the question of the Cape Verde Islands, we shall find collateral evidence to prove their discovery in 1460. In the Torre do Tombo, which is rich in documents of the period, not a single one is found bearing reference to the Cape Verde Islands anterior to December 3rd, 1460,* under which date they *are* mentioned, and it is in the highest degree improbable that, had they been discovered at the period given by Cadamosto, so many years would have been allowed to pass without their being colonized by so energetic a colonizer as Prince Henry, whereas in 1461 we find the colonization proceeding with considerable rapidity. Moreover Faria y Souza distinctly gives 1460 as the date of the discovery. On the 3rd of December, 1460, King Affonso V., being at Evora, made a donation to his brother Fernando, Prince Henry's adopted son and heir, of the islands hitherto discovered—to possess them in like manner as he had received them from Prince Henry. These islands, some of which have names now unknown, are recounted in the order of the groups, the last being the Cape Verde group, of which the following islands are mentioned :—S. Jacobe (Santiago); S. Filippe (Fogo); De las Moyaes (Maio); S. Christovao, supposed to be Boavista, a name apparently falsely given by Cadamosto years afterwards ; Ilha Lana, most probably the Ilha do Sal, which from its proximity to Boavista could scarcely fail of being, as here, mentioned next to it. The remainder of the Cape Verde Islands were soon after discovered by some mariners in the service of Prince Fernando, when they received their collective name from the cape off which they lay. The King ceded them to that Prince on the 19th of September, 1462. The first colonized was Santiago.

* Torre do Tombo, liv. i., of Affonsi V., fol. 61.

CHAPTER XVII.

THE DEATH OF PRINCE HENRY.

1457—1460.

AFTER the death of the hapless Duke of Coimbra at Alfarrobeira, the agitations which had had their source in private
intrigue were laid aside, and gave place to opportunities for
concentrated national action. King Affonso, energetic and
warlike, occupied himself with those conquests on the north
coast of Africa which gained for him the surname of "the
African."

In 1454 Constantinople fell into the hands of the Turks,
and the Pope summoned all the princes of Europe to a general
crusade against the infidels. In 1457 a special. legate, the
Bishop of Silves, was sent to Affonso by Calixtus III. with
the bull of the Crusade. The King entered warmly into
the plan, and made great preparations for it, offering to
supply twelve thousand men yearly. He also struck, with
the view of making Portuguese money of more value in
the foreign countries through which his march would lie,
a new piece of gold money, which had a cross on one side,
and was called a *cruzado*. In no country had the spirit of
chivalrous enthusiasm for the defence of the faith been preserved with less diminution from its ancient loyalty than in
Portugal; but the zeal which animated King Affonso was
manifestly inadequate, with his limited exchequer, to contend against the Turks, unless the Pope's appeal were
warmly responded to by other sovereigns. Such, however,

was not the case, and after the death of Calixtus III., in 1458, the crusade came to an end.

In this position of affairs, the object which the King had proposed to himself of fighting against the infidels, and the avoidance of those financial losses which his people apprehended, became reconciled in the notion of directing his forces against Africa. His first thought was to attack Tangier, but remembering its strength, and how much it had cost the Portuguese on a former occasion, he fixed on Alcaçar Seguer, or Alcaçar the Little, to distinguish it from Alcaçar Quivir, or the Great, as the place to be attacked.

On the 30th of September, 1458, Affonso sailed from Setuval with a fleet of ninety sail, and on the 3rd of October landed near Sagres, where Prince Henry gave him a magnificent reception. When the contingents from Mondego, Oporto, and elsewhere, joined the royal fleet at Lagos, they mustered in all two hundred and twenty sail, and on the 16th of October twenty-five thousand men disembarked, though not without some opposition and loss, off Alcaçar. The artillery and implements for the siege were promptly landed, and that same evening the order was given to invest the town. A portion of the ramparts was soon broken down, and at midnight Prince Henry, having constructed a battery in a favourable position, brought to bear a large piece of ordnance, a few shots from which made a considerable breach in the wall. The Moors, who, it must be acknowledged, had hitherto offered a brave and troublesome resistance, were overcome with fear at this result, and sent to propose terms of surrender. Prince Henry replied that " the King's object was the service of God, and not to take their goods or force a ransom from them. All that he required was that they should withdraw with their wives and children and effects from the town, but leaving behind them all their Christian prisoners." They begged for time to reflect, which was prudently refused, with a threat that if the town had to be taken by main force, all would be put to the sword. On this the Moors submitted, and sent the King hostages for the suspension of the conflict.

On the morrow they withdrew from the city unmolested, under the Prince's warrant for their safety. The Portuguese entered in triumph, the Mosque was consecrated, and thanks were offered for the conquest. Duarte de Menezes was appointed Governor of the place, and the King, who then assumed the title of Lord of Alcaçar, withdrew by sea to Ceuta.

In a short time the King of Fez brought a large force to lay siege to Alcaçar. Affonso had at first intended to march from Ceuta to the assistance of the place, but soon found that it was necessary to raise more men in Portugal, if he was effectually to relieve the besieged. A letter was shot into the town to tell the Governor his plans, and an answer was shot back, saying that Menezes was failing in provisions and stores. This letter, which was written in French, unfortunately fell into the Moorish camp, and the King of Fez, availing himself of the condition of the Portuguese, offered favourable terms if the Governor would surrender. Dom Duarte not only refused, but to show how little he feared the Moors, had the boldness to offer the King his scaling-ladders, if he chose to accept them. After some further attempts, the King of Fez withdrew for the purpose of raising fresh troops, and on the 13th of November returned with thirty thousand cavalry and a vast force of infantry and artillery. The siege had now lasted fifty-three days, when, on the 2nd of January, 1459, the Moors were obliged to retire with immense loss. When Dom Duarte saw that the siege was about to be raised, he sent a message to the King, recommending him to try a little longer before he quite gave it up.

King Affonso now perceiving the advantage which would result from this place having a mole for the mooring of small craft, sent out twenty-six vessels, laden with materials, masons, and labourers. Dom Duarte commenced the construction of the mole on the 12th of March, and it was finished by the end of July, in spite of the continued hindrances offered by the Moors to the progress of the work. Affonso V. had already, on the 15th of September, 1448, transferred entirely to Prince Henry the trade in Guinea in the old acceptation of the word, in which Arguin was included, for he had decreed that no ships should sail beyond Cape Boyador without the Prince's permission, and the transgressor of this prohibition should forfeit his ship to the Prince; but that all ships sailing with his permission should pay him tribute, of the fifth or tenth part of their freight. On the 7th of June, 1454, Affonso granted to the Order of Christ, for the discoveries made and to be made at their expense, entire spiritual jurisdiction in Guzulla (Gozola), Guinea, Nubia, and Ethiopia, with all its accustomed rights, and in the same manner in which it was exercised at the house of their Order at Thomar; and on the 26th of December, 1458, Prince Henry signed a decree "in my town" (Villa do Iffante), stipulating that the Order of Christ should receive tribute of the twentieth, instead of the tithe, of all merchandise from Guinea, whether slaves, gold, or whatsoever it might be, and the remainder should fall to whomsoever held the dominion, as the Prince then held it, by royal prerogative. This record is preserved in the collection of Pedro Alvarez, Pt. III., fol. 17-18.

We have no public act of Prince Henry to record between his return from Alcaçar and his death on Thursday, the 13th of November, 1460, with the exception of the already-mentioned donation, on the 18th September, 1460, of the ecclesiastical revenues of Porto Santo and Madeira to the Order of Christ, and of the temporality to the King and his successors.

We have already seen that he carried into effect the promises which had been made on his behalf to Nomimansa, the King of the Barbaçins, by his faithful navigator, Diogo Gomez. It would seem that that loyal servant was about his master's person at the time of his death, inasmuch as, by the King's command, he remained constantly near the Prince's remains till they were removed from Lagos to their last resting-place in Batalha. It is therefore a satisfaction to be able to give the old sailor's own account of the matter in his own language.

"In the year of our Lord 1460," he says, "Prince Henry fell ill in his town on Cape St. Vincent, and of that sickness he died on Thursday, the 13th of November, of the same year. And the same night on which he died, they carried him to the church of St. Mary in Lagos, where he was buried with all honour. At that time King Affonso was in Evora, and he, together with all his people, mourned greatly over the death of so great a Prince, when they considered all the expeditions which he had set on foot, and all the results which he had obtained from the land of Guinea, as well as how much he had laid out in continuous warlike armaments at sea against the Saracens in the cause of the Christian faith.

"At the close of the year King Affonso ordered me to be sent for, for, by the King's command, I had remained constantly in Lagos near the body of the Prince, giving out whatever was necessary to the priests, who were occupied in constant vigils and in Divine service in the church. And the King ordered that I should look and examine if the body of the Prince was decomposed, for it was his wish to remove his remains to the most beautiful monastery called Santa Maria de Batalha, which his father, King João I., had built for the Order of Friars Preachers. When I approached the body of the deceased, I found it dry and sound, except the tip of the nose, and I found him clothed in a rough shirt of horse-hair. Well doth the Church sing 'Thou shalt not give thine holy one to see corruption.' That my Lord the Infant had remained a virgin till his death, and what and how many good things he had done in his life, it would be a long story for me to relate.

"The King then issued a command that his brother Dom Fernando, Duke of Beja, and the bishops and nobles should go and convey the body to the aforesaid monastery of Batalha, where the King would await its arrival.

THE TOMB OF PRINCE HENRY.
IN THE MONASTERY OF BATALHA.

"And the Prince's body was placed in a large and most beautiful chapel which King João his father had built, and where lie the bodies of the King and his Queen Philippa, the Prince's mother, together with his five brothers, the memory of all of whom is worthy of praise for evermore. There may they rest in holy peace. Amen." On the face of the tomb, on the south side of the Founder's Chapel,* which contains the mortal remains of Prince Henry, and which is in a line with those of his brothers, Dom Pedro, Dom João, and the Constant Prince, are three escutcheons. On the first are sculptured Prince Henry's own arms; on the second the cross, device, and motto of the Order of the Garter, the riband of which had been conferred on him by King Henry VI., in 1442-3,† and on the last the cross of the military order of Christ. Over the tomb is a recumbent statue of the Prince in full armour, with a kind of turban bound round the head. This is protected by a sort of canopy worked in minute sculpture. On the frieze of the tomb, intertwined with ilex boughs, is the Prince's well-known motto, "Talent de bien faire," and

* The following remarks from the pen of our late distinguished ecclesiologist, Dr. Mason Neile, will give some notion of the beauty of the noble specimen of Christian architecture which King João erected at Batalha, and in which he and his family are entombed. He says, "The traveller who is a man of taste will be more than delighted to observe the manner in which this unique temple is being restored, so that in a few years it will have recovered its ancient purity, not to say splendour, and which, for its exquisite workmanship, its unrivalled cloisters, its marvellous Founder's Chapel, its nave, aisles, chapter-house, and Capella Imperfeita, is perhaps the most striking edifice in Christendom. In a few years its exterior, as well as interior, will be little short of perfection; and if Dom Ferdinand were a person endued with as much wealth as he is with taste, there might be some hopes that the present generation would not pass away without seeing finished the truly wonderful Capella Imperfeita, the very parts of which are replete with all that man's ingenuity can imagine, and his skill execute. It were worth all the trouble of a trip to Portugal for any one to come to Batalha to revel in the inexhaustible beauty of this superb monument of the taste of bygone days." It is not unlikely, from the friendly intercourse that existed between Portugal and England, that Dom Manoel conceived the idea of imitating Henry VII.'s chapel in the Capella Imperfeita.

† His Excellency the Count de Lavradio has informed me that he has traced the identical collar of Prince Henry to its present holder as a Knight of the Order, the Earl of Clarendon.

below the frieze, in a single line, the following inscription :
" Aqui jaz o muito alto e muito honrado senhor o Ifante dom
anrique governador da ordem da cavallaria de no.......om
Joham e rainha philipa, que aquy jazem nesta capella cuias
almas deos por sua mercee aja o qual se finou em......na
era de mil e.........." The first of the gaps here
marked has arisen from a fault in the stone. The other
two, which should contain the date, seem to show that the
tomb was prepared during the Prince's lifetime, and that,
after his death, the day, month, and year of his decease were
neglected to be inserted.

The following is Azurara's description of Prince Henry :—
" He was large of frame and brawny, and stout and strong
of limb. His naturally fair complexion had by constant
toil and exposure become dark. The expression of his face
at first sight inspired fear in those who were not accustomed
to him, and when he was angry, which rarely happened, his
look was very formidable. Stout of heart and keen in
intellect, he was extraordinarily ambitious of achieving great
deeds. Neither luxury nor avarice ever found a home with
him. In the former respect he was so temperate that after
his early youth he abstained from wine altogether, while
the whole of his life was reputed to have been passed in
inviolate chastity. As for his generosity, the household of
no other* uncrowned Prince formed so large and excellent
a training school for the young nobility of the country.
All the worthies of the kingdom, and still more foreigners

* Note by the Vicomte de Santarem. "This remark quoted from Azurara,
who was the Prince's cotemporary, shows the error into which Fr. de Luiz de
Souza fell when, in his Historia de S. Domingos, lib. 6, fol. 331, he said that the
Infant was elected King of Cyprus, which mistake was repeated by José Soares
de Sylva in his memoirs of King João I.; and, if the words of Azurara were not
sufficient to prove this, it might be shown by dates and historical facts. In fact,
the kingdom of Cyprus, which Richard I. took from the Greeks in 1191, was
afterwards granted by him to Guy de Lusignan, whose posterity held the crown
till 1487, and as Prince Henry was born in 1394 and died in 1460, he could not
have been elected to a kingdom which was governed by a legitimate royal line.
Moreover, in the list of the kings of Cyprus, the name of Prince Henry does not
occur. It may be supposed that Fr. Luiz de Souza confounded Henry Prince
of Galilea, son of James I., King of Cyprus, with Prince Henry of Portugal."

of renown, found a general welcome in his house, and there were frequently assembled in it men of various nations, the diversity of whose habits presented a curious spectacle. None left that house without some proof of the Prince's generosity. His self-discipline was unsurpassed; all his days were spent in hard work, and it would not readily be believed how often he passed the night without sleep, so that by dint of unflagging industry he conquered what seemed to be impossibilities to other men. His wisdom and thoughtfulness, excellent memory, calm bearing, and courteous language, gave great dignity to his address.

" He was constant in adversity, and humble in prosperity, and it was impossible for any subject of any rank to show more obedience and reverence to the sovereign. This was especially noticeable in his conduct to his nephew Don Affonso, even at the beginning of his reign. He never entertained hatred or ill-will towards any, however serious the offence they might have committed against him. So great was his benignity in this respect, that the wiseacres said that he was deficient in retributive justice, although in other matters he was very impartial. No stronger example of this could be shown than his forgiveness of some of his soldiers who deserted him in the attack on Tangier, when he was in the utmost danger. He was devoted to the public interests of the kingdom, and took great pleasure in trying new plans for the general welfare at his own expense. He gloried in feats of arms against the enemies of the Faith, but earnestly sought peace with all Christians. He was universally beloved, for he did good to all and injured none. He never failed to show due respect to every person, however humble, without lowering his own dignity. A foul or indecent word was never known to issue from his lips.

" He was very obedient to all the commands of Holy Church, and attended all its offices with great devotion, and they were celebrated with as much solemnity and ceremony in his own chapel as they could be in any cathedral church.

He held all sacred things in profound reverence, and took delight in showing honour and kindness to all who ministered in them. Nearly one half of the year he passed in fasting, and the hands of the poor never went empty away from his presence. His heart never knew what fear was, except the fear of committing sin. Assuredly," continues Azurara, " I know not where to look for a Prince that shall bear comparison with this one."

Such was the exalted character of the man whom we honour as the *originator of continuous modern discovery*. In the prefatory chapter to this work, where the Prince's purpose was spoken of, a passing allusion was made to his dignity as the son of a King, and there was an especial object in the mention of that reality. All modern discovery found its origin in one great event—the rise of the powers which bordered on the Atlantic; and this rise, although slow, was identical with the strengthening of the respective monarchies. At the close of the middle ages, the Kings were, in all these countries, the real centres of their nations, whilst in the " Roman Empire " many contending claims existed, but no general government. This difference had long been in favour of the East as far as commerce and navigation were concerned. But now the balance began to turn to the other side. The Hanseatic confederacy, powerful as it might be, was but a confederacy; and Venice, however magnificent, was but a city. The really modern states of Western Europe had the germs of quite another force and power within them.

The first discoveries of the Portuguese were originated by that exuberant regal power which was free to leave the paternal realms, and to extend itself beyond the Mediterranean in wars against the infidels. This movement also received a new intensity by the emigration of the able seamen of Italy, Germany, and the Netherlands to the rising states along the Atlantic. Under the liberal inducements of Prince Henry, men of these three nations held prominent positions in the early naval exploits of the Portuguese. But not Portugal only rose by their talents; the newly united kingdoms of Castile and Aragon, England, and France received with avidity the offers of service of the most gifted men of those nations which had held the sway of the sea.

It is a notable fact, and one that greatly redounds to the honour of Italy, that the three Powers, which at this day possess almost all America, owe their first discoveries to the Italians: Spain to Columbus, a Genoese; England, to the Cabots, Venetians; and France, to Verazzano, a Florentine: a circumstance which sufficiently proves, that in those times no nation was equal to the Italians in point of maritime knowledge and extensive experience in navigation.

It is, however, remarkable, that the Italians, with all their knowledge and experience, have not been able to acquire one inch of ground for themselves in America, a failure which may be ascribed to the penurious mercantile spirit of the Italian republics, to their mutual animosities and petty wars, and to their contracted and selfish policy.

Indeed, it may be said that it was principally to the efforts of Italians and Hanseatics that the dominion of the waters was lost to Italy and the Hanse Towns, and passed to the nations of the West. Nor can this be deplored or ascribed to ingratitude; the new regal powers, such as Portugal, disposed of better means to carry out extensive plans of discovery, to make the first and necessary sacrifices, and to pursue one purpose with that unremitting earnestness which is so seldom found in republics. Nor were they inapt pupils in the practical development of nautical knowledge. Cadamosto, himself a Venetian, and well acquainted with the progress of navigation in the Mediterranean, declares that the caravels of Portugal were the best sailing ships afloat. " Sendo le caravelle di Portugallo i migliori navigli che vadano sopra il mare di vella."

Furthermore, their geographical situation along the Atlantic made them also, beyond comparison, fitter for these endeavours than the old masters of what are merely inland waters compared with the mighty oceanic seas.

Nevertheless for the prosecution of these endeavours the knowledge of the latter was of the utmost value.

During the long period in which Prince Henry was continuing his maritime explorations he did not cease to cultivate the science of cartography. In this he was warmly seconded by his nephew King Affonso V. We have, unfortunately, nothing to show as the result of the cartographical labours of the geographer Mestre Jayme, whom the Prince had procured from Majorca, to superintend his school of navigation and astronomy at Sagres, whither he had also brought together the most able Arab and Jewish mathematicians that he could obtain from Marocco or the Peninsula; but we have already seen what good service had been practically rendered by the Venetian Cadamosto and the Genoese Antonio de Nolli, whose discoveries gave extension to the grant recently conferred on Portugal by a Bull of Pope Nicholas V., dated January 8, 1454, of all Guinea beyond Capes Non and Boyador as far as a certain large river reputed to be the Nile (Senegal) which they had then reached. The discovery that beyond Cape Verde the coasts trended eastward, inspired the King with new energy, for he assumed therefrom that it would soon lead to India. He thought it possible that in that direction the meridian of Tunis, and perhaps even that of Alexandria, had been already passed. He gave names to rivers, gulfs, capes, and harbours in the new discovery, and sent to Venice draughts of maps on which these were laid down, with a commission for the construction of a mappemonde on which they should be pourtrayed.

It was to the Venetian Fra Mauro of the Camaldolese Convent of San Miguel de Murano, that this commission was entrusted. King Affonso V. spared no expense, and Fra Mauro paid the draughtsmen from twelve to fifteen sous a day, while from 1457 to 1459 he himself gave all possible pains ·to perfecting his task. The practised draughtsman Andrea Bianco was called to take a part in its execution. At length this magnificent specimen of mediæval cartography was completed, and by desire of the King despatched to Portugal, in charge of the noble Venetian Stefano Trevigiano, on the 24th April, 1459.   In the same year, on the 20th of October, the drawings and writings, and a copy of the mappemonde, were enclosed in a chest and sent to the abbot of the convent, from which it would seem that Fra Mauro was then dead. It is to be presumed that while elaborating the mappemonde for King Affonso he made at the same time a copy which he intended to leave to the convent.   In the convent library still exists the register of Receipts and Expenditure of the convent, written by the Abbot Maffei Gerard, afterwards in 1466 Patriarch of Venice, and in 1489 Cardinal.   In that register is a note of the current cost of the map.*   (See Count Carli, tom. 9 of his works, page 9, and tom. 13, part 3, page 212, and the extract from M. Villoison's letter to him.)

It is on this map in especial, which preceded by forty years the periplus of the Cape of Good Hope by Vasco da Gama, that we see more clearly laid down the southern extremity of Africa, under the name of " Cavo di Diab." We find delineated a triangular island on which, north-east of Cavo di Diab (our Cape of Good Hope), are inscribed the names of " Soffala " and " Xengibar." This southern extremity is separated from the continent by a narrow strait.   An inscription on Cape Diab states that in 1420 an Indian junk from the East doubled the Cape in search of the islands of men and women (separately inhabited by each), and after a sail of two thousand miles in forty days, during which they saw nothing but sea and sky, they turned back, and in seventy days' sailing reached Cavo di Diab, where the sailors found on the shore an egg as big as a barrel, which they recognised as that of the bird Crocho, doubtless the roc or rukh of Marco Polo, a native bird of Madagascar.

It has been already seen that the Arabs who traded

---

* There is a vellum copy of this planisphere in the British Museum made in 1804 by Mr. William Frazer, but according to Dr. Vincent it is not perfectly accurate.

on the east coast of Africa were prevented, by the force of the current, from venturing southward of the Cape, afterwards named by the Portuguese the Cabo dos Corrientes. It could only, therefore, be by communication with the natives, or from some daring expedition such as that recounted by Fra Mauro, that the form of the southern extremity of Africa could have been learned. The Indian junk, after being carried westward by the Great Lagulhas stream might, after passing forty days in the Atlantic, return by the southern connecting current, which, reinforced by the west wind in more southern latitudes (between 378 and 40°), brings back a portion of the waters of the Atlantic eastward into the Indian Ocean.   (See "Humboldt's Geographie du Nouveau Continent," page 344.)

It is more remarkable that the Camaldolese geographer makes no mention of the sources from which he derived his information.   He does not even mention the names of the most renowned voyagers, not even that of his own country-man, Cadamosto, whose recent discoveries he was made acquainted with by direct communication.   The unfortunate Doge, Francesco Foscarini, states in a letter that " when he considered the success of Cadamosto's voyage, and witnessed the plan and commencement of Mauro's work, he trusted that Prince Henry would therein find new inducements to continue his explorations."   But the sums expended by the Prince on his maritime expeditions were so large, that not only were his own revenues exhausted, as well as the profits derived from commerce with the African coast, but he died heavily involved in debts, which were partly paid by his nephew and adopted son, Dom Fernando, and partly by Don Manuel, the son of Fernando, while Duke of Beja.   The Duke of Braganza, Dom Fernando I., in a declaration or codicil, dated 8th of November, 1449, declares that Prince Henry owed him, in 1448, nineteen thousand three hundred and ninety-four and a-half golden crowns, somewhat under £70,000, for the payment of which he had pledged his lands and goods, and in his will the Duke states that this debt

STATUE OF PRINCE HENRY.

OVER THE SIDE GATE OF THE MONASTERY
AT BELEM.

was further increased by sixteen thousand and eighty-four golden crowns, nearly £60,000 more.

But we have already seen that the Prince did not confine his expenditure or his patronage to the development of geographical knowledge. Having already in 1431 purchased residences for the University of Lisbon, which had previously been obliged to rent its house-room, he, by a deed dated 25th March, 1448, established the chair of Theology in that University, and subsequently confirmed it by a charter dated from the Villa do Infante, at Sagres, the 22nd of September, 1460. He ordered that every Christmas-day twelve silver marks should be given to the lecturer in that science out of the tithes of the island of Madeira. These important services gained for him the honourable designation of *Protector of the studies of Portugal,* in like manner as the maritime expeditions won for him the epithet of *the Navigator.*

His great nephew, the King Dom Manuel, had a statue of him placed over the centre column of the side gate of the church of Belem, in memory of his having been the founder of the little chapel of Restello for the service of sailors in the harbour, which chapel had stood upon the site of that magnificent church. On 24th of July, 1840, in the reign of Dona Maria II., at the instance of His Excellency the Viscount, now Marquis, de Sá da Bandeira, then Secretary of State of the Navy and Colonies, a monument to Prince Henry, prepared in 1839, was finally erected at Sagres, a representation of which, from a drawing most kindly sent to the author by His Excellency, is here given.

The monument consists of one piece of marble, twelve palms and a half high, embedded in the wall over the inner gate of the principal entrance of the fort of Sagres. On the upper part of the monument is sculptured, as in the drawing, in semi-relief, the escutcheon of the Prince, with an armillary sphere on the right, and a ship in full sail on the left. The lower part of the monument contains two panels with an inscription on the one below the sphere in

Latin, and another on the one below the ship in Portuguese. The two inscriptions are as follows :—

AETERN · SACRUM.
HOC · LOCO.
MAGNUS · HENRICUS · JOAN. I. PORTUGAL · REG. FILIUS
UT · TRANSMARINAS · OCCIDENTAL · AFRICÆ · REGIONES
ANTEA · HOMINIBUS · IMPERVIAS · PATEFACERET
INDEQUE · AD · REMOTISSIMAS · ORIENTIS · PLAGAS
AFRICA · CIRCUMNAVIGATA
TANDEM PERVENIRI · POSSET
REGIAM · SUÆ · HABITATIONIS · DOMUM
COSMOGRAPHIÆ · SCHOLAM · CELEBRATISSIMAM
ASTRONOMICAM · SPECULAM · AMPLISSIMAQUE · NAVALIA.
PROPRIIS · SUMPTIBUS · CONSTRUI · FECIT
MAXIMOQUE · REIPUBLICÆ · LITTERARUM · RELIGIONIS
TOTIUSQUE · HUMANI · GENERIS · BONO
AD · EXTREMUM · VITÆ · SPIRITUM
INCREDIBILI · PLANE · VIRTUTE · ET · CONSTANTIA
CONSERVAVIT · FOVIT · ET · AUXIT.
OBIIT · MAXIMUS · PRINCEPS
POSTQUAM · SUIS · NAVIGATIONIBUS · AB · ÆQUINOCTIAL · AD · VIII.
VERSUS · SEPTEMTRIONEM · GRADUM
PERVENIT
QUAMPLURESQUE · ATLANTICI · MARIS · INSULAS · DETEXIT
ET · COLONIS · AB · LUSITANIA · DEDUCTIS
FREQUENTAVIT
XIII . · DIE · NOVEMBR. · AN. · DOM. · M.CDLX.
MARIA · II. PORTUGAL · ET . ALGARB · REGINA.
EJUS · CONSANGUINEA
POST · CCCLXXIX · ANNOS
H. M. P. J.
CURANTE · REI · NAVALIS · ADMINISTRO
VICE · COMITE · DE · SA · DA · BANDEIRA
M.DCCCXXXIX.

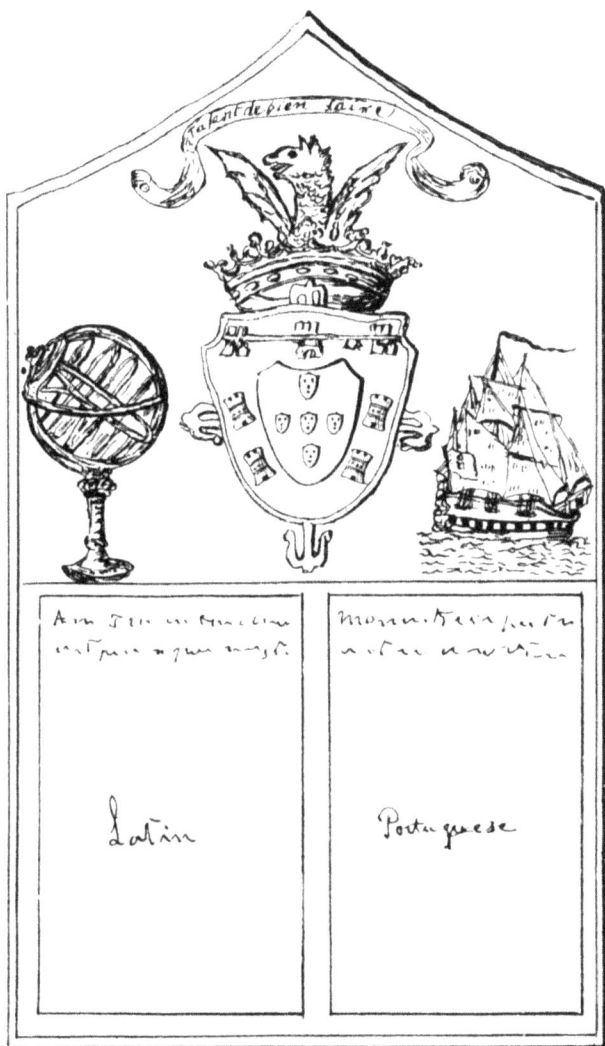

MONUMENT TO PRINCE HENRY AT SAGRES.

Monum · consagrado · á eternidade · o grande ·
Infante · d. henrique · filho · de · el-rei · de · portugal ·
d · joão · I. tendo · emprehendido · descobrir · as · regiões ·
até · então · desconhecidas · de · africa · occidental ·
e · abrir · assim · caminho · para · se · chegar · por · meio ·
da · circumnavegação · africana · até as · partes · mais ·
remotas · do · oriente · fundou · nestes · lugares · á · sua ·
custa · o · palacio · da · sua · habitação · a · famosa ·
escola · de · cosmografia · o · observatorio ·
astronomico · e · as officinas · de · construcção .
naval · conservando · promovendo · e · augmentando ·
tudo · isto · até · ao · termo · da · sua · vida · com ·
admiravel · esforco · e · constancia · e · com ·
grandissima · utilidade · do · reino · das · letras ·
da · relegião · e · de · todo · o · genero · humano · falleceo ·
este · grande · principe · depois · de · ter · chegado ·
com · suas · navegações · ate · o · 8° · gr · de · latitude ·
septemtr · e · de · ter · descoberto · e · povoado · de ·
gente · portuguezza · muitas · ilhas · do · atlantico ·
aos · xiii. · dias · de · novembro · de · 1460 · d · maria · II. ·
rainha · de · portugal · e · dos · algarves · mandou ·
levantar · este · monumento · á · memoria · do ·
illustre · principe · seu · consanguineo · aos · 379 ·
annos · depois · do · seu · fallecimento · sendo ·
ministro · dos · negocios · da · marinha · e ·
ultramar · o · visconde · de · sá · da · bandeira ·
1839.

The following is a translation :—

SACRED FOR EVER.

IN THIS PLACE

the Great Prince Henry, son of John I., King of Portugal, having undertaken to discover the previously unknown regions of West Africa, and also to open a way by the circumnavigation of Africa to the remotest parts of the East, established at his own cost his Royal Palace, the famous School of Cosmography, the Astronomical Observatory, and the Naval Arsenal, preserving, improving, and enlarging the same till the close of his life with admirable energy and perseverance, and to the greatest benefit of the kingdom, of literature, of religion, and of the whole human race. After reaching by his expeditions the eighth degree of north latitude, and discovering and planting Portuguese Colonies in many islands of the Atlantic, this great Prince died on the 13th of November, 1460. Three hundred and seventy-nine years after his death, Maria II., Queen of Portugal and the Algarves, commanded that this monument should be erected to the memory of the illustrious Prince, her kinsman, the Viscount de Sá da Bandeira being Minister of Marine. 1839.

To the kindness of His Excellency the Marquis de Sá da Bandeira, I am also indebted for the accompanying plan of the promontory of Sagres, which was taken at the time by Captain Lourenço Germack Possollo, to whose able management the erection of the monument was entrusted.

On this plan will be seen the site of the present small fort, which was erected in 1793, and the traces of the few ancient walls and ruins that remain. The hard granite rock of which the promontory consists is hollowed out at its base into a natural arch, and there are holes worn through to the surface, through which in time of storms from the south-west, the sea drives the air with terrific force, and expels to a considerable height any objects which may be in the way. On some occasions the sea water is driven through these holes in great quantity, and falls down on the surface of the earth in the form of rain. This salt-water shower, which will sometimes extend to a distance of nearly two miles, goes far to destroy the very few traces of vegetation which are to be found on this desolate and sterile spot.

CHAPTER XVIII.

THE STORMY CAPE.

1460—1487.

THE death of Prince Henry próduced the effect that might have been expected. The progress of discovery received for the time a check when the presiding genius was removed from the scene of action. In the main the tendencies of King Affonso were rather towards conquest in Mauritania, and the support of his pretensions to the throne of Castile, than to the prosecution of discoveries on the west coast of Africa. Nevertheless the "talent de bien faire" had left behind it its impress in its example and its benefits, and we are not without something to record in the way of discovery, between the death of the Prince in 1460, and that of his nephew, King Affonso V., in 1481. Indeed in the year following the death of the Prince, the King was induced, by the great traffic in gold and negroes at the island of Arguin, to build a fort there to insure its safety. Its construction and commandership were committed to Soeiro Mendez, a gentleman of his household, to whom and to his heirs the King, by deed of July 26th, 1464, made a grant of the governorship-in-chief of the fortress.

Cadamosto had reached the Rio Grande, and from his pen we have an account of the exploration of more than six hundred miles yet further south by a gentleman of the King's household, named Pedro de Cintra, whom the King had sent out in command of two armed caravels. The narrative was dictated to Cadamosto by a young Portuguese, who had been his secretary in his own two voyages, and who, after accompanying Pedro de Cintra, returned to his former master, who still resided at Lagos. The date of the voyage is not given, but it was either in 1461 or 1462, since it occurred between the death of Prince Henry, at the close of 1460, and Cadamosto's departure from the Peninsula at the beginning of 1463. De Cintra first went to the two large inhabited islands, discovered by Cadamosto in his second voyage, at the mouth of the Rio Grande, on one of which they landed. In the miserable straw-thatched hovels in the interior they found some wooden figures, which led them to think that the blacks were idolaters, but as they were unable to hold any conversation with them, they returned to the ship and proceeded on their voyage. After sailing forty miles, they reached the mouth of a large river, about three or four miles in breadth, called Beseque, from the name of the chief who lived at its mouth. A hundred and forty miles further on they came to a cape, which they called Cape Verga. The hills were lofty, and eighty miles beyond they came to another cape, which the sailors all agreed was the highest they had ever seen. It was covered with beautiful green trees, and had at its summit a point shaped like a diamond. In honour of Prince Henry, and in remembrance of his residence at Cape Sagres, the Portuguese gave it the name of " Cape Sagres of Guinea." The people worshipped wooden images in the shape of men, to which at mealtimes they offered food. They were tawny rather than black, and had figures branded on their faces and bodies. They had no clothes, but simply wore pieces of the bark of trees in front of them. They had no arms, for they had no iron in their country. They lived on rice, honey, and vegetables, such as beans and kidney-beans, of a finer and larger kind than those of Europe. They had also beef and goats' flesh, but in no great abundance. Near the cape were two little islands, one about six miles distant, the other eight, but too small to be inhabited. They were thickly covered with trees. Those who lived on this river * used very large canoes, each carrying from thirty to forty men, who rowed standing, without row-locks. They had their ears pierced with holes all round, and wore in them a variety of gold rings. Both the men and women had the cartilage of their noses pierced and a ring passed through it, like the buffaloes in Italy; but these they took off when they ate.

About forty miles beyond Cape Sagres they found another river, which they called the San Vicente, about four miles broad at the mouth, and some five miles further they came to another river, called Rio Verde, yet broader at the mouth than the San Vicente. The country and coast were very mountainous, but there was good anchorage everywhere. Four-and-twenty miles from this Cape was another, which they called Cape Ledo, or "Joyous," on account of the beauty and verdure of the country. Further on was a lofty mountain range extending fifty miles, covered with fine trees, at the end of which, at about eight miles out at sea, were three little islands, the largest about ten or twelve miles in circumference. These they called the Selvagens, and the mountain they called Sierra Leona, on account of the roaring of the thunder which is constantly being heard on its cloud-enveloped summit.

Thirty miles beyond Sierra Leona they found a large river, three miles broad at its mouth. They called it Rio Roxo, or Red River, because passing through a red soil, it assumed that colour. Beyond was a cape, also of red colour, which they named Cabo Roxo; and about eight miles out to sea, an uninhabited island, which for the same reason they called

---

* The original text is exceedingly faulty, as for instance where above it is stated that the natives were marked with fire the Italian expression is "com ferro affocata," with heated *iron*, whereas immediately afterwards it is stated that there was no iron in the country. Again, the two islets just mentioned are declared to be "one distant from the other six miles, the other eight," a piece of Hibernicism for which one is unprepared. So here reference is made to "this river" when no river whatever has been named. The river alluded to must be the Pongas, at the mouth of which Cape Sagres is situated.

Ilha Roxa. From this island (which is about ten miles from the Rio Roxo) the north star seemed to be about the height of a man above the sea. Beyond Cabo Roxo they discovered a gulf, into which flowed a river, and this they named Santa Maria das Neves,* " St. Mary of the Snows." They saw it on the 2nd of July, the visitation of the Blessed Virgin. On the other side of the river was a point, and opposite that, a little way out at sea, a small island. The gulf was full of sandbanks, running ten or twelve miles along the coast. The sea broke here with great violence, and there was a very powerful current, both at the ebb and flow of the tide. They called this island Ilha dos Bancos, on account of· these sandbanks.

Twenty-four miles beyond this island is a great cape, called Cabo de Santa Anna, because it was discovered on St. Anne's day, the 26th of July. Sixty miles beyond they found another river, which they called Rio das Palmas, on account of the many palms which grew on its banks; but its mouth, though of considerable breadth, was full of sand-banks, which made it very dangerous. This was the cha-racter of the coast the whole distance between Cabo de Santa Anna and this river. About sixty miles further they discovered another small river, which they called Rio dos Fumos, because when they discovered it they could see nothing on land but smoke. Four-and-twenty miles beyond this river, they discovered a cape jutting out into the sea, which they called Cabo del Monte, because beyond it they saw a very lofty mountain. Coasting thence for sixty miles, they saw another small cape, not very high, but similarly capped by a hill. This they called Cabo Mesurado. Here they observed a great number of fires, lighted by the blacks in consequence of their getting sight of the ships, the like of which they had never seen before. Sixteen miles beyond this cape, there was a wood of fine trees,

---

* There would appear to be some blunder here, as the feast of St. Mary of the Snows is on the 5th of August, which would not accord with the chronology of the voyage.

reaching down to the sea. This they called the Bosque de Santa Maria, or St. Mary's Grove.

The caravels came to anchor beyond this wood, but no sooner had they arrived than some little canoes, with two or three naked men in each, came towards them, some of them having the remains of what seemed to be human teeth hanging on their necks. One of them they captured in order to bring him into communication with other blacks in Portugal, that they might gain information respecting his country, but nothing of importance could be gathered from him. He was subsequently sent back to his own country with clothes and other presents. Cadamosto informs us that no other ship had returned from that coast up to the period of his departure from the Peninsula, on the 1st of February, 1463.

On the 12th of June, 1466, the King granted privileges to the colonists with respect to the Guinea trade, which were abused by them to an extent that caused the King by a new charter to restrict the use of these privileges to the limits of his original grant.

In 1469, King Affonso V. rented the trade of the African coast to Fernam Gomez, for five hundred cruzados a year, for five years, reserving the ivory-trade only to the crown, and stipulating for the discovery of a hundred leagues of coast annually. This stipulated exploration was to commence at Sierra Leona, the point reached by Pedro de Cintra and Soeiro da Costa, who were the latest previous discoverers. The latter, who had already distinguished himself as one of the first explorers from Lagos, subsequently discovered the river which received his name, but which is now known as the Great Bassam or Assinie River. The explorers selected by Fernam Gomez were João de Santarem, and Pedro de Escobar, both knights of the King's household. The pilots were Martin Fernandez and Alvaro Esteves, the latter having at that time the highest repute as a navigator in the whole kingdom.

In January, 1471, they discovered the coast afterwards

named La Mina, where so large a trade in gold-dust was carried on, and in the same year crossed the line and extended their explorations even as far as Cape St. Catherine, thirty-seven leagues beyond Cape Lopo Gonsalves.

Fernam Gomez acquired great wealth by this traffic, so that he was able to render good service to the King in his wars in Marocco. When his contract expired in 1474, the King conferred on him a coat-of-arms argent, three negroes' heads collared or, and with rings in their noses and ears. He also gave him the surname of Mina, in commemoration of his important discovery.

The last of the explorers, during the reign of King Affonso V., was a knight of his household named Sequeira, who discovered Cape St. Catherine, two degrees south of the equator.

On the death of Affonso V., his son and successor, João II., entered with zeal into the views of his predecessors and of his uncle Prince Henry. Before he came to the throne, a part of his revenues had been derived from the African trade, and the fisheries connected therewith, so that he had every inducement to prosecute its extension. With this view he not only ordered the completion of the Fort of Arguin, which had been commenced years before, but resolved on the construction of another, on a larger scale, at S. Jorge da Mina. The gold traffic had at first been carried on at a place called Saama, discovered in 1472, by João de Santarem and Pedro de Escover, in the service of Fernam Gomez, already mentioned; but San Jorge de Mina was now selected for its superior convenience.

That the fort might be constructed the most expeditiously, both for preventing objections and saving his people from exposure to the dangers of the climate, the King took the precaution to have the stones cut and fashioned in Portugal. With these, and bricks, and wood, and other needful materials, he loaded ten caravels and two smaller craft. He sent out also provisions sufficient for six hundred men, one hundred of whom were officers to superintend the work. The command of this fleet was given to Diogo de Azambuja.

It set sail on the 11th December, 1481, and after stopping to conclude a favourable treaty with Bezeguiche, the lord of the harbour and court which bore his name, they reached La Mina on the 19th of January, 1482. On the following morning they suspended the banner of Portugal from the bough of a lofty tree, at the foot of which they erected an altar, and the whole company assisted at the first mass that was celebrated in Guinea, and prayed for the conversion of the natives from idolatry, and the perpetual prosperity of the church which they intended to erect upon the spot.

By good luck they found there a small Portuguese vessel, the captain of which, João Bernardes, was engaged in traffic with the natives, and him they made interpreter between Caramansa, the chief of the place, and Azambuja. The interview took place with the greatest ostentation possible on both sides, a kind of rivalry in which, as may be supposed, the negro prince had a very sorry chance of producing any very imposing effect. Azambuja appeared in a tunic of brocade, with a collar of gold and precious stones, and his captains were all in holiday attire, while Caramansa, who was no less ambitious of making a good display, was habited, like the rest of his people, in the best vestments with which nature had provided them. With their skins anointed and glistening till their native blackness was made blacker still, they considered their toilette perfect, although their only garment was an apron of monkeys' skin or palm leaves. To this extreme simplicity, however, Caramansa himself was in so far an exception that his arms and legs were adorned with bracelets and rings of gold, and round his neck a collar from which hung small bells, and some sprigs of gold were twisted into his beard, so that the curls were straightened by the weight.

Azambuja then addressed the chieftain in the name of King João, commending to him the Christian religion, which if he would recognise and be baptized, the King would regard him as a brother, and make with him an alliance, offensive and defensive, against their common enemies, and enter into a treaty for the interchange of the products of their respective countries. With this view he proposed, with the chieftain's permission, to found a permanent establishment in his country which should serve as a place of security against their enemies, as a refuge to the Portuguese who visited the coast, and also as a storehouse for their merchandise. Caramansa, who was very shrewd for a negro, after some hesitation, gave his consent. On the following day Azambuja put the work in hand, but no sooner was it commenced than the negroes showed signs of an intention to interrupt it. Fortunately mischief was prevented by Azambuja's learning that this arose from displeasure that the requisite presents had not as yet been offered to the chieftain. The oversight was soon remedied, and the work was set about with so much activity that in twenty days the fort was in a condition to repel an attack. Azambuja also built a church on the site, where on his arrival he had erected an altar. Both the church and the fort were dedicated to St. George. In the former, a daily mass was established in perpetuity for the soul of Prince Henry, and to the latter the King conceded the privileges of a municipality. Azambuja took up his abode there, with a garrison of sixty men, and sent back the rest to Portugal with gold and slaves and other articles of merchandise. By a charter of King John II., dated 17th March, 1485, Diogo Azambuja received in recognition of his great services in the wars, and especially in the construction of the fortress of San Jorge da Mina, the permission to add a castle to his arms in commemoration of the fact.

Hitherto the Portuguese in making their explorations had contented themselves by setting up crosses by way of taking formal possession of any country; but these crosses soon disappeared, and the object in setting them up was frustrated. They would also carve on trees the motto of Prince Henry,

" Talent de bien faire," together with the name which they gave to the newly-discovered land. In the reign of King João, however, they began to erect stone pillars surmounted by a cross. These pillars, which were designed by the King, were fourteen or fifteen hands high, with the royal arms sculptured in front, and on the sides were inscribed the names of the King and of the discoverer, as well as the date of the discovery, in Latin and Portuguese. These pillars were called Padrãos.

In 1484, Diogo Cam, a knight of the King's household, carried out with him one of these stone pillars, and passing Cape St. Catherine, the last point discovered in the reign of King Affonso, reached the mouth of a large river, on the south side of which he set up the pillar, and accordingly called the river the Rio do Padrão. The natives called it Zaire. It was afterwards named the Congo, from the country through which it flowed. Diogo Cam ascended the river to a little distance, and fell in with a great number of natives, who were very peacefully inclined, but although he had interpreters of several of the African languages, none of them could make themselves understood. He accordingly determined to take some of the natives back with him to Portugal, that they might learn the Portuguese language and act as interpreters for the future. This was easily managed, and without any violence, by sending Portuguese hostages to the King of Congo, with a promise that in fifteen months the negroes should be restored to their country. He took with him four of the natives, and on the voyage they learned enough Portuguese to enable them to give a fair account of their own country and of those which lay to the south of it. The King João was greatly gratified, and treated the negroes with much kindness and even munificence, and when Diogo Cam took them back the following year, the King charged them with many presents for their own sovereign, accompanied by the earnest desire that he and his people would embrace the Christian religion. Up to the year 1485, John II. used the title of King of Portugal and the Algarves on this side the sea and beyond the sea in Africa,* but in this year he added thereto that of Lord of Guinea.†

In this remarkable voyage Diogo Cam was accompanied by Martin Behaim, the inventor of the application of the astrolabe to navigation, and to whom has been erroneously attributed the first idea of the discovery of America.

A singular train of collateral events places Behaim in curious juxtaposition and comparison with the great Columbus, whose glory he never wished to disparage, although others have attempted to do so for him.‡ Born in the same year, the two men died in the same month. Behaim, though a native of Nuremberg, took up his residence with his wife in a remote island in the Atlantic, Fayal, of which his father-in-law, Jobst de Huerter, was the Captain Donatary: Columbus, a native of Genoa, married the daughter of that Perestrello to whom, as we have already seen, Prince Henry gave the commandership of Porto Santo. Like Behaim, he lived with his wife on her family property in that singularly analogous position, so calculated to develop the ardent desire of each for geographical discovery. Both these illustrious men were at Lisbon at the same time, and both engaged in nautical projects. The same physicians of King João II., Mestre Rodrigo and Mestre Josef, who were entrusted by Diogo Ortiz, Bishop of Ceuta, to examine the project of Columbus for sailing to Cipango by the west, worked with Martin Behaim in the construction of an astrolabe adapted to the purposes of navigation. Another link between Columbus and Behaim was the tutor of the latter, the celebrated Regiomontanus (Johann Müller, a native of Koenigsberg in

* This arose from the name of Algarb being given by the Moors to the Prince of Fez, while the southernmost province of Portugal bore the same name.

† Ruy de Pina. Chron. cap. 19 of the Ined. de Hist. Port., published by the Royal Acad. of Sciences, tom. 2, page 65. Joao P. Ribeiro. Dissert. Chronol. e Critica, tom. 2, page 207, and Garcia de Resende. Chron de D. João 2, cap. 56.

‡ Chief among these is M. Murr, in a memoir originally written in German and translated into French with the title of "Notice sur le Chevalier Martin Behaim avec la description de son Globe Terrestre, traduite par H. J. Jansen." It is inserted at the end of Amoretti's edition of Pigufetta's Voyage Round the World published Paris, An 9.

Franconia). In 1463 he dedicated to Toscanelli (whose letter to Columbus is so famous in the history of the discovery of America) his treatise on the Quadrature of the Circle, in which he refuted the pretended solution of that problem by the Cardinal Nicolas de Cusa. Dissatisfied with the astronomical tables of Affonso the Wise, known as the Alphonsine Tables, but which he maliciously called the Alphonsine Dream, Regiomontanus published at Nuremberg his famous astronomical Ephemerides, calculated prospectively for the years 1475 to 1506, and which were used on the coasts of Africa, America, and India in the first great voyages of discovery of Bartholomew Dias, Gama, and Columbus. (See Humboldt, Examen Critique, tom. i. p. 274.)

But the most prominent material that has been employed for detraction from the fame of Columbus in favour of Martin Behaim, was the famous globe made by the latter in 1492, and still existing in the possession of his descendants in their ancient mansion in Nuremberg. All sorts of claims have been set up by the Nurembergers on behalf of their distinguished countryman on the asserted evidence of this globe. Hartmann Schedel, in the famous Nuremberg Chronicle, published in 1493, had happened to speak of Behaim and Cam having crossed the equator and reached the other hemisphere, and this suggestive declaration seems to have supplied the Nurembergers with the idea that long before Columbus or Magellan sailed in those seas, Behaim had discovered not only America, but the straits of Magellan. The best refutation of these assertions is Behaim's globe itself, copies of which are given in the elegant Life of Behaim, by Dr. F. W. Ghillany, published in Nuremberg, 1853, and in the magnificent atlas prepared by the Vicomte de Santarem, and elsewhere. Even letters by Behaim himself, found in the archives of Nuremberg, have been referred to in ratification of the same claims ; but their futility is proved by the date of the letters themselves (1486) plainly pointing to the voyage with Diogo Cam, the limits of which are clearly defined, and from which Behaim returned in April or May of that year.

There is on Behaim's globe a legend of much importance to this part of our narrative. Below the Ilhas do Principe and S. Thomé is the following statement:—" These islands were discovered by the ships of the King of Portugal in 1484. We found them all deserts, nothing but woods and birds. The King of Portugal sends to them every year those who are condemned to death, both men and women, to cultivate the land and sustain themselves with its produce, so that they may be inhabited by Portuguese. It is spring there when it is winter in Europe, and the birds and beasts are all different from ours. There is a great abundance of amber there, called in Portugal *algalia*," by which I presume he means civet. Now Barros and others make these islands to have been discovered in the time of Affonso [before 1481]. Galvão says 1471 or 1472, but I have not found Galvão generally trustworthy for dates. De Barros' expression is :—" There were also discovered, by command of King Affonso, the islands of S. Thomé, Annobon, and Principe, and others, of which we do not now speak particularly, because we do not know when or by what captains they were discovered; but we do know, by common report, that more was discovered in that King's reign than we have been able to write down." It is of course impossible therefore to say, under such circumstances, whether in Behaim's voyage in 1484 these islands were for the first or second time discovered. It is, however, generally believed, and with high probability, that João de Santarem and Pero de Escobar, both knights of the King's household, went out in 1470, on account of Fernam Gomes, to explore the coast beyond Cape Palmas, and took with them, for their pilots, Martin Fernandez of Lisbon, and Alvaro Esteves of Lagos, and that, in spite of the calms, south winds, and northward currents common in that gulf, they managed to run along the whole of the coast of the kingdom of Benin, and on the 21st of December, St. Thomas's Day, sighted a lofty island covered with wood, to which they gave the name of that apostle. On the 1st of January, 1471, they are supposed to have come upon a smaller island, to which they gave the name of Anno Bom, or Good Year, in memory of the happy omen that it was discovered on New Year's Day. And in truth a good year it was, for in that same month of January they made the first traffic in gold on the Gold Coast, in the village of Sama, between Cape Three Points and La Mina, whither they were carried by the currents and breezes from the south, after having sighted the terra firma of Cape Lopo Gonsalves. In this same voyage they discovered the Ilha do Principe, but it is not known on what day. It was, probably, in the passage from Cape Lopo Gonsalves to the Gold Coast, in 1471; and as they originally gave the island the name of Santo Antão, or Saint Anthony, we may infer that it was discovered on the 17th of January, which is the day of that saint's commemoration. It afterwards received the name of Ilha do Principe, because the King's eldest son had assigned to him, as his appanage, the duty on the sugars grown in the island. Whether the Ilha Formosa, or Beautiful Island, discovered by Fernam do Pó, a gentleman of the King's household, whose name it afterwards received, was discovered in this voyage, or as some have supposed in 1486, when, as we shall presently see, João Affonso de Aveiro was sent by King João II. on an especial mission to the King of Benin, and in which voyage the first African pepper was brought to Portugal, we possess no evidence to show. However all this may have been, it would seem that the islands were now, in Cam's voyage, for the first time brought under the notice of the Government and turned to any account. But there is another point in connection with these islands, which demands consideration. It will have been noticed that in previous voyages, when islands at a distance from the mainland had been discovered, it had been through the vessels being driven on them by storms; as, for example, the discovery of Porto Santo by Zarco, and of the Cape Verde Islands by Antonio de Nolle and Diogo Gomez, but in the present case we have islands, one, S. Thomé, more than fifty, the other, Annobon, more than eighty leagues distant from the mainland, discovered without the interference of any storm whatever, of which we are informed. The reasonable inference seems to be that the navigators used their newly-improved nautical instruments to good purpose, and were able to leave the coast with impunity, which their predecessors were not in the position to do, for want of being able to take the altitude. From Behaim's globe we derive the following statement :—

" In the year 1484, King John of Portugal fitted out two caravels, well provided with men, provisions, and munitions of war for three years, and he ordered that after passing the Straits, they should proceed southward and eastward as far as they possibly could. The vessels were laden with all sorts of merchandise for barter. There were also taken out eighteen horses with their harness for presents to the several kings, one for each, as we might find it expedient. We also took all sorts of spices to show the natives what we went in search of. We sailed from Lisbon straight to Madeira, where the Portuguese sugar grows. Passing the Canaries, we found some Moorish chiefs, with whom we interchanged presents, and afterwards came to the kingdom of Gambia, where the malaguette grows, eight hundred leagues distant from Portugal. Thence we passed twelve hundred leagues to the dominions of the King of Furfur, where grows the pepper called Portuguese pepper.* Far beyond that country we found the *casca de canella* [or cinnamon], where, having then sailed a distance of two thousand three hundred leagues, we turned back and reached Lisbon in the nineteenth month from our departure."

This statement is confirmatory of the dates derived from Barros. As Diogo de Azambuja reached La Mina on the 19th of January, 1482, and remained there two years and seven months, he would be back in Lisbon at the end of August or beginning of September, 1484, and as Diogo Cam did not start till his return, if he left in October or

---

* The designation of the " Grain Coast " is derived from the " Grains of Paradise," " Guinea Grains " or malaguette pepper, which is there produced.

November the addition of the nineteen months above stated by Behaim would make their return to Lisbon fall in May or June of 1486, as stated by Barros. It must, however, be confessed that the cinnamon mentioned by Behaim is not to be found on the west coast of Africa, and must have been confounded by him with some other aromatic tree that grew on that coast.

Diogo Cam did not forget his promise to return with his charges to Congo within the fifteen months. When he reached the Congo River he was received with great welcome by the natives, and by their King. He then proceeded further south, and planted two pillars surmounted by crosses, one named St. Augustine, in 15° 50' south, and the other at a point which they called the Manga das Areas, or Sleeve of Sands, in 22°, now called Cape Cross by the English. The cross is still in good preservation, only part of one of the arms being gone. (See a letter by William Messem, in *Nautical Magazine* for 1855, p. 211.) It is here that the country of the Cimbebas terminates, and that of the Hottentots begins. Cam thus traversed more than two hundred leagues beyond the Congo, landing occasionally, and taking some of the natives for the sake of the language.

On his return he was received by the King of Congo with marked affection, and had the happiness of inspiring him with a great desire to receive instruction in the tenets of the Christian religion. For this purpose, he not only requested that priests might be sent out from Portugal, but he himself despatched one of his own subjects, named Caçuta, with some youths to urge this request. On their arrival, the King and Queen stood sponsors for Caçuta, who received the King's name of João for his Christian name, with the surname of Silva, from his other godfather Ayres da Silva, the King's chamberlain. The whole of the little embassy were baptized before their return to Africa, in the year 1490, and thus originated the diffusion of Christianity in those benighted countries.

The expedition which took them back consisted of three

ships under the command of Gonzalo de Sousa, but this commander died at Cape Verde, and was succeeded by his nephew Ruy de Sousa. On their arrival in Congo, they were warmly received by an aged uncle of the King named Mani Sono, who very shortly received baptism, and was named Manuel. His son was also baptized, and took the name of Antonio. This was the first baptism that was administered in those heathen countries. It took place on Easter-day, the 3rd of April, 1491. Twenty-five thousand men were present at the ceremony. The King was fifty leagues away at the time, but when he heard of it he testified his approval by bestowing on his uncle a large increase of territory, and he ordered the idols to be destroyed throughout his dominions. Indeed, so zealous was he for the maintenance of reverence for everything sacred, that on one occasion when some of his people made a disturbance at the door of the church which the Portuguese had constructed of boughs, he would have had them put to death but for the intercession of the priests. The King's residence was at Ambasse Congo, about twenty leagues from the sea-coast, where he received Ruy de Sousa. When at two leagues from the city he was met by a chieftain, accompanied by a great host of men formed in procession, who to the noise of trumpets and kettle-drums, barbarously constructed, sang the praises of the King of Portugal, three or four singing a verse, and the whole body joining in the chorus. The King sat on a throne of ivory, raised on a lofty wooden platform, so that he could be seen from all sides. From his waist upwards, his black and glittering skin was uncovered. Below that he wore a piece of damask which had been given him by Diogo Cam. On his left arm was a bracelet of copper, and from the shoulder hung a dressed horse's tail, which was a symbol of royalty. He had a cap on his head resembling a mitre made of palm leaves so skilfully that it had the appearance of stamped velvet. Ruy de Sousa made his obeisance to him in the Portuguese fashion, which the King returned in his; that is, he put his right hand on the ground as if to take up dust; he then passed his hand first over Sousa's breast, and then over his own, which was the greatest courtesy he could show him. He not only gave permission to build a church, but ordered one of his chieftains to provide materials and labourers, so that no time might be lost. The first stone was laid on the 3rd of May, and the work proceeded so rapidly that the church was completed on the 1st of June. It was dedicated to the Holy Cross, and afterwards became the Cathedral Church of a bishopric. The King himself received baptism in presence of a hundred thousand men, who were brought together both by curiosity and the preparations for a war with some rebels, who had done great mischief in his territory. He took the name of João, and the Queen that of Leonora, from the Portuguese sovereigns. After 'the ceremony he proceeded to the battle, and with more than eighty thousand men in the field, won an easy victory over the rebels. When they returned the King's eldest son was baptized, and took the name of Affonso.

The King's second son, however, named Panso Aquitimo, not only rejected the Christian religion, but excited others to do the same. One great ground of dissatisfaction was that the Church forbad them to have more than one wife, and at this the King himself took offence, and relaxed from his original fervour, even so far as to leave the crown to his second son to the prejudice of the eldest. At the death of the old King, however, Affonzo recovered his right by force, and, firm to the religion of his adoption, zealously developed the Christian faith throughout his dominions, and sent his children and grand-children to Portugal to be educated, and two of these young princes afterwards received consecration as Bishops.

In the course of a century from this time, the Portuguese having become well established in Congo, we find one of their countrymen, Duarte Lopes, going on a mission from the King of Congo to Pope Sixtus V. and Philip II., King of Spain and Portugal, for the purpose of representing the deplorable condition of Christianity in the country at that time, and begging for missionaries. Lopes then related to Felipe Pigafetta, the account of his observations while in Africa during the years 1578 to 1587 ; and this narrative, under the title of " Relatione del Reame di Congo," was published by Pigafetta at Rome, in 1591, 4°. This rare work is accompanied by two maps, of one of which a reduction is annexed, and from which, as well as from the account which I shall proceed to extract from the text, it will be seen that the two great equatorial lakes, Victoria Nyanza and Albert Nyanza, with their probable southern feeder, lake Tanganyika, the positive existence of which has only been made known to us in recent years by our noble explorers, Burton and Speke, and Sir Samuel Baker, were actually laid down and described from information gathered in Africa by a Portuguese three hundred years ago. But though so laid down and described, these three great facts, of such vital importance to the question of the discovery of the sources of the Nile, slept and remained unrepeated by geographers during all those centuries, until our brave adventurers unfolded the truth from absolute personal observation.

The single fact of the map exhibiting, as none of its predecessors or successors had done, these three important lakes so recently discovered, would be sufficient to justify us in hoping for enlightenment on points which have not yet been established by satisfactory modern observation. But, vague and strange as its delineation will appear to eyes accustomed to neater and more systematic cartography, it contains several other items of information which I can point out as wanting in subsequent maps, until they had become matters of fact substantiated by recent explorations.

To begin at the north, it is not improbable that in the Lago Chinanda we have Clapperton's Lake Chad, although considerably north of the true position, and the Lago de Nubia may well be the Liba Lake; but of these I speak with much hesitation. I can with far greater confidence call attention to the fact that on this map for the first time

MAP
(DATE OF 1578-1587)
Showing the
EQUATORIAL LAKES
VICTORIA & ALBERT NYANZA,
the latter fed by
LAKE TANGANYIKA,
AS DELINEATED IN
DUARTE LOPEZ'
"CONGO"
Edited by F. Pigafetta.
ROME 1591.

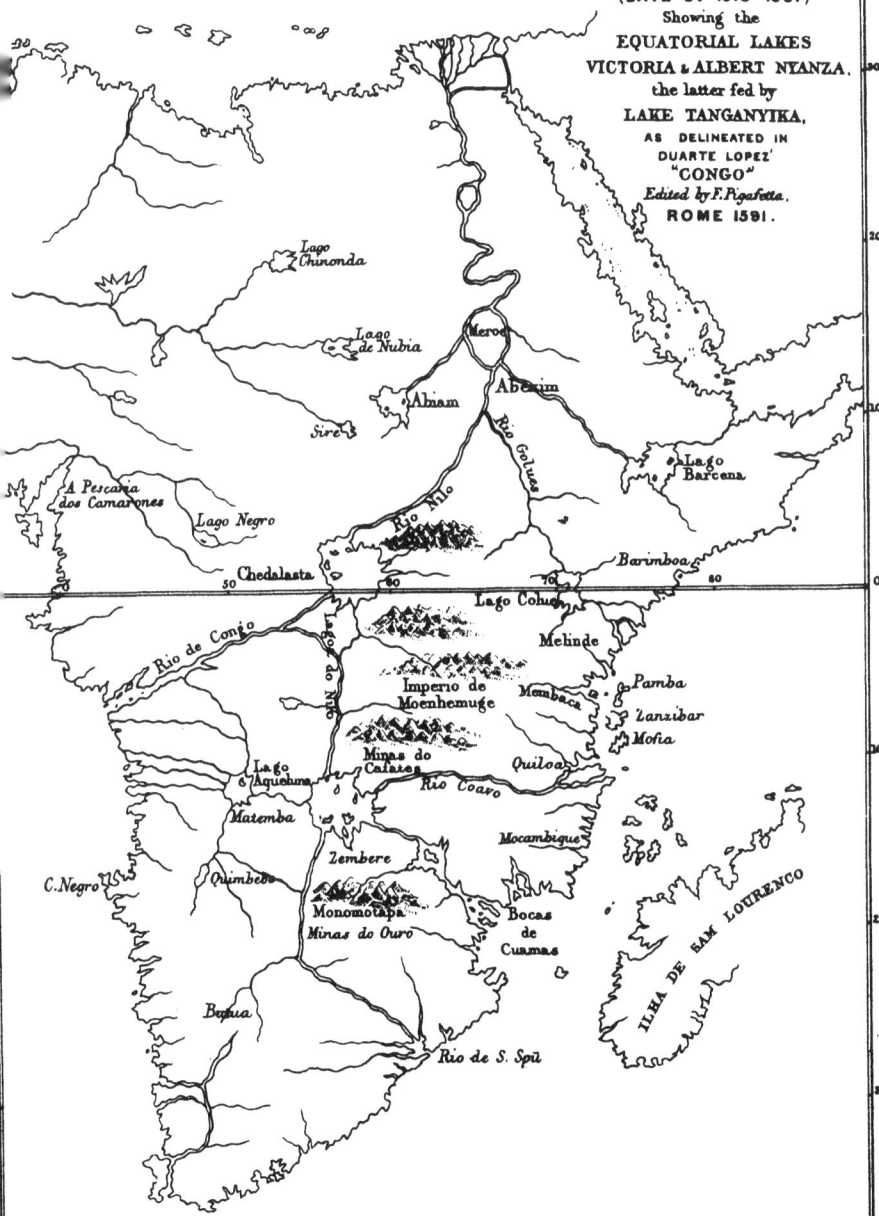

Lago Chinonda

Lago de Nubia

Meroe

Abnam

Abexim

Sire

Rio Goias

Lago Barcena

A Pescaria dos Camarones

Lago Negro

Rio Nlo

Chedalasta

Barimboa

Lago Colue

Melinde

Rio de Conjo

Imperio de Moenhemuge

Membaca

Pamba

Zanzibar

Mofia

Minas do Calaie

Quiloa

Rio Coavo

Lago Aquelma

Matemba

Mocambique

C. Negro

Quimbebe

Zembere

Monomotapa

Minas do Ouro

Bocas de Cuamas

ILHA DE SAM LOURENZO

Butua

Rio de S. Spu

is laid down the great empire of Monomoezi, or Uniamuezi, occupying in a remarkably striking manner a position between the easternmost of the two equatorial lakes and another vast lake to the south-west, exactly corresponding with the true position of that country between the Victoria Nyanza and Lake Tanganyika. In the north-east is the Lago Barcena corresponding with Lake Dembea, with an affluent of the White Nile issuing from it, —a fact by no means unworthy of notice, even though the indistinctness of the delineation leaves us in doubt whether the Atbara or Bar-el-Azreh may be intended : moreover, the name of Barsena still survives in another affluent of the White Nile. Nor is it without significance that north-westward of the Lake Colue, which answers to the Victoria Nyanza, there occurs the word Barimboa, closely expressing Baringo, the name of the water north-west of that great lake.

If we travel further south, we find near to each other the names of Matemba and Quimbebe, suggestive of an indistinct piece of information respecting Kabebe, the court of the great Sovereign of Matiamvo, to whom the King of Casembe was a tributary. Yet further south, on the Tropic of Capricorn, we find the word Butua representing on its proper position the country of the Bechuanas. We have here a sufficient amount of approximately correct information as established by recent exploration, to justify us in inquiring what further the author of the map can tell us with reference to the important subject of the tide. Unfortunately we get not the slightest recognition of *two* great lakes south of these on the Equator. One only is spoken of, but I propose to show that the two great lakes of Tanganyika and Livingstone's Nyassa have been confused into one, doubtless through the information being procured from various sources. The following is the statement in the work which the map was made to illustrate :—

" The Nile does not rise in the country of Bel Gian, *i.e.* Prester John (the Emperor of Abyssinia), nor in the Mountains of the Moon, nor, as Ptolemy writes, from two lakes lying east and west of each other, with about four hundred and fifty miles between them. For in the latitude in which he places these two lakes lies the kingdom of Congo and Angola on the west; and on the east are the empire of Monomotapa and the kingdom of Sofala, the distance from sea to sea being twelve hundred miles. In this region Lopes stated that there was only one lake, on the confines of Angola and Mono-motapa. It is one hundred and ninety-five miles in diameter, as he learned from the people of Angola on the west, and those of Sofala and Monomotapa on the east; and while they give us a full account of this, they mention no other lakes, whence we may con-clude that there is no other in that latitude. It is true that there are two lakes, not lying east and west, but north and south of each other, and about four hundred miles apart. Some of the natives think that the Nile, issuing from the first lake, flows underground and again appears; but Lopes denies this. The first lake is in 12° S. lat., and like a shell, and surrounded by very lofty moun-tains, the highest of which on the east are called Cafates, and on both sides are mountains from which saltpetre and silver are dug. The Nile flows thence four hundred miles due north, and enters another very great lake, which the natives call a sea. It is larger than the first, for it is two hundred and twenty miles across, and lies under the equinoctial line. Respecting this lake very certain information is given by the Anzichi, near Congo. They say that there are people on it who sail in great ships, and who write and have weights and measures, such as they have not in Congo. Their houses were built of stone and lime, and equalled those of the Portuguese, whence it might be inferred that Prester John was not far off. From this second lake the Nile flows seven hundred miles to the island of Meroe, and receives other rivers, the principal of which is the River Colues, so named because it issues from a lake of that name on the borders of Melinde, and when the Nile reaches Meroe it divides into two branches, and embraces a high land named Meroe, to the right of which, on the east, is a river named Abagni that rises in the Lake Bracina and crosses the empire of Prester John till it reaches that island."

Now if there be any value in this statement at all, coin-ciding as it does with considerable accuracy with what we

now know of the relative positions of the two equatorial
lakes and Tanganyika, it is impossible to avoid identifying
the latter lake with that here described as the headwater of
the Nile, which confirms the suggestion recently put forth
by our distinguished geographer, Mr. Findlay, that the
waters of the Lake Tanganyika fall into the Albert Nyanza.
(See Transactions of Royal Geographical Society Meeting
of June 3rd, 1867.) At the same time, the latitude of
12° S., and the placing the lake described on the confines of
Angola and Monomotapa, plainly indicate the Lake Nyassa
of Livingstone; but, clearly, it is quite possible for a
certain amount of accurate information to have been derived
from the natives with respect to both these lakes, though
from want of completeness in the information, confusion
may easily have arisen.

While, however, the teaching of the doctrines of Christianity was thus successful in Congo, it was far otherwise in
the kingdom of Benin, which lay between Congo and the Fort
of St. Jorge da Mina. At about the same time that Diogo
Cam was returning for the first time under such propitious
circumstances from Congo, one João Affonso de Aveiro was
commissioned by the King of Benin to convey an ambassador
to the King of Portugal, with a request that he would send
missionaries to teach his people the Christian religion. His
real object, however, was much more to strengthen his
hands against his enemies than to secure the blessings of
Christianity. The mission accordingly languished, and the
unwholesomeness of the locality caused many deaths,
amongst the earliest to succumb being Aveiro himself. The
negro ambassador, however, had informed King João that
eastward of Benin, some three hundred and fifty leagues in
the interior, lived a powerful monarch named Ogane, who
held both temporal and spiritual dominion over all the
neighbouring kings, and that the King of Benin on his own
elevation to the throne sent him an embassy with rich
presents, and received from him the investiture and insignia
of sovereignty. These latter consisted of a staff and cap of

shining brass by way of sceptre and crown, with a cross of brass. Without this ceremony the kings were not held to be legitimized. The ambassadors never saw this monarch during the whole term of their stay at his court. Only on the day of audience he showed one of his feet, which they kissed with reverence as something holy. At their departure a cross of brass was thrown over the neck of each in the name of the King, and this liberated the wearer from all slavery, and was to him as an ennobling order of chivalry.

The story tallied so remarkably with the accounts of Prester John which had been brought to the peninsula by Abyssinian priests, that the King was seized with an ardent desire to get enlightened upon this subject, for he plainly saw how immensely his double object of spreading Christianity and extending his commerce by opening the road to the Indies would be furthered by an alliance with such a sovereign. It was, as has been shown in a previous chapter, the idea of the geographers of the time that the sources of the Senegal and the Nile were very near to each other. The King therefore gave orders that as soon as the fortress at the mouth of the Senegal was completed the ascent of the river should be made as far as its source; but he little foresaw the difficulties of such an undertaking. He nevertheless determined that both by sea and land the attempt should be made to reach the country of Prester John.

By sea he sent, in August, 1486, two vessels of fifty tons respectively, under the command of Bartholomeu Dias and João Infante. A smaller craft which carried the provisions was commanded by Pedro Dias, Bartholomeu's brother. Of this voyage, however, we shall speak more fully after that we have described the measures which the King adopted with the view of finding, if possible, the country of Prester John by land. The first persons whom he sent out with this object were Father Antonio de Lisboa and one Pedro de Montarryo; but when they reached Jerusalem they found that without knowing Arabic it would be useless to continue their voyage, and therefore they returned.

On the 7th of May, 1487, however, the King despatched two men who were not wanting in that respect, viz., Pedro de Covilham and Affonso de Payva. They went by Naples and Rhodes to Alexandria and Cairo, and so to Aden, where they separated with an agreement to meet at a certain time at Cairo. They left Lisbon for Naples, where, says Alvarez, their bills of exchange were paid by the son of Cosmo de Medicis; and from Naples they sailed to the island of Rhodes. Then crossing over to Alexandria, they travelled to Cairo as merchants, and proceeding with the caravan to Tor on the Red Sea, at the foot of Mount Sinai, gained some information relative to the trade with Calicut. Thence they sailed to Aden, where they parted; Covilham directed his course towards India, and Payva towards Suakem in Abyssinia, appointing Cairo as the future place of their rendezvous.

At Aden, Covilham embarked in a Moorish ship for Cananor, on the Malabar coast, and after some stay in that city, went to Calicut and Goa, being the first of his countrymen who had sailed on the Indian Ocean. He then passed over to Sofala, on the eastern coast of Africa, and examined its gold mines, where he procured some intelligence of the island of St. Lawrence, called by the Moors the Island of the Moon, now known as Madagascar.

Covilham had now, according to Alvarez, heard of cloves and cinnamon, and seen pepper and ginger; he therefore resolved to venture no further until the valuable information he possessed was conveyed to Portugal. With this idea he returned to Egypt; but found on his arrival at Cairo, where he met with messengers from King João, that Payva had died a short time before. The names of these messengers were Rabbi Abraham of Beja, and Joseph of Lamego; the latter immediately returned with letters from Covilham, containing, among other curious facts, the following remarkable report:—" *That the ships which sailed down the coast of Guinea might be sure of reaching the termination of the continent, by persisting in a course to the south; and that when they should arrive in the eastern ocean, their best direction must be to inquire for Sofala, and the Island of the Moon"* (*Madagascar*). Rabbi Abraham and his companion, having already visited the city of Baghdad and the island of Ormuz, had made themselves acquainted with many particulars respecting the spice trade. This alone was sufficient to recommend them to the patronage of João II., and they accordingly were employed by him to seek Covilham and Payva at Cairo, with additional directions to go to Ormuz and the coast of Persia, in order to improve their commercial information.

Covilham eagerly embraced this opportunity to visit Ormuz, and having attended Abraham to the Gulf of Persia, they returned together to Aden, whence the latter hastened to give King João an account of their tour, and Covilham embarked for Abyssinia to complete that part of his voyage which the death of Payva had hitherto frustrated.

Crossing the Straits of Babelmandeb, he landed in the dominions of the Negus. That prince took him with him to Shoa, the residence of the court, where he met with a very favourable reception. He at length became so necessary to the prince, that he was compelled to spend the remainder of his life in Abyssinia. He married in that country, and from occupying highly important posts, amassed a considerable fortune. It is stated by Alvarez, that when, in 1525, the Portuguese embassy under Don Rodriguez de Lima arrived in Abyssinia, Covilham shed tears of joy at the sight of his fellow-countrymen. He passed thirty-three years of his life in Abyssinia, and died there. From his letter to King João, already quoted, it will be seen that to him is to be assigned the honour of the theoretical discovery of the Cape of Good Hope, as that of the practical discovery will presently be shown to belong to Dias and Da Gama.

Meanwhile, in the year 1488, the King had fitted out a considerable armament with the view of founding another station at the mouth of the Senegal, similar to that of San

Jorge da Mina, but this project met with very different success. It so happened that the Prince of the Jaloffs, a man whose vicious habit of life made the cares of ruling irksome, had to a certain extent abandoned the government to his uterine brother, named Bemoi, and in so doing had slighted the claims of his two brothers, the sons of the late king. Bemoi, who was a man of talent and energy, strengthened himself against the princes, his rivals, by forming a close alliance with the Portuguese, to whom he never failed to show every possible attention and kindness. All went on well till the death of the King, who was assassinated at the instigation of his brothers. Bemoi now found himself engaged in open warfare, and naturally appealed for help to his allies. King João promised him every help if only he would become a Christian and be baptized, and for this purpose sent out ambassadors with presents and accompanied by missionaries. Bemoi promised to do what was required of him, but objected that it was highly inexpedient, during a civil war, to make a change which would naturally alienate even many of his own partizans, but he engaged that, if he should obtain quiet possession of the kingdom, he would not only embrace Christianity, but would make the whole nation do the same. A year thus passed, during which the commerce was seriously interrupted by the war, and the Portuguese merchants complained to King João, who, finding that Bemoi did not embrace Christianity, ordered all his subjects under heavy penalties to leave him and return to Portugal. Bemoi became alarmed, and sent a nephew of his in company with the Portuguese, with a collar of gold and a hundred picked slaves as a present to the King, in the hope of securing his assistance. There was not time, however, for him to receive the answer, for he was beaten and with difficulty escaped to the fortress of Arguin, whence he embarked for Portugal, with twenty-five of his most faithful adherents.

When the King heard of his arrival he had him conducted to the Palace of Palmella, where he was treated with the greatest magnificence until he should make his public entry into Lisbon. On that occasion his passage through the streets was an ovation, and he was received with the greatest pomp, both by the King and Queen at separate palaces, each surrounded by a numerous court of ladies and grandees. For a long time Bemoi had been receiving instruction in the tenets of Christianity; so that the King's anxiety was gratified by his spontaneous request, that he and his companions might be admitted by baptism into the Christian Church. He was baptized in the Queen's palace, by the Bishop of Ceuta, on the 3rd of December, 1489, and received the King's name of João. On the following day the King dubbed him knight, and gave him for arms, gules a cross or between the five escutcheons of Portugal.

Meanwhile the King equipped twenty caravels, well provided with men, and provisions, and munitions of war, and everything requisite for the construction of a fortress, together with a number of missionaries for the conversion of the heathen. Unhappily for the fulfilment of the King's desires, the command was entrusted to Pedro Vaz da Cunha, a man of brutal nature, who, in a moment of spleen at finding the foundations of the new fortress laid in an unhealthy position, in which it would be his duty for some time to reside, stabbed Bemoi to death upon an empty pretence that he had plotted treason against him. Not only the negroes, but the Portuguese themselves were horrified at this act of baseness, which caused the King much pain. He contented himself, however, with leaving Da Cunha to his remorse, which would probably be but a trivial punishment to so heartless a coward.

But it is time we revert to that most important expedition of which Bartholomeu Dias was the commander, and which, as we stated on page 338, set sail for the south in 1486. It was fitting that a Dias should be the first to accomplish the great task which it had been the ruling desire of the life of Prince Henry to see effected. It was a family of daring navigators. João Dias had been one of the first who had doubled Cape Boyador, and Diniz Dias was the first to pass the Senegal and reach Cape Verde. The expedition of Bartholomeu started about the end of August, and made directly for the south. Passing the Manga das Areas, where Diogo Cam had placed his furthest pillar, they reached a bay to which they gave the name of Angra dos Ilheos. Here Dias erected a pillar, which was broken some seventy years ago. The point is now called Dias Point or Pedestal Point. From seaward is seen what looks like two conical shaped islands, on the highest of which stood the cross. These hillocks stand out dark from the surrounding sand, and probably gave rise from their tint to the name of Serra Parda, or the Dark Hills, in which Barros places this monument. Proceeding southward, Dias reached another point, where he was delayed five days in struggling against the weather, and the frequent tacks that he had to make induced him to call it Angra das Voltas, or Cape of the Turns or Tacks. It is still called Cape Voltas, and forms the south point of Orange River. From this they were driven before the wind, for thirteen days, due south, with half-reefed sails, and were surprised to find a striking change in the temperature, the cold increasing greatly as they advanced. When the wind abated, Dias, not doubting that the coast still ran north and south, as it had done hitherto, steered in an easterly direction with the view of striking it, but finding that no land made its appearance, he altered his course for the north, and came upon a bay where were a number of cowherds tending their kine, who were greatly alarmed at the sight of the Portuguese, and drove their cattle inland. Dias gave the bay the name of Angra dos Vaqueiros, or the Bay of Cowherds. It is the present Flesh Bay, near Gauritz River.

It is a fact specially worthy of notice that in this voyage an entirely different system was adopted with respect to the natives than had prevailed hitherto. Instead of capturing the negroes that they chanced to find on the coast, they had orders to leave on the shore at intervals negroes and negresses well dressed and well affected towards Portugal, to gather information respecting Prester John, to speak in praise of the Portuguese from experience of kindnesses received, and to infuse a desire to contract alliances with them. In accordance with these instructions two negroes had been restored at Angra do Salto (the Bay of the Capture) so called from Diogo Cam having captured them at this place. They had left also a negress at Angra dos Ilheos (Angra Pequeña), and another at Angra das Voltas. An unfortunate event, however, occurred which neutralised the effect of this well-intended plan. In proceeding eastward from Flesh Bay, Dias reached another bay, to which he gave the name of San Bras, where he put in to take water. In doing this he met with determined opposition from the natives, who threw stones at his men. They were thus compelled to resort to their own weapons in self-defence, and an unfortunate shot from an arblast struck one of the Caffres dead, and thus the favourable impressions which had been looked for from a pacific system of procedure were nullified by an act of violence which they would gladly have avoided. Continuing east, Dias reached a small island in Algoa Bay, on which he set up another pillar with its cross, and the name of Santa Cruz, which he gave to the rock, still survives; and as they found two springs in it, many called it the Penedo das Fontes. This was the first land beyond the Cape which was trodden by European feet, and here they set on shore another negress.

The crews now began to complain, for they were worn out with fatigue, and alarmed at the heavy seas through which they were passing. With one voice they protested against proceeding further. Dias, however, was most anxious to prosecute the voyage. By way of compromise he proposed that they should sail on in the same direction for two or three days, and if they then found no reason for proceeding further, he promised they should return. This was acceded to. At the end of that time they reached

a river some twenty-five leagues beyond the island of Santa
Cruz, and as João Infante, the captain of the second ship,
the *S. Pantaleon*, was the first to land, they called the
river the Rio do Infante. It was the river now known as
the Great Fish River.

Here the remonstrances and complaints of the crews compelled Dias to turn back. When he reached the little island
of Santa Cruz, and bade farewell to the cross which he had
there erected, it was with grief as intense as if he were
leaving his child in the wilderness with no hope of ever
seeing him again. The recollection of all the dangers that
he and his men had gone through in that long voyage, and
the reflection that they were to terminate thus fruitlessly,
caused him the keenest sorrow. He was, in fact, unconscious of what he had accomplished. But his eyes were
soon to be opened. As he sailed onwards to the west of
Santa Cruz he at length came in sight of that remarkable
cape which had been hidden from the eyes of man for so
many centuries. In remembrance of the perils they had
encountered in passing that tempestuous point, he gave to it
the name of Cabo Tormentoso, or Stormy Cape, but when
he reached Portugal and made his report to the King,
João II., foreseeing the realization of the long-coveted
passage to India, gave it the enduring name of Cape of
Good Hope.

The one grand discovery which had been the object of
Prince Henry's unceasing desire was now effected. The
joy of the homeward voyage was, however, marred by a
most painful incident. Dias had, by way of precaution, left
behind him, off the coast of Guinea, the small vessel containing the supplies of provisions. He now went in search
of it, it being nine months since they had parted company.
When they reached it, they found three men only surviving
out of the nine that had been left, and one of these, named
Fernando Colaço, a scrivener from Lumiar, near Lisbon,
was so weakened by illness that he died of joy when he saw
his companions. The cause of the loss had been that, while

the Portuguese were holding friendly communication with
the negroes, the latter were seized with a covetous desire to
possess some of the articles which were being bartered, and
as a short means of obtaining them killed the owners. Not
to return empty-handed, Dias put in at St. Jorge da Mina,
and received from the commander, João Fogaza, the gold
which he had taken in barter. He then proceeded to
Lisbon, which he reached in December, 1487, after an
absence of sixteen months and seventeen days.

In that voyage he had discovered three hundred and fifty
leagues of coast, which was almost as much as Diogo Cam
had discovered in his two voyages. In the seven hundred
and fifty leagues explored by these two captains, six
Padrãos, or pillars, had been set up. The first, called S.
Jorge, at the river Zaire or Congo; the second, called St.
Augustine, at the Cape Negro; the third, which was Diogo
Cam's last, at the Manga dos Areas, or Sleeve of Sand
(Cape Cross); the fourth, called Santiago, which was Dias's
first, at Sierra Parda (Dias or Pedestal Point); the fifth,
called San Felipe, at the Cape of Good Hope; and the
sixth, Santa Cruz, at the island of that name. This great
and memorable discovery was the last that was made in the
reign of King João II.

CHAPTER XIX.

RESULTS WESTWARD.

1470—1507.

" It was in Portugal," said Ferdinand Columbus, the son and biographer of the most illustrious navigator that the world has seen,—" it was in Portugal that the admiral began to surmise, that, if the Portuguese sailed so far south, one might also sail westward, and find lands in that direction." The period of Christopher Columbus' sojourn in Portugal was from 1470 to the close of 1484, during which time he made several voyages to the coast of Guinea in the Portuguese service. While at Lisbon he married Felipa Moñiz de Perestrello, daughter of that Bartholomeu Perestrello to whom we have already seen that Prince Henry had granted the commandership of the island of Porto Santo.*

---

* Prince Henry had originally, with the consent of the king his father, conferred this grant on Perestrello for his lifetime only, but subsequently, on the 1st of November, 1446, gave it in perpetuity to him and his successors, and the grant was afterwards confirmed by King Affonso his nephew. On the death of Perestrello the Prince gave it, with the consent of his widow Isabel Moñiz, to Pedro Correa da Cunha, a gentleman of his own household, who had married a daughter of Perestrello, to be held by him during the minority of his wife's brother, also named Bartholomeu Perestrello. Pedro Correa subsequently contracted with Bartholomeu's mother and uncle, who were also his guardians, for the concession of the governorship for a certain sum of money. This was done with the permission of the Prince, who issued a warrant to that effect dated Lagos, May 17th, 1458, which was confirmed by King Affonso V. at Cintra on the 17th of August, 1459 ; but the governorship subsequently reverted to Bartholomeu Perestrello, son of the first grantee, as is shown by the confirmation of it made to him by King Affonso V , on the 15th of March, 1473, and still existing in the Torre do Tombo.

For some time Columbus and his wife lived at Porto
Santo with the widow of Perestrello, who, observing the
interest he took in nautical matters, spoke much to him of
her husband's expeditions, and handed over to him the
papers, journals, maps, and nautical instruments which
Perestrello had left behind him. *

" It was not only," says Ferdinand Columbus (see *Vida*,
cap. 8), " this opinion of certain philosophers, that the greater
part of our globe is dry land that stimulated the admiral ; he
learned also from many pilots, experienced in the western
voyages to the Azores and the island of Madeira, facts and
signs which convinced him that there was an unknown land
towards the west.   Martin Vicente, pilot of the King of
Portugal, told him that at a distance of four hundred and fifty
leagues from Cape St. Vincent, he had taken from the water
a piece of wood sculptured very artistically, but not with an
iron instrument.   This wood had been driven across by the
west wind, which made the sailors believe, that certainly
there were on that side some islands not yet discovered.
Pedro Correa, brother-in-law to the admiral, told him, that
near the island of Madeira he had found a similar piece of
sculptured wood, and coming from the same western direction.   He also said that the King of Portugâl had received
information of large canes having been taken up from the
water in these parts, which between one knot and another
would hold nine bottles of wine, and Herrera (Dec. 1, lib. i.
cap. 2) declares that the King had preserved these canes,
and caused them to be shown to Columbus.   The colonists
of the Azores related, that when the wind blew from the

* Las Casas, in his History of the Indies, tells us distinctly that Columbus
derived much information from Perestrello's maps and papers, and adds that
" in order to acquaint himself practically with the method pursued by the
Portuguese in navigating to the coast of Guinea, he sailed several times with
them as if he had been one of them."   Las Casas says that he learned this from
the admiral's son Diego, adding that " some time before his famous voyage
Columbus resided in Madeira, where news of fresh discoveries was constantly
arriving, and this," he says, " appeared to have been the occasion of Christopher
Columbus coming to Spain, and the beginning of the discovery of this great
world (America)."

west, the sea threw up, especially in the islands of Graciosa and Fayal, pines of a foreign species. Others related, that in the island of Flores they found one day on the shore two corpses of men, whose physiognomy and features differed entirely from those of our coasts. Herrera, perhaps from the MSS. of Las Casas, says, that the corpses had broad faces, different from those of Christians. The transport of these objects was attributed to the action of the west winds. The true cause, however, was the great current of the Gulf, or Florida stream. The west and north-west winds only increase the ordinary rapidity of the ocean current, prolong its action towards the east, as far as the Bay of Biscay, and mix the waters of the Gulf stream with those of the currents of Davis Straits and of North Africa. The same eastward oceanic movement, which in the fifteenth century carried bamboos and pines upon the shores of the Azores and Porto Santo, deposits annually on Ireland, the Hebrides, and Norway, the seeds of tropical plants, and the remains of cargoes of ships which had been wrecked in the West Indies.[*]

While availing himself of these sources of information, Columbus studied with deep and careful attention the works of such geographical authors as supplied suggestions of the feasibility of a short western passage to India. Amongst these, the " Imago Mundi " of Cardinal Pierre d'Ailly (Petrus de Aliaco) was his favourite, and it is probable that from it he culled all he knew of the opinions of Aristotle, Strabo, and Seneca, respecting the facility of reaching India by a western route. Columbus' own copy of this work is now in the cathedral of Seville, and forms one of the most precious items in the valuable library, originally collected by his son Ferdinand, and bequeathed to the cathedral on condition of its being constantly preserved for public use. It contains many marginal notes in his own handwriting, but of comparatively little importance.

The fondness of Columbus for the works of Pierre d'Ailly, a Frenchman, has caused a recent French writer, M. Margry,

---

[*] Humboldt, Examen Critique, vol. ii. p. 246—251.

to put forth the empty pretension that the discovery of
America was due to the influence of French teaching, whereas,
not only was the "Imago Mundi" itself a compilation from
ancient authors, but the first edition was not printed till
many years after Columbus had devoted himself to the purpose which ended in his great discovery, for his famous
correspondence with Toscanelli, of which I shall presently
speak, occurred in 1474. M. Margry, indeed, *asserts*, but
without giving his authority, that in the Columbian Library
at Seville are D'Ailly's treatises *printed at Nuremberg in
1472.* This is in contravention of all the bibliographers—
Panzer, Ebert, Hain, Serna Santander, Lambinet, and Jean
de Launoy.

The earliest date assigned to the first edition of the " Imago
Mundi," is *about* 1480 by Serna Santander, 1483(?) by Lambinet, while Jean de Launoy, in his " Regii Navarræ Gymnasii
Parisiensis Historia," Parisiis, 1677, tom. ii. page 478, distinctly gives it the date of 1490. Humboldt, who had
Columbus' copy in his hands, and who, as the subject was
especially his own, cannot be suspected of sleeping over such
an important point, adopts De Launoy's date of 1490, while
Lambinet gives the queried date of 1483 from actual colla-
tion with another work printed in that year, at Louvain, in
the very identical type, by John of Westphalia. In the recently published second volume of the " Ensayo de una
bibliotheca de libros españoles raros," por Don Bartolomé
Gallardo, is a list of the books in the Columbian Library,
but D'Ailly's " Imago Mundi" is not therein mentioned,
although his " Quæstiones," printed much later by Jean
Petit at Paris, a far less important book, is inserted. The
omission is to be regretted, as we might have hoped for
some illustrative comments from the author.

But perhaps it may be suggested that Columbus may
have possessed, or seen, a *manuscript* copy of Pierre d'Ailly
at a yet earlier period. We will willingly suppose it for the
sake of the argument; but even then the reasoning will
fail, for I find that the very portion of the " Imago Mundi,"
writen in 1410, which is assumed to have supplied the

inspiration for the discovery of America, and which Columbus quoted in his letter to Ferdinand and Isabella from Haiti in 1498, is *taken by Pierre d'Ailly, without acknowledgment, almost word for word, from the " Opus Majus" of Roger Bacon* written in 1267, a hundred and forty-three years before, as will be seen at page 183 of that work, printed Londini, 1733, fol. See Humboldt, Examen Critique, tom. i. pp. 64-70.

Unfortunately Roger Bacon was not a Frenchman, but there remains for M. Margry the consolatory fact that no Englishman is likely to avail himself of the circumstance which I have just enunciated, to claim for his countrymen the honour of having inspired Columbus with the idea which led to the discovery of America, although, by M. Margry's process of reasoning, he might do so if he would. True, Roger Bacon had been a student in the University of Paris; but this fact did not communicate the character of French inspiration to the ancient authors whose statements he quotes. True also (but this is a circumstance either unknown or unnoticed by M. Margry), Ferdinand Columbus tells us that his father was principally influenced in his belief of the smallness of the space between Spain and Asia, by the opinion of the Arab astronomer Al Fergani, or Alfragan, to that effect; and it is further true, that Alfragan is treated of by Pierre d'Ailly, in his "Mapa Mundi." This is a separate work from the "Imago Mundi," although it happens to have been printed with it, at a period which we have shown to be posterior to Columbus' correspondence with Toscanelli, in 1474. It follows, therefore, that either; 1st, the great explorer obtained his knowledge of Alfragan's opinion through one of the Arabo-Latin translations to which he seems to have had recourse during his cosmographical studies in Portugal and Spain (see Humboldt, Examen Critique, tom. i. p. 83), in which case French influence is eliminated; or 2ndly, he derived it from a manuscript of Pierre d'Ailly before 1474, which there is no evidence to show; or 3rdly, he derived it from his printed copy of Pierre d'Ailly, in which case the influence of Alfragan on his mind could not have been

primarily suggestive, but only corroborative of conclusions
to which he had come several years before that book was
printed. And in either of the two latter cases, the information supplied by Alfragan would not become French because
adduced by a Frenchman, unless we introduce into serious
history a principle analogous to the old conventional English
blunder of giving to the toys manufactured in Nuremberg the
name of "Dutch toys," because imported through Holland.

The suggestions derived from these works were corroborated by the narratives of Marco Polo and Sir John
Mandeville,. whose reports of the vast extent of Asia eastward led to the reasonable inference that the westward
passage to the eastern confines of that continent could not
demand any considerable length of time. The natural
inclination of Columbus for nautical enterprise being thus
fostered by the works that he studied, and by the animating
accounts of recent adventurers, as well as by the glorious
prospects which the broad expanse of the unknown world
opened up to his view, we find that in the year 1474 his
ideas had formed for themselves a determined channel, and
his grand project of discovery was established in his mind
as a thing to be done, and done by himself. The combined
enthusiasm and tenacity of purpose which distinguished his
character, caused him to regard his theory, when once
formed, as a matter of such undeniable certainty, that no
doubts, opposition, or disappointment, could divert him
from the pursuit of it.

It so happened that while Columbus was at Lisbon, a
correspondence was being carried on between Fernando
Martinez, a prebendary of that place, and the learned Paolo
Toscanelli of Florence, respecting the commerce of the
Portuguese to the coast of Guinea and the navigation of the
ocean to the westward. This came to the knowledge of
Columbus, who forthwith despatched by an Italian then at
Lisbon a letter to Toscanelli, informing him of his project.
He received an answer in Latin, in which, to demonstrate
his approbation of the design of Columbus, Toscanelli sent

him a chart, the most important features of which were laid down from the descriptions of Marco Polo. The coasts of Asia were drawn at a moderate distance from the opposite coasts of Europe and Africa, and the islands of Cipango, Antilla, &c., of whose riches such astonishing accounts had been given by this traveller, were placed at convenient spaces between the two continents.

While all these exciting accounts must have conspired to fan the flame of his ambition, one of the noblest points in the character of Columbus had to be put to the test by the difficulty of carrying his project into effect. The political position of Portugal, engrossed as it was with its wars with Spain, rendered the thoughts of an application for an expensive fleet of discovery for the time worse than useless, and several years elapsed before a fair opportunity presented itself for making the proposition.

At length, as we have already seen, about the year 1480, Martin Behaim rendered the astrolabe useful for the purposes of navigation, and shortly afterwards Columbus submitted to the King of Portugal his proposition of a voyage of discovery westward. The King at first received him discouragingly, but was at length induced to refer the proposition to a council consisting of the great mathematicians and geographers, Roderigo and Josef, and Cazadilla, Bishop of Ceuta, the king's confessor, who treated the question as an extravagant absurdity.

The King, not satisfied with their judgment, then convoked a second council, consisting of a large number of the most learned men in the kingdom; but their deliberations only confirmed the verdict of the first junta, and a general sentence of condemnation was passed upon the proposition. As the King still seemed inclined to make a trial of the scheme of Columbus, some of his councillors, who were enemies of the Genoese, and at the same time loth to offend the King, suggested a plan which suited their own views, but which was as short-sighted as it was dishonest. Their design was to procure from Columbus a detailed

account of his plan that it might be submitted to the
council, and then, under the false pretext of conveying
provisions to the Cape Verde Islands, to despatch a caravel
on the voyage of discovery. King João, deviating from his
general character for prudence and generosity, yielded to
their insidious advice, and their plan was acted upon; but
the caravel which was sent out, after keeping on its westward course for some days, encountered a storm, and the
crew, possessing none of the lofty motives of Columbus to
support their resolution, returned to Lisbon, ridiculing the
scheme in excuse of their cowardice. So indignant was
Columbus at this unworthy manœuvre, that he resolved to
offer his services to some other country, and towards the
end of 1484 he left Lisbon secretly with his son Diego.

It is not difficult to understand why the King of Portugal
should have hesitated to accept the proposition of Columbus.
Nearly seventy years of continued effort on the part of the
Portuguese to realise the great conception of Prince Henry,
afforded substantial proof of their conviction of the soundness of that conception. Many years before Columbus
proposed to reach India by the sea, Prince Henry had
finished a life which had been spent in aiming at the same
result by another route. That route, therefore, though by
no means free from great dangers, was identified with their
hopes in the future as well as their predilections in the past.
What wonder that they refused to resign a course so
hopeful, comparatively so simple, and so essentially their
own, in favour of a project replete with danger, and which
they regarded as the chimera of a visionary?

The learned and careful Muñoz states his opinion that
Columbus went immediately from Portugal to Genoa, and
made a personal proposition to that government, but met
with a contemptuous refusal. Great obscurity, however,
hangs over his history during the first year after his leaving
Portugal, but from calculations based on his own statements, it would seem that it was in 1485 that he made his
first application to the court of Spain. It is well known

that the lively interest which the worthy prior of the Franciscan convent of Santa Maria de Rabida, Fray Juan Perez de Marchena, took in his guest, and his anticipated influence with his friend Fernando de Talavera, prior of the monastery of Prado, and confessor to the Queen, was the cause that first induced Columbus in the spring of 1486 to venture to the Spanish court in the hope of gaining a favourable audience. On reaching Cordova, however, he had the mortification to find that Talavera regarded his design as preposterous. The court also was engrossed with the war at Granada, so that all hope of gaining attention to his novel and expensive proposition was out of the question. At length, at the close of 1486, Mendoza, archbishop of Toledo, and grand cardinal of Spain, became impressed with the high importance of the scheme as set forth by the earnest and lucid reasoning of Columbus. He adopted his cause, and became his staunch protector and friend. Through his means an audience with the sovereign was procured, and it was resolved to submit the proposition to the judgment of the *literati* of the country. But here again Columbus found himself in a painful predicament. He was to be examined at Salamanca by a council of ecclesiastics, whose ignorance of cosmography and blind conclusions from misinterpreted texts of Scripture stood in strong opposition to his arguments, and he began to find himself in danger of being convicted not only of error, but of heresy. Fortunately one learned man of the number, Diego de Deza, tutor to Prince Juan, and afterwards Archbishop of Seville, appreciated the lucid arguments of the adventurer, and aiding him with his own powers of language and erudition, gained for him not only a hearing, but even approval from some of the most learned of the council. At length, in 1491, after a succession of vexatious delays, Talavera, the chief of the council, was commanded to inform Columbus that the cares and expenses of the war precluded the possibility of their Highnesses engaging in any new enterprises, but that, when it was concluded, there would be both the will and the opportunity to consider the subject further. Regarding this as nothing better than a courteous evasion of his application, Columbus retired, wearied and disappointed, from the court, and were it not that an attachment which he had formed at Cordova made him reluctant to leave Spain, it is probable that he would have gone to France, under the inducement of an inviting letter from that quarter.

The interval till 1492 was spent in a succession of appeals to the Spanish court, and in contending against all the vexatious variety of obstacles that ignorance, envy, or a pusillanimous economy could suggest.

At length, having overcome all obstacles, he set sail with a fleet of three ships on the 3rd of August, 1492, on his unprecedented and perilous voyage. The ordinary difficulties which might be expected to occur in so novel and precarious an adventure were seriously aggravated by the alarming discovery of the variation of the needle, as well as by the mutinous behaviour of his crew; and his life was upon the point of being sacrificed to their impatience, when the fortunate appearance of land, on the morning of the 12th October, converted their indignation into compunction, and their despondency into unbounded joy.

In this first voyage the discovery was made of the islands of St. Salvador, Santa Maria de la Concepcion, Exuma, Isabella, Cuba, Bohio, the Archipelago off the south coast of Cuba, called by Columbus the Jardin del Rey, or King's Garden, the islands of St. Catherine and Hispaniola. On this latter Columbus erected the fortress of La Navidad, and established a colony. He set sail on his return voyage on the 16th January, 1493, and, after suffering severely from a storm and a wearisome struggle with the trade winds, reached the island of St. Mary's on the 18th of February. Scarcely had he and his tempest-tost crew commenced their thanksgivings for their safe return to the abode of civilised men, when the governor of the island, acting under the general orders of the King of Portugal, surrounded them and took them all prisoners. This reception of the admiral on his return to the old world is well described by Washington Irving, as an earnest of the crosses and troubles with which he was to be requited through life for one of the greatest benefits that ever man had conferred upon his fellow-beings. He was at length liberated, with an apology, invited to the court, and received most graciously by the King and Queen, but not without evident manifestations of jealousy and chagrin on the part of some of the courtiers, and propositions to take away his life. The magnanimity of the King prevented this injustice, and leaving Portugal in safety, on the 13th of March, Columbus arrived on the 15th, at the little port of Palos, from whence he had sailed on the 3rd of August in the preceding year. His reception in Spain was such as the grandeur and dignity of his unrivalled achievement deserved, and his entrance into Barcelona was scarcely inferior to a Roman triumph.

The description of his voyage, which he had addressed to the Spanish sovereigns through their treasurer, caused so much excitement, that numerous editions of it were issued in the same year (1493) from the various great printing cities of Europe; and the narrative, embodied in *ottava rima* by the Florentine poet, Giuliano Dati, was sung about the streets to announce to the Italians the astounding news of the discovery of a new world.[*]

It is not my duty here to lead the reader through details of the explorations made by Columbus in his four voyages. It has been my purpose to show the correctness of my assertion in the first chapter, that " while this vast achievement of Columbus was the link that united the old world with the new, the explorations instituted by Prince Henry of Portugal were in truth the anvil on which that link was forged." It was an event in which all humanity was con-

---

[*] Believing at the time that the copy of this extremely scarce and curious poem, then recently purchased by the British Museum, was unique, I reprinted it as an appendix to the Introduction to my " Select Letters of Columbus," printed for the Hakluyt Society in 1847.

cerned, but one which was recompensed with the basest
ingratitude even from those most closely and beneficially
interested in it.

The seductive adulation of the court and the people shown
for the moment to Columbus, did not divert his thoughts
from the preparations for a second expedition. A stay of
six months sufficed to make all ready for this purpose,
during which period a papal bull was obtained which fixed
the famous line of demarcation, determining the right of
the Spanish and Portuguese to discovered lands; which
line was drawn from the north to the south pole, at a
hundred leagues west of the Azores and Cape de Verde
Islands; the discoveries to the westward were to belong to
Spain, and those to the eastward to Portugal. It may be
well here to remark that the success of Columbus in obtaining a second armament gave rise to a malignant feeling
towards him on the part of Juan Rodriguez Fonseca, Bishop
of Badajos, who had treated him as a visionary, which
eventually led to such disgraceful ill-usage of the admiral,
as will remain a stain upon the character of Spain while
the name of Columbus exists in the memory of man.

On the 25th of September, 1493, Columbus sailed westward, taking his departure from Cadiz with a fleet of three
large ships of heavy burthen, and fourteen caravels, and
after a pleasant voyage reached the island of Dominica on
the 2nd of November. In this voyage he discovered the
Caribbee Islands, Jamaica, an archipelago named by
Columbus the Queen's Gardens, and supposed to be the
Morant Keys, Evangelista, or the Isle of Pines, and the
island of Mona.

He sailed with his fleet finally for Spain on the 28th of
April, 1496, and after working his way for nearly two
months against the whole current of the trade winds,—during
which provisions became so reduced, that there was talk of
killing, and even eating, the Indian prisoners,—he reached
the bay of Cadiz on the 11th of June. The emaciated state
of the crew when they disembarked, presenting so mournful

a contrast with the joyous and triumphant appearance
which they were expected to make, produced a very dis-
couraging impression upon the opinions of the public, and
reflected a corresponding depression upon the spirits of
Columbus himself. He was reassured, however, by the
receipt of a gracious letter from the sovereigns inviting
him to the court; a letter the more gratifying to him that
he had feared he was fallen into disgrace. He was received
with distinguished favour, and had a verbal concession of
his request to be furnished with eight ships for a third
voyage. He was doomed, however, to have his patience
severely tried by the delay which occurred in the performance of this promise, which was partly attributable to
the engrossing character of the public events of the day,
and partly to the machinations of his inveterate enemy, the
Bishop Fonseca.

It was not till the 30th of May, 1498, that he set sail
from San Lucar, with six of the eight vessels promised, the
other two having being despatched to Hispaniola with provisions in the beginning of the year. When off Ferro, he
despatched three of his six vessels to the same island, with
a store of fresh supplies for the colony, while with his remaining three he steered for the Cape Verde Islands, which
he reached on the 27th of June. On the 5th of July, he left
Boavista, and proceeded southward and westward. In the
course of this voyage the crews suffered intensely from the
heat, having at one time reached the fifth degree of north
latitude, but at length land was descried on the 31st of
July,—a most providential occurrence, as but one cask of
water remained in the ship. The island they came to
formed an addition to his discoveries ; and as the first land
which appeared consisted of three mountains, united at their
base, he christened the island, from the name of the Trinity,
La Trinidad. It was in this voyage that he discovered Terra
Firma, and the islands of Margarita and Cubagua. On
reaching Hispaniola, to which he was drawn by his anxiety
on account of the infant colony, he had the mortification to

find that his authority had suffered considerable diminution, and that the colony was in a state of organized rebellion. He had scarcely, by his active and at the same time politic conduct, brought matters to a state of comparative tranquillity, when a new storm gathered round him from the quarter of the Spanish court. The hatred of his ancient enemies availed itself of the clamour raised against him by some of the rebels who had recently returned to Spain, and charges of tyranny, cruelty, and ambition were heaped unsparingly upon him. The King and Queen, wearied with reiterated complaints, at length resolved to send a judge to inquire into his conduct,—injudiciously authorising him to seize the governorship in the place of Columbus, should the accusations brought against him prove to be valid. The person chosen was Don Francisco de Bobadilla, whose character and qualifications for the office are best demonstrated by the fact, that, on the day after his arrival in Hispaniola, he seized upon the government before he had investigated the conduct of Columbus, who was then absent; he also took up his residence in his house, and took possession of all his property, public and private, even to his most secret papers. A summons to appear before the new governor was despatched to Columbus, who was at Fort Concepcion; and in the interval between the despatch of the summons and his arrival, his brother (Don Diego) was seized, thrown into irons, and confined on board of a caravel, without any reason being assigned for his imprisonment. No sooner did the admiral himself arrive, than he likewise was put in chains, and thrown into confinement. The habitual reverence due to his venerable person and exalted character, made each bystander shrink from the task of fixing the fetters on him, till one of his own domestics, described by Las Casas as "a graceless and shameless cook," filled up the measure of ingratitude that he seemed doomed to experience, by riveting the irons, not merely without compunction, but with alacrity. In this shackled condition he was conveyed, in the early part of October, from prison to the ship that was to convey him home; and when Andreas Martin, the master of the caravel, touched with respect for the years and great merit of Columbus, and deeply moved at this unworthy treatment, proposed to take off his irons, he declined the offered benefit, with the following magnanimous reply: " Since the king has commanded that I should obey his governor, he shall find me as obedient to this, as I have been to all his other orders; nothing but his command shall release me. If twelve years' hardship and fatigue; if continual dangers and frequent famine; if the ocean first opened, and five times passed and repassed, to add a new world, abounding with wealth, to the Spanish monarchy; and if an infirm and premature old age, brought on by these services, deserve these chains as a reward, it is very fit I should wear them to Spain, and keep them by me as memorials to the end of my life." This in truth he did, for he always kept them hung on the walls of his chamber, and desired that when he died they might be buried with him.

His arrival in Spain in this painful and degraded condition produced so general a sensation of indignation and astonishment, that a warm manifestation in his favour was the immediate consequence. A letter, written by him to Doña Juana de la Torre, a lady of the court, detailing the wrongs he had suffered, was read to Queen Isabella, whose generous mind was filled with sympathy and indignation at the recital. The sovereigns immediately commanded that he should be set at liberty, and ordered two thousand ducats to be advanced for the purpose of bringing him to court with all distinction and an honourable retinue. His reception at the Alhambra was gracious and flattering in the highest degree; the strongest indignation was expressed against Bobadilla, with an assurance that he should be immediately dismissed from his command, while ample restitution and rewards were promised to Columbus, and he had every sanction for indulging the fondest hopes of returning in honour and triumph to St. Domingo. But here a grievous disappointment awaited

him; his re-appointment was postponed from time to time
with various plausible excuses. Though Bobadilla was dismissed, it was deemed desirable to refill his place for two
years, by some prudent and talented officer, who should be
able to put a stop to all remaining faction in the colony, and
thus prepare the way for Columbus to enjoy the rights and
dignities of his government both peacefully and beneficially
to the crown.

The newly selected governor was Nicolas de Ovando, who,
though described by Las Casas as a man of prudence,
justice, and humanity, certainly betrayed a want both of
generosity and justice in his subsequent transactions with
Columbus. It is possible that the delay manifested by the
sovereigns in redeeming their promise might have continued
until the death of Columbus, had not a fresh stimulant to
the cupidity of Ferdinand been suggested by a new project
of discovering a strait, of the existence of which Columbus
felt persuaded, from his own observations, and which would
connect the New World which he had discovered with the
wealthy shores of the East. His enthusiasm on the subject
was heightened by an emulous consideration of the recent
achievements of Vasco de Gama and Cabral, the former of
whom had in 1497 found a maritime passage to India by the
Cape, and the latter in 1500 had discovered for Portugal
the vast and opulent empire of Brazil. The prospect of a
more direct and safe route to India than that discovered by
De Gama, at length gained Columbus the accomplishment
of his wish for another armament; and finally, on the 9th
of May, 1502, he sailed from Cadiz on his fourth and last
voyage of discovery.

It is painful to read the description given of the splendour
of the fleet with which Ovando left Spain to assume the
government of Hispaniola, and to contrast it with the slender
and inexpensive armament granted to Columbus for the purpose of exploring an unknown strait into an unknown ocean,
the traversing of whose unmeasured breadth would complete
the circumnavigation of the globe. Ovando's fleet consisted

of thirty sail, five of them from ninety to one hundred and
fifty tons burden, twenty-four caravels of from thirty to
ninety tons, and one bark of twenty-five tons; and the
number of souls amounted to about two thousand five
hundred. The heroic and injured man, to whose unparalleled
combination of noble qualities the very dignity which called
for all this state was indebted for its existence, had now, in
the decline of his years and strength, and stripped both of
honour and emolument, to venture forth with four caravels,
—the largest of seventy, and the smallest of fifty tons burthen,—accompanied by one hundred and fifty men, on one
of the most toilsome and perilous enterprises of which the
mind can form a conception.

On the 20th of May he reached the Grand Canary, and
starting from thence on the 25th, took his departure for the
west. Favoured by the trade winds, he made a gentle and
easy passage, and reached one of the Caribbee Islands, called
by the natives Mantinino (in all probability Martinique), on
the 15th of June. After staying three days at this island,
he steered northwards, touched at Dominica, and from
thence directed his course, contrary to his own original intention and the commands of the sovereigns, to St. Domingo.
His reason was that his principal vessel sailed so ill as to
delay the progress of the fleet, which he feared might be an
obstacle to the safety and success of the enterprise, and held
this as a sufficient motive for infringing the orders he had
received. On his arrival at San Domingo, he found the
ships which had brought out Ovando ready to put to sea on
their return to Spain. He immediately sent to the governor
to explain that his intention in calling at the island was to
procure a vessel in exchange for one of his caravels, which
was very defective, and further begged permission for his
squadron to take shelter in the harbour from a hurricane,
which, from his acquaintance with the prognostics of the
weather, he had foreseen was rapidly approaching. This
request was ungraciously refused; upon which Columbus,
though denied shelter for himself, endeavoured to avert the

danger of the fleet, which was about to sail, and sent back immediately to the governor to entreat that he would not allow it to put to sea for some days. His predictions and requests were treated with equal contempt, and Columbus had not only to suffer these insulting refusals and the risk of life for himself and squadron, but the loud murmurings of his own crew that they had sailed with a commander whose position exposed them to such treatment. All that he could do was to draw his ships up as close as possible to the shore, and seek the securest anchorage that chance might present him with. Meanwhile the weather appeared fair and tranquil, and the fleet of Bobadilla put boldly out to sea. The predicted storm came on the next night with terrific fury, and all the ships belonging to the governor's fleet, with the exception of one, were either lost, or put back to San Domingo in a shattered condition. The only vessel that escaped was the one which had been freighted with some four thousand gold pieces, rescued from the pillage of Columbus's fortune. Bobadilla, Roldan, and a number of the most inveterate enemies of the admiral, perished in this tremendous hurricane, while his own fleet, though separated and considerably damaged by the storm, all arrived safe at last at Port Hermoso, to the west of San Domingo. He repaired his vessels at Port Hermoso, but had scarcely left the harbour before another storm drove him into Port Brazil. But we must not follow him through the remainder of this unhappy voyage, the toils and perils of which were aggravated to Columbus by extreme bodily suffering, and which closes by his reaching Jamaica, where he would in all probability have perished, but for the activity and zeal of the faithful and devoted Diego Mendez.* When at length, through the agency of Mendez, two ships arrived from Hispaniola to the assistance of the admiral, he was enabled, on the 28th of June, 1504, to leave his wrecked vessels

* The highly interesting description of that brave man's exploits on behalf of Columbus, has been quoted by Navarrete from his will, and is translated in my " Select Letters of Columbus," printed for the Hakluyt Society, 1847.

behind him, and start with revived hopes for San Domingo, which he reached on the 13th of August.

On the 12th of September, 1504, he set sail for Spain; the same tempestuous weather, which had all along tended to make this his last voyage the most disastrous, did not forsake him now. The ship in which he came home sprung her mainmast in four places in one tempest, and in a subsequent storm the foremast was sprung, and finally, on the 7th of November, he arrived, in a vessel as shattered as his own broken and care-worn frame, in the welcome harbour of San Lucar.

It is impossible to read, without the deepest sympathy, the occasional murmurings and half-suppressed complaints which are uttered in the course of the veteran navigator's touching letter to the sovereigns describing this voyage. These murmurings and complaints were wrung from the manly spirit of Columbus by sickness and sorrow, and though reduced almost to the brink of despair by the injustice of the King, yet do we find nothing harsh or disrespectful in his language to the sovereign. A curious contrast is presented to us. The gift of a world could not move the monarch to gratitude; the infliction of chains, as a recompense for that gift, could not provoke the subject to disloyalty. The same great heart which through more than twenty wearisome years of disappointment and chagrin gave him strength to beg and to buffet his way to glory, still taught him to bear with majestic meekness the conversion of that glory into unmerited shame.

The two years which intervened between this period and his death, present a picture of black ingratitude on the part of the crown to this distinguished benefactor of the kingdom, which it is truly painful to contemplate. We behold an extraordinary man, the discoverer of a second hemisphere, reduced by his very success to so low a state of poverty that in his prematurely infirm old age he is compelled to subsist by borrowing, and to plead, in the apologetic language of a culprit, for the rights of which the very sovereign whom he has benefited has deprived him. The death of the benignant and high-minded Isabella, in 1505, gave a finishing blow to his hope of obtaining redress, and we find him thus writing, subsequently to this period, to his old and faithful friend, Diego de Deza:—"It appears that His Majesty does not think fit to fulfil that which he, with the Queen, who is now in glory, promised me by word and seal. For me to contend for the contrary, would be to contend with the wind. I have done all that I could do; I leave the rest to God, whom I have ever found propitious to me in my necessities." The selfish and cold-hearted Ferdinand beheld his illustrious and loyal servant sink, without relief, under bodily infirmity, and the paralysing sickness of hope deferred; and at length, on the 20th of May, 1506, the generous heart which had done so much without reward, and suffered so much without upbraiding, found rest in a world where neither gratitude nor justice is either asked or withheld.

His body was in the first instance buried at Valladolid, in the parish church of Santa Maria de la Antigua, but was transferred in 1513 to the Cartuja de las Cuevas, near Seville, where a monument was erected over his grave with the memorable inscription:—

"A Castilla y a Leon
Nuevo Mundo Dió Colon."

In the year 1536, both his body, and that of his son Diego, who had been likewise buried in the Cartuja, were transported to St. Domingo, and deposited in the cathedral of that city. From hence they were removed to Havanna in 1795, on the cession of Hispaniola to the French, and the ashes of the immortal discoverer now quietly repose in the cathedral church of that city. A tardy tribute has been at length paid to his memory by his fellow-citizens of Genoa, and the first stone of a monument in commemoration of his achievements was laid in that city on the 27th of September, 1846.

But injustice, unhappily, was not buried with Columbus in the tomb. It was but one twelvemonth after his death

that an attempt was made, and only too successfully, to
name the new world which he had discovered, after another
who was not only his inferior, but his pupil in the school of
maritime enterprise.   In an obscure corner of Lorraine, at
the little cathedral town of St. Dié, a cluster of learned
priests, who had there established a printing press under the
auspices of René II., Duke of Lorraine, suggested to give to
the newly-discovered continent the name of the Florentine,
Amerigo Vespucci, whose nautical career did not commence
till after Columbus had returned from his second voyage to
the western hemisphere.   The first time that the name of
Amerigo came into notice was in the year 1504, when Johann
Ottmar published at Augsburg the " Mundus Novus," a
description of Vespucci's third voyage, now extremely rare,
embodied in a letter addressed by Vespucci himself to
Lorenzo di Pier Francesco de' Medici.   In this voyage,
which occupied from May, 1501, to September, 1502, he was
in the service of Portugal, and explored the coasts of South
America as far as beyond the fifty-second degree.   But it
was not till May, 1507, when Columbus had been a twelvemonth dead, that the world was informed of four voyages
professed to have been made by Vespucci, of which the one
just mentioned was only the third, the two former having
been made, as he states, in the service of Spain.   As the
first of these was asserted to have taken place between May
20th, 1497, and October, 1499 [say 1498], and, if correct,
would involve the discovery by him not only of the north
coasts of South America, but a large extent of the coast of
North America also, and that in priority of the claims both
of Cabot and Columbus as regards the discovery of the
American continent, it has been a matter of keen interest
to many to examine minutely the correctness of Vespucci's
claim to having made this voyage.   This is not the place to
enter into the complicated arguments in which this question
is involved, but the reader may be interested to know some of
the leading positions of the dispute on both sides.   For this
purpose he must be invited to travel back with us a few years.

There was a Spaniard who had accompanied Columbus in his second voyage in 1493, named Alonza de Ojeda, small of stature, but of indomitable energy, courage, and perseverance. He was charged by Columbus to explore the gold mines in Hispaniola, and was not only successful in bringing back samples of gold, but also in capturing a formidable cacique, named Caonabo, who was doing his utmost to drive away the Spaniards from his territories. Ojeda subsequently quarrelled with Columbus, and when he returned to Spain in 1498, he was provided by the Bishop Fonseca, Columbus' enemy, with a fragment of the map which the admiral had sent to Ferdinand and Isabella, showing the discoveries which he had made in his last voyage. With this assistance Ojeda set sail for South America, accompanied by the pilot, Juan de la Cosa, who had accompanied Columbus in his first great voyage in 1492, and of whom Columbus complained that, " being a clever man, he went about saying that he knew more than he did," and also by Amerigo Vespucci.

They set sail on the 20th of May, 1499, with four vessels, and after a passage of twenty-seven days came in sight of the continent, two hundred leagues east of the Oronoco. At the end of June, they landed on the shores of Surinam, in six degrees of north latitude, and proceeding west saw the mouths of the Essequibo and Oronoco. Passing the Boca del Drago of Trinidad, they coasted westward till they reached the Cape, which Ojeda named the Capo de la Vela in Granada.

It was in this voyage that was discovered the Gulf, to which Ojeda gave the name of Venezuela, or Little Venice, on account of the cabins built on piles over the water,* a mode of life which brought to his mind the water-city of the Adriatic. From the American coast Ojeda went to the Caribbee Islands, and on the 5th of September reached Yaguimo, in Hispaniola, where he raised a revolt against

---

* In the same manner as, in ages long gone by, the recently discovered lake dwellings in Switzerland were constructed, and as they still are in Borneo and elsewhere.

the authority of Columbus. His plans, however, were frustrated by Roldan and Escobar, the delegates of Columbus, and he was compelled to withdraw from the island. On the 5th of February, 1500, he returned, carrying with him to Cadiz an extraordinary number of slaves, from which he realized an enormous sum of money.

At the beginning of December, 1499, the same year in which Ojeda had set sail on his last voyage, another companion of Columbus, in his first voyage, Vicente Yañez Pinzon, sailed from Palos, was the first to cross the line on the American side of the Atlantic, and on the 20th of January, 1500, discovered Cape St. Augustine, to which he gave the name of Cabo Santa Maria de la Consolacion, whence returning northward he followed the westerly trending coast, and so discovered the mouth of the Amazon, which he named Paricura.

Within a month after his departure from Palos, he was followed from the same port and on the same route by Diego de Lepe, who was the first to discover, at the mouth of the Oronoco, by means of a closed vessel, which only opened when it reached the bottom of the water, that, at a depth of eight fathoms and a half, the two lowest fathoms were salt water, but all above was fresh. Lepe also made the observation that beyond Cape St. Augustine, which he doubled, as well as Pinzon, the coast of Brazil trended south-west, which may have first given the idea of the pyramidal conformation of South America.

In October of that same year, 1500, Rodrigo de Bastidas, guided by the counsels of Juan de la Cosa, sailed from Cadiz, and coasting the Terra Firma of South America, reached the Gulf of Uraba and the Puerto del Retrete (Puerto Escribanos), in the isthmus of Panama, seventeen miles east of the Puerto de Bastimentos, where in 1510 Diego de Nicuesa founded the once celebrated, but now destroyed, town of Nombre de Dios. He returned to Europe in September, 1502, at which time Juan de la Cosa was in Seville, where a courier arrived with the news of the

discoveries of Bastidas, and announcing that he had landed
in Portugal and had brought with him several Indian
slaves. Juan de la Cosa forthwith repaired to Lisbon to
ascertain the truth of the report. In 1504 he was himself
employed by Queen Isabella, who gave him four vessels for
the purpose of making fresh explorations. His expedition
was so successful that he was able to hand over to the crown
491,708 maravedis as the royalty of one-fifth on the amount
of gold which he had brought back.

Now, in Vespucci's own accounts of his first two voyages,
there exists so much indistinctness and uncertainty that it
is only by correlation with the dates of departure, the number of ships mentioned, and casual coincidences of description with the voyages above described, that any approximation to conviction could be arrived at with respect to the
reality and identity of the more certain of the two, the
second. The Baron von Humboldt seems to have discovered here and there points of coincidence with all, but
there can be now no doubt that Vespucci's voyage in 1499
was identical with that of Ojeda. The establishment of this
identity is of value, because in treating of the first and most
important of the asserted voyages of Vespucci, it will be
necessary, for reasons which will make themselves apparent,
to call the reader's attention to the second.

It is not till the 18th of July, 1500, that we find a date
given to any letter of Vespucci's descriptive of any maritime explorations of his, and that date appertains to a letter
addressed to Lorenzo di Pier Francisco de' Medici, and
descriptive of his asserted *second* voyage. Had any letter
addressed by him to either of the illustrious men to whom
he subsequently wrote been indited immediately after his
return from his first voyage, it would have been a strong
point in his favour, which is now unfortunately wanting.
Such a deficiency is the more remarkable that a voyage of
the kind, involving priority in the discovery of the continent
which Columbus had gone in quest of, was a fact so stupendous that silence on such a subject baffles our comprehension.

And while we have to rely for the fact solely on Vespucci's own word, unsupported by one single assertion of any contemporaneous witness, we do find distinct corroborative evidence of the reality of his so-called second voyage described by him in his letter, dated 18th July, 1500, evidence which is exceedingly damaging to the probability of the first having ever taken place.

In the process instituted by the Procurator-General against the heirs of Columbus, Alonzo de Ojeda bore witness that in the expedition which he undertook to the coast of Paria " *after the admiral* (Columbus) in 1499, he took with him Juan de la Cosa, pilot, Morigo Vespuche, and other pilots." This ambiguous sentence may or may not place Vespucci in the category of pilots. I incline⁻to think not, but that he was the astronomer of the expedition, his part being as he himself says " per ajutare a discoprire, " to aid in making discoveries. But the witness of Ojeda is here quoted for the purpose of showing that Vespucci was with him in his visit to the coast of Paria in June and July, 1499, and yet so far from Vespucci having made him aware that he had seen that coast before, he (Ojeda) distinctly declared that he himself was the *first man* who came thither to explore *after the admiral.* Silence on the part of Vespucci respecting so great a previous discovery might under any circumstances be regarded as incredible; under *such* circumstances, well-nigh impossible. Yet not even once does Vespucci himself claim, in so many plain words, that he had first discovered the continent, but leaves the fact to be derived from what he relates. Equally difficult of comprehension is it that, in the aforesaid process, no trace or mention whatever of such a discovery, made by a man so conspicuous at that time in the service of Spain, should have appeared, when every report that was inimical to the priority of the discovery of the Terra Firma by Columbus was seized, not only with avidity, but with malice, *the period of that discovery being the principal object of the process.*

It is further remarkable that whereas Ojeda in this well

attested voyage of 1499, a voyage perfectly recognised by
Columbus and established by documentary evidence, speaks
of houses built on piles, like those in Venice, which made
him give the country the name of Venezuela, or little Venice,
we find in the asserted previous voyage to the same shores
by Vespucci in 1497, exactly the same observation, and the
comparison of the houses built on the water with Venice.
The occurrence of the same observation and idea to two
persons at two different times is sufficiently remarkable, but
it is yet more so that in 1499 Ojeda should be left in ignorance that his companion had observed the same curious fact
in 1497. Then the entire omission in the "four voyages"
of the names of those with whom he sailed, cannot fail
to have a mischievous effect on the judgment of most,
especially when brought into correlation with the facts
already adduced. Indeed, so great was the indistinctness
and complication connected with the letters of Vespucci,
that the Baron von Humboldt, who devoted years to the
examination of the history of American discovery, found it
difficult to decide to his own satisfaction to which of the
different Spanish and Portuguese expeditions the navigator
was successively attached.

In spite, however, of all these causes of mistrust, there have
been powerful and hearty defenders of the correctness of
Vespucci's statements just as we have received them. The
latest of them—and the Florentine navigator could not
have had a more earnest or more conscientious advocate—
is his Excellency Senhor F. A. de Varnhagen, envoy from
the Court of Brazil to Lima. In the asserted first voyage,
he makes Vespucci to have passed from the gulf of Honduras
round the peninsula of Yucatan, to Vera Cruz and Tampico,
and thus to have sailed between the west point of Cuba and
the mainland, verifying its insularity, and also to have
explored the coasts of Florida. In arriving at this conclusion,
my valued friend Senhor de Varnhagen, like all who have
ventured on exploring the complications of this perilous first
voyage, has been forced into an entanglement. He brings

Vespucci into the port of Tampico or Panuco, in compliance with the text which places the port alluded to in 23° north latitude ; but from this port the navigators, according to the text, sailed eight hundred and seventy (or, as Senhor de Varnhagen suggests, three hundred and seventy) leagues constantly towards the north-west (" tuttavia verso el maestrale)." The man who would sail three hundred and seventy leagues north-west from Tampico must do so upon wheels across dry land. The text goes on to say that after they had been out thirteen months, being very fatigued, and their ship in bad condition, they put into *the finest harbour in the world*, when they met with great kindness from the inhabitants, with whose aid they repaired their vessels, &c.

From this statement, combined with another of Vespucci's, that he had traversed one quarter of the globe's circumference, from Lisbon to beyond the fiftieth degree of south latitude, Senhor de Varnhagen is tempted to adapt the Bay of Chesapeake to this description of " the finest harbour in the world," which would necessitate, contrary to his own sense of correctness, the retention of the eight hundred and seventy instead of three hundred and seventy leagues. But in order to make their passage from this harbour in harmony with the text, which places a group of islands at one hundred leagues, or seven days' journey E.N.E. therefrom, Senhor de Varnhagen confesses that a bay in the east coast of Florida would meet the necessities of the case better than the Bay of Chesapeake, since he had decided that the islands referred to, and which bore the name of Iti, were the Bermudas. Unfortunately, however, we shall seek in vain for the finest harbour in the world on the east coast of Florida.

Manifestly, it would be simply ungenerous and vexatious to withhold large latitude to the computations of an investigator so patient, laborious, and conscientious as Senhor de Varnhagen, when brought to bear upon a text so indistinct and unmanageable as this description of Vespucci's assumed

first voyage.   My learned and highly honoured friend himself acknowledges that, at the part we have referred to, the
text is incomplete and obscure.   It is this very incomplete-
ness and obscurity which has made this voyage a torment to
every one who has attempted to reconcile it with known
facts, either geographical or historical.

Upon such insecure grounds for a decision, one feels but
little inclination to withdraw from John and Sebastian Cabot,
the honour of having been the first to discover the continent
of America since the times of the early Scandinavian expedition.   It is a fact, about which there is no doubt, that on
the 24th of June, 1497, the Cabots discovered the coast of
North America, "with Bristol men, in a ship called the
Matthew," and explored from Hudson's Bay to the southern
part of Virginia.   This was a year before Columbus landed
on the Terra Firma of South America.*   Yet none the less,
for ever and for aye, must Columbus be esteemed the discoverer of America.

Vespucci, it is clear, was not without his merit.   His
voyage with Ojeda, already described, together with his previous voyage, if it really took place, won for him sufficient
renown to induce the King of Portugal to invite him to his

* Not to interrupt inopportunely the statement of the deeds and deserts of
Columbus, I resort, though unwillingly, to a note to insert here, in its chrono-
logical position, another important voyage.   In the year 1500 Gaspar Cortereal
(whose father, João Vaz Cortereal, was governor of Tercera, and is stated by
Father Cordeiro to have discovered the Terra de Bacalhaos or Land of Codfish,
now called Newfoundland, so early as 1463, nearly thirty years before the great
success of Columbus) sailed from Lisbon with two ships, and steering northward
from the Azores discovered the land since known as Canada, and gave the
Portuguese name of Terra do Labrador to the country still known thereby.
That country was frequently designated by geographers in the following century
"Corterealis," after his name.   Again in May, 1501, he made another voyage north
with two ships, and reached Greenland, but a storm there separated the ships,
and his consort only returned.   Cortereal was never more heard of.   His brother
Miguel, who went in search of him in 1502, met with an exactly similar fate;
and in 1503 King Manuel sent out two ships expressly with the object of learning
what had befallen them, but in vain.   Lastly, a third brother, Vasco Eannes
de Cortereal, prepared to follow their traces, but the king would not give his
sanction to the last survivor of this courageous family thus placing himself in a
peril which seemed to have a fatality for the race.

service. He possessed nautical and astronomical knowledge, which he had turned to good account. With respect to his third voyage (the first made for Portugal), which itself has been the subject of severe dispute, Navarrete, who was by no means prejudiced in his favour, concedes that "it may be concluded from documents found in the archives of the Casa de Contratacion, at Seville, that Vespucci did navigate along the coasts of Brazil, that he had seen Cape St. Augustine, and fixed its latitude at 8° south." The purpose of the expedition to which he was attached was a double one : first, to examine the country discovered by Cabral, and next to seek a westward route to the Moluccas ; and it is only justice to add that, but for the inclemency of the weather, and the uncertainty of the ships holding out, he was in a fair chance of anticipating both Magalhaens and Balboa in reaching the Pacific. The expedition consisted of three ships, which touched at Cape Verde, and there met Cabral on his homeward voyage, which will be hereafter described. From thence the course they sailed was S.W. & ¼ S., and sixty-seven days, of which forty-four were very stormy, brought them to land in five degrees of south latitude. They cast anchor on the 17th of August, off Cape Saint Roque, which they then so named from having sighted it on the festival of that saint the day before, and in the name of the King of Portugal they took possession of the country.

On the 18th they again landed to take in water, and saw a great number of natives on a neighbouring mountain, but from which they had not the courage to descend. They therefore left some bells and small mirrors on the beach and went on board, when the natives came down and showed their admiration of the things which had been left behind. On the morrow smoke was observed at different points along the coast, which they regarded as an invitation, and two of the crew offered to go amongst them, and take with them some of their small articles of traffic. Permission was granted, on condition that they returned in five days. Seven days passed without their return, during which, from time to time, a few Indians would make their appearance, but with looks that betokened no good intention.

On the 26th of August, the crews again landed, and the Indians sent their women among the sailors, but when one of the latter approached, the women immediately surrounded him, and one of the women, with a bludgeon, broke open his head and stretched him dead on the spot. They then carried him away to the mountain, while the Indians came forward defiantly and discharged a great quantity of arrows. With difficulty the Portuguese reached their boats. Four cannon shot dispersed the natives, but when they returned to the mountain they began cutting up the body of their victim into pieces, which they first showed and then roasted. This left little hope as to the fate of the two first. The crews clamoured for vengeance, but the commander thought it more prudent to pursue his voyage. They proceeded E.S.E. till they reached Cape St. Augustine, which they then so named in honour of the day, the 28th August. Rounding the cape, they followed the coast, often anchoring and communicating with the inhabitants. On the 4th of October, they discovered the mouth of the San Francisco, and on the 1st of November, the Bahia de todos os Santos.

Sailing still south, they found themselves, on the 3rd of April, beyond the fifty-second degree of south latitude. On that day a tempest arose of such severity that they were in great dread. They were obliged to take in all sail and to run under bare poles. The nights became very long, that of the 7th of April lasting fifteen hours. On that day, in the midst of the storm, a new land made its appearance, the coast of which they followed for nearly twenty leagues; but it was quite wild, and they saw no inhabitants and no port. But what with the intensity of the cold and the thickness of the fog, they were scarcely in a condition to take note of anything. In this state of things they resolved to turn their prow homewards. Senhor de Varnhagen is of opinion that this wild uninhabited country was the island of Georgia, lat. 34° 30' S., long. 37° W., the description of which by

Captain Cook fully corroborates the inference. The fleet first made its way to Sierra Leone, thence to the Azores, and finally reached Lisbon on the 7th of September, 1502. It was the description of this grand voyage made in the service of Portugal, accompanied by his own bold expression, that he had explored regions which he might be permitted to call a New World, that first brought the name of Vespucci into prominent relief in 1504. In itself the voyage was a great and noble achievement, eminently and justly calculated to inspire enthusiastic admiration of the qualities developed in its performance. The ground of complaint is not to be found in the admiration of those qualities, but in the injustice to another's fame to which that admiration afterwards became the stepping-stone. Meanwhile we will proceed to lay before the reader, as briefly as we may, the details of Vespucci's fourth voyage.

The fourth voyage of Vespucci was undertaken with great hopes on his own part of important consequences. Before starting on it he announced his intention to proceed to the coast by way of the south, and when he should have reached his destination " to do many things for the glory of God, the service of his country, and *the perpetual memory of his own name*." His thoughts, like those of Columbus, were constantly directed to finding the rich islands of the east by the coasts of the new country opposite to Africa. Moreover, information from India directed the attention of the Portuguese government to the port of Malacca, and it was resolved to send out a small fleet, and Vespucci was offered the command of one of the ships.

At first two ships only seem to have been thought of, but at length six were fitted out. If we were to accept Vespucci's own statement, they started on the 10th of May, 1503, but Senhor de Varnhagen judiciously suggests from internal evidence that the 10th of June, the date given by Damião de Goes to the departure of the expedition of Gonzalo Coelho, with which Vespucci's was identical, was the correct one. After a stay of thirteen days in the

harbour of Santiago in the Cape Verde Islands, the commander of the expedition sailed south-east, making for
Sierra Leone; but the weather was bad, and the wind contrary, and after four days he altered his course and steered
south-west.

On the 10th August, when in three degrees S. latitude, they
saw on the horizon the island now known by the name of
Fernando de Noronha, and on a rock near this island the
principal vessel, of three hundred tons, was wrecked. The
crew was happily saved, but everything else went to the
bottom. Vespucci was then four leagues from the island,
and received orders to take his ship in search of a harbour.
He obeyed, but soon lost sight of the other vessels. He
fell in with one after the lapse of eight days, and the two
together returned and took in water at the island, and thence
made for Bahia, which evidently had been already discovered, as the King's instructions had indicated this
harbour as the point of rendezvous in case of separation.
In seventeen days they reached Bahia, and remained there
two months and four days, in hope of the arrival of the
other three ships, but in vain. They then proceeded south,
and after several communications with the inhabitants,
put into a harbour, where they found a great quantity of
dye-wood (*Brazil* wood), with which they loaded their
ships.

They remained five months in this harbour, and established there a little factory, which they fortified with twelve
cannon, and garrisoned with four-and-twenty armed men.
It is to Senhor de Varnhagen's researches that we owe the
identification of this port, in which, so soon after the discovery of Brazil, a factory for facilitating the commerce of
the dye-wood there found, and from which the country took
its name, was established. It was the port of Cape Frio.
In 1854 Senhor de Varnhagen discovered in the Torre do
Tombo the "*Llyvro*" of Duarte Fernandez, and published it
for the first time in note 13, p. 427, *et seq.*, of his "Historia
Geral do Brazil." Rio de Janeiro, 1854. In this work it is

shown that in the year 1511, the ship *Bretoa*, commanded by Christovão Pires, went to load dye-wood at the port of Cape Frio, where on an island in the harbour was a factory, with its factor, &c. After a passage of seventy-seven days, the two ships reached Lisbon on the 18th of June, 1504. The other vessels were not arrived, and when Vespucci wrote his account, dated the 4th of September of that year, he believed that they were all lost. Senhor de Varnhagen seems to have reasons for supposing that in pursuing their course towards Malacca, they reached the La Plata River, and that that river, as well as the cape at its mouth, then received from them the name of Santa Maria.

In February, 1505, Vespucci again entered the service of the King of Spain, and by a patent of the 24th of April, of that year, we find him naturalised as a Castilian. His occupation was to attend to the equipment and provisioning of ships destined for the Indies, for which he received an annual salary of thirty thousand maravedis. On the 22nd of March, 1508, the post of Pilot-Major of the kingdom was created for him, with a considerable salary attached, and in August of the same year, a royal letter was issued to him, which was to be read and proclaimed through all the towns and villages of Spain, in which he was charged to examine pilots on the use of the astrolabe and quadrant, to prove their acquaintance with the theory as well as the practice of navigation, to give them certificates, to receive payment from them for instruction, and to preside over the compilation of a sailing instruction-book to be named " Padron Real," which should receive constant corrections from information brought by pilots coming from the Indies, which information they were bound to supply to the Casa de Contratacion at Seville. This post he held for five years, and died at Seville on the 22nd of February, 1512, having just reached his sixty-first year.

There can be little doubt that the conferment of this honourable and comfortable post was led to by the reputation that had accrued to his name by the publication in

France, Germany, and Italy, of the suggestion to give to the new world in honour of him, that name which has ever since been attached to it, the name of America.

But let us trace the history of this name. When Vespucci was at Seville in 1501, we find from a statement in a letter describing his third voyage, addressed to his old schoolfellow, Pietro Soderini, Gonfaloniere of Florence, that one Giuliano Giocondi, then resident at Lisbon, was sent to him by Dom Manoel, King of Portugal, to seduce him from the service of the King of Spain, in which mission Giocondi was successful. Another letter describing the same voyage, but addressed by Vespucci to Lorenzo di Pier Francesco de' Medici, was translated from Italian into Latin by another member of the Giocondi family. This was no less than the celebrated Fra Giovanni Giocondi, of Verona, who had gained great renown as an architect at Venice, but was, at the time we speak of, engaged in the service of Louis XII., and built the bridge of Notre Dame at Paris, which is at present standing, together with, as some had supposed, the *petit pont* in continuation, crossing the southern branch of the Seine.*

Now at the time that Fra Giocondi was thus engaged in Paris, a young man of great talent, named Mathias Ringmann, a native of Schlestadt, on the eastern side of the Vosges mountains, was also pursuing his studies in the French capital at the college of Cardinal Lemoine. Ringmann is better known in the literary world by the pseudonym of Philesius, with the adjunct of Vosgesigena, in allusion to his birth in the Vosges mountains. He was a great proficient in Latin versification, and when he returned to his native Alsace, he found the fiercest literary rivalry existing between two contiguous parties of students, the one recognised as the Suevi or Swabians, the other as the

---

* This supposition was drawn from the following couplet by Sanazzaro :—
    "Jocundus geminum imposuit tibi, Sequana, pontem,
    Hunc tu jure potes dicere Pontificem."
But it has since been disproved.

Rheni or Rhine-men. Among the latter Ringmann soon distinguished himself by the gracefulness, no less than the wit, of his versification. At the University of Fribourg, the party of the Swabians found a talented but discreditable supporter in a dissolute professor, named Jacob Locher, better known as Philomusus.

A pique occasioned by some able verses of Ringmann in defence of his own party, induced Locher to resort to a mode of retribution of the most brutal and disgraceful character. At the close of the year 1505, Ringmann, who was at the time but twenty-two, a beardless young man of inoffensive manners and far from strong in frame, happened to be on a visit to the Syndic Zasius at Fribourg. Locher, having heard that on a certain day Ringmann intended to proceed on his way through the Black Forest, secured the assistance of eight armed confederates, and awaited the arrival of his victim by the wall of the Carthusian Convent about two miles from Fribourg, which lay on his road. Totally free from suspicion, Ringmann came as was expected, and was forthwith seized by this troop of cowards, who untrussed him, and inflicted on his bare body a severe and ignominious flagellation. *This whipped and weakly youth was the originator of the name which now belongs to the whole of the vast western world.*

For reasons which I shall now proceed to describe, there is great room for supposing that, when in Paris, Ringmann had made the acquaintance of Fra Giovanni Giocondi. From Paris he carried back with him to Alsace that admiration for Vespucci and his achievements which no one in Paris, of whom we have as yet heard, was so likely to have instilled into him as Giocondi; and in August, 1505, he became the editor, at Strasburg, of an edition of Giocondi's translation of Vespucci's above-mentioned letter, of which seven other editions are extant, but only one with a date, viz., that published by Johann Ottmar, at Augsburg, in 1504. In this edition of 1505 there are not only a set of verses by Ringmann, in laudation of Vespucci's discoveries

in his so-called third voyage, but there is also a Latin
epistle on the same subject to one Jacobus Brunus, whom
he addresses as his Achates and also as his "second self."
We thus find even at this early period an intellectual and
earnest advocate of the glory of Vespucci existing in Alsace.

A short distance beyond the line which separated that
province from Lorraine, stood the small cathedral city of
St. Dié, on the banks of the Meurthe, within the dominions
of René II., Duke of Lorraine, a prince who greatly distinguished himself by his encouragement of the arts and
of literature. The Duke's secretary was Walter Lud, one of
the Canons of the Cathedral of St. Dié. A zealous friend
of literature, this worthy priest established a gymnasium or
college at St. Dié under the Duke's auspices, and, what is
still more remarkable, he there set up also a printing press.
Ringmann became professor of Latin at the College, and
corrector of the press in the printing office ; and in 1504
another important personage joined this little confraternity.
This was Martin Waldseemüller, or, as he is better known
by his Græco-latinized pseudonym, Hylacomylus, a native
and student of Fribourg, who, going in the vintage season
of that year, in conformity with an annual habit of his, to
eat grapes in Lorraine, became so charmed with the society
of his learned friends at St. Dié, that he made up his mind
to take up his abode there, and became the teacher of
geography at the college. On the 25th of April, 1507, *a
year after the death of Columbus*, this latter member of the
clique produced, from the St. Dié printing-press, a little work
entitled "Cosmographiæ Introductio," to which was ap-
pended a Latin translation of Vespucci's four voyages as
described by himself, and addressed to Duke René of
Lorraine, though it can be shown by the contents to have
been really intended for Soderini. In the same year appeared at Strasburg a work, now of great rarity and possibly
unique, by Walter Lud himself, entitled "Speculi orbis
Succinctiss. sed neque pœnitenda neque inelegans Declaratio et Canon," which throws much light on difficulties

presented by Waldseemüller's publication.* Not only was it from that publication that the world was, for the first time, made aware of four voyages made to America by Vespucci, and one of them involving absolute priority in the discovery of the continent of America, but in the text which preceded the narrative of those voyages, the name of America was now, for the first time, suggested for the newly discovered western world.

Before we proceed to notice the form of that suggestion, we naturally pause with the reader to enquire how came these letters, hitherto unknown to the world, to make their appearance now for the first time at St. Dié? I say "for the first time," because while these letters, which are in Latin, bear a date (1507), the oldest Italian edition bears neither date nor place of imprint, and although, by the paper and type, it may be recognised as of nearly the same period as the Latin, there is no reason, with which I am acquainted, for believing that it was printed before the latter. From Walter Lud's "Speculum" we find that the letters were sent from Portugal to Duke René in French, and from the French translated at Lud's request into Latin by another Canon of St. Dié, named Jean Basin de Sandacourt. From this we must infer that the French version of the letters of Vespucci, intended for King René (and which was probably in manuscript—for no copy in type has ever been heard of), was prepared in Lisbon under the eye of Vespucci himself. But whence the connection between King René and Vespucci? That question has never been clearly answered, but I think I can offer a solution. It is a fact not entirely without significance that, immediately after the sentence from Walter Lud's work, which speaks of the letters coming from Portugal, we find the following remark, "and the

---

* To my great good fortune this valuable work, which consists only of four leaves, was purchased by the British Museum two years ago, when I was engaged on an examination of this very subject for a "Memoir on a Mappemonde by Leonardo da Vinci" in the collections at Windsor, being the earliest map hitherto known containing the name of America, printed in the Archæologia.

booksellers carry about a certain epigram of our Philesius
(Ringmann) in a little book of Vespucci's, translated from
Italian into Latin by Giocondi of Verona, the architect from
Venice." We have seen the connection of the Giocondi
with Vespucci. We have seen also the connection of Ringmann with the work of Fra Giovanni Giocondi. Here lies,
as I submit, the probable solution of the enigma. The
interest taken by Ringmann in the glory of Vespucci has
been clearly demonstrated. He has infused into the little
circle of St. Dié a similar interest, and, certes, the question
of a claim to the glory of having discovered a new world
and of a right to confer on it a name, is one which might
excite an interest in the most phlegmatic. But these men
are possessed of a printing press, and we can imagine the
keenness of their pleasure in having the opportunity to set
forth a subject which would throw so bright a reflection on
the obscurity of their secluded valley. Well might Pico de
Mirandola express his surprise that so learned a cluster of
men should exist among those wild rocks. (See his letter
to the Editor of the Ptolemy of 1513.) One of the members of that little circle is private secretary to the Prince of
the Duchy, and that Prince is remarkable for taking a pride
in connecting his name with the spread of knowledge and
refinement. Vespucci has a French translation made from
the original narrative, which he had drawn up in his own
Hispanicized Italian for Soderini, and this French translation
he sends to the Duke René, for whom there is, I think,
little doubt that it was made expressly at Ringmann's suggestion. It is reasonable to suppose that Vespucci, who
was no good linguist, was too ignorant of French to revise
the translation, and hence the explanation of what has
hitherto been to so many, and myself amongst the number,
an inexplicable enigma.* This account of the four voyages

---

* That he did not revise the French translation is proved by the fact that in
the fourth voyage there is a blunder which he could not have allowed to pass,
had he seen it. In naming the Bahia de todos os Santos (Bay of all Saints),
Vespucci had in his half Spanish, half Italian original, written " Bahia di tueti i

thus sent in French to Duke René is prefaced by an address,
in which Vespucci reminds the personage to whom he is
writing that " in their youth they had been friends, and had
together learned the elements of grammar from the writer's
uncle, Fra Georgio Antonio Vespucci." Now, we know
from the antiquary Giuliano Ricci, that Soderini, to whom
Vespucci's original Italian letters were addressed, *had been*
his schoolfellow, which Duke René could not easily have
been. We have, therefore, only to entertain the hypothesis
which I have now suggested, viz., that the French translation was made by suggestion from Ringmann for Duke
René without revision by Vespucci, to find the explanation
of this puzzle, about which so much has been written without any satisfactory conclusion. It may further be stated,
by way of showing more fully the likelihood of communication between Ringmann and Vespucci, that the former had
already made two journeys into Italy in connection with the
subject of an edition of Ptolemy, which was intended to be
prepared at St. Dié, and for which Giovanni Francesco Pico
de Mirandola made him a present of a Greek MS. of that
geographer. These journeys would naturally bring him into
contact with friends of Vespucci, whose praises he so zealously proclaimed, that Pico de Mirandola himself states
that, in consequence thereof, he had testified his sympathy
by adding to the hymn to Christ some verses in honour of
Vespucci's Lusitanian voyage. We have evidence that in
1508 the preparation of this edition of Ptolemy was going
on in full vigour, but Duke René died in December of that
year, the printing-press at St. Dié was broken up, and Ringmann went home to Schlestadt, where he died in 1511 at the
early age of twenty-nine. It is probable that his withdrawal may have caused the delay in the production of that
really valuable work till 1513. When it did appear, it

sancti," the first word being Spanish and Portuguese, the rest Italian. In the
original script, however, he wrote the " h " in the first word so like a " d " as to
mislead both the printer of the Italian and the French translator. The result was
that the word " Bay" was converted into " Abbey," and appeared as such in Italian,
French, and Latin, both in books and on maps for very many years afterwards.

contained a new map, entitled, " Tabula Terre Nove," by Hylacomylus, on which, strange to say, the name of America does not appear, but on the contrary there is inserted on the very *continent* of South America the following legend:—" Hæc terra cum adjacentibus insulis inventa est per Columbum Januensem ex mandatis Regis Castilliæ." As far as we have been hitherto able to trace the motives and meaning of the suggestion of the name of America, it seems that this sentence stands in direct contradiction of the only basis on which the suggestor could have pretended to give this honour to Vespucci.

But now at length we come to the mode of suggesting the name of America. In the " Cosmographiæ Introductio " of Hylacomylus, occur the following words : " And the fourth part of the world having been discovered by Americus, may well be called Amerige, which is as much as to say, the land of Americus or America."*

And a few pages later he says : But now these parts are more extensively explored, and as will be seen in the following letters, another fourth part has been discovered by Americus Vesputius, which I see no just reason why any one should forbid to be named Amerige, which is as much as to say, the land of Americus or America, from its discoverer Americus, who is a man of shrewd intellect; for Europe and Asia have both of them taken a feminine form of name from the names of women."†

In September of the same year appeared a re-issue at St. Dié of this same book, and in 1509 a new edition of it was issued from the printing-press of Johann Grüninger, of Strasburg. Now in this very same year, 1509, the name of America, thus proposed two years before, appears as if it

---

* " Et quarta orbis pars quam quia Americus invenit, Amerigen quasi Americi terram, sive Americam nuncupare licet."

† " Nunc vero et hæc partes sunt latius lustratæ, et alia quarta pars per Americum Vesputium, ut in sequentibus audietur, inventa est, quam non video cur quis jure vetet ab Americo inventore, sagacis ingenii viro, Amerigen quasi Americi terram sive Americam dicendam, cum et Europa et Asia a mulieribus sua sortita sint nomina. Ejus situm et gentium mores et bis binis Americi navigationibus quæ sequuntur liquide intelligi dant."

were already accepted as a well-known denomination in an anonymous work, entitled "Globus Mundi," printed also at Strasburg in that year. This was three years before the death of Vespucci. But although the work is anonymous, the colophon has supplied me with the means of associating the adopter of the suggestion with the suggester himself. The colophon runs thus : " Ex Argentina ultima Augusti, 1509. J. Grüniger *(sic)* imprimebat, Adelpho Castigatore." Now this Adelphus was a physician, a native of Mühlingen, near Strasburg, who afterwards established himself in the latter city. But the just-mentioned re-issue in 1509, of the " Cosmographiæ Introductio," containing the suggestion of the name of America, appeared from the press of this same Johann Grüninger, with the following words in the colophon : " Johanne Adelpho Mulicho, Argentinensi, Castigatore." Mulicho simply means native of Mühlingen. The coincidence clearly brings the suggester and the adopter of the suggestion into remarkably close proximity.

The first place in which we find the name of America, used a little further a-field, is in a letter dated Vienna, 1512, from Joachim Vadianus to Rudolphus Agricola, and inserted in the Pomponius Mela of 1518, edited by the former. The expression used is " America discovered by Vesputius."* But although this Vadianus, whose real name was Joachim Watt, writes from Vienna in 1512, I find that he was a native of St. Gall, whence in 1508, being then twenty-four years old, he went to the High School at Vienna. His learned disputations and verses gained him the chair of the Professorship of the Liberal Arts at that school, and he subsequently studied medicine, of which faculty he obtained the doctorate. This attachment to the study of medicine recalls to my mind a fact which awakens a suspicion that he may have been a personal friend of John Adelphus, just referred to, and if so, of the little confraternity of St. Dié. Before Adelphus established himself in Strasburg, he had practised as a physician at Schaffhausen, and this at the time when Joachim

* " Americam a Vespuccio repertam."

Watt was a young man, still resident at St. Gall, which is distant from Schaffhausen seventy English miles, a distance which would offer very little hindrance to Swiss intercommunication. Whether this suspicion be worth anything or no, I advance it as a possible clue to yet further researches which may show the process by which this spurious appellation of America became adopted, through the efforts of a small cluster of men in an obscure corner of France.

The earliest engraved map of the new world yet known as bearing the name of America, is a mappe-monde by Appianus, bearing the date of 1520, annexed to the edition by Camers, of the Polyhistoria of Julius Solinus (Viennæ Austr.), 1520, and a second time to the edition of Pomponius Mela by Vadianus, printed at Basle in 1522. The earliest manuscript map hitherto found bearing that name, is in a most precious collection of drawings by the hand of Leonardo da Vinci, now in Her Majesty's collections at Windsor, to which, from an examination of its contents, I have assigned the date of 1513-14.

I have thus endeavoured to unravel the intricate story of a great and irreparable injustice. No one can deny to Vespucci the credit of possessing courage, perseverance, and a practical acquaintance with the art of navigation ; but he had never been the commander of an expedition, and had it not been for the great initiatory achievement of Columbus, we have no reason to suppose that we should ever have heard his name.

" To say the truth," as has been well remarked by the illustrious Baron von Humboldt, " Vespucci shone only by reflection from an age of glory. When compared with Columbus, Sebastian Cabot, Bartholomeu Dias, and Da Gama, his place is an inferior one. The majesty of great memories seems concentrated in the name of Christopher Columbus. It is the originality of his vast idea, the largeness and fertility of his genius, and the courage which bore up against a long series of misfortunes, which have exalted the Admiral high above all his cotemporaries."

CHAPTER XX.

MEANWHILE great things had been doing in the East. The
grand discovery of Bartholomeu Dias was not to remain
fruitless, although it may fairly be wondered at that so long
an interval should have been allowed to elapse between that
discovery in 1487 and the realisation of its advantages by
Vasco da Gama ten years later. Some have even added to
the reasonable inquiry, an unreasonable insinuation that the
success of Columbus proved to be the effective stimulus to
the second important expedition. No chimera was ever
more untenable when examined by the light of facts and
dates. Indeed the interval of five years between the two
grand discoveries of Columbus and Da Gama is in itself
sufficient to show that we must look elsewhere for an explanation of the delay. It will be remembered that before
Dias had returned at the close of 1487, Payva and Covilham
had been sent by land to Eastern Africa, and that from
Cairo, in 1490, Covilham had sent home word to the King
confirmatory of the fact that India was to be reached by the
south of Africa. It happened, however, that in this same
year, 1490, King John was seized with an illness so severe
that his life was in the utmost jeopardy. This was supposed
to have been caused by his drinking the water of a fountain
near Evora, which was thought to have been poisoned, inasmuch as two Portuguese gentlemen who had drunk of it,
died. Through great care, and the pure air and tranquillity

of his pleasure-palaces of Santarem and Almerino, the King recovered; but though his life was saved, the vigour of his constitution was irreparably impaired. Shortly after this partial restoration to health, the King, by order of his physicians, stayed at Santarem during the summer months for the sake of bathing in the Tagus, when one day he sent for his son Affonso to join him in the bath. The Prince at first excused himself, but afterwards reflecting that such excuse was unbefitting the reverence due to a message from the King his father, mounted his horse and rode quickly to repair his fault. The King had, however, entered the bath, and Affonso proposed to his companion, João de Meneses, a race on horseback. In the midst of the race a young man crossing the path startled the Prince's horse, which reared up, fell, and rolled over him. The injuries that he received were such that he died that same night. The Prince was in his seventeenth year, and by his death the succession, which for three centuries and a half had continued without interruption in the male line, fell to the collateral line of the Dukes of Viséu, a fearful blow to the King's peace of mind.

In 1492 the King again fell dangerously ill, and in addition to the general infirmity of his frame, black spots showed themselves on his body, which confirmed the belief that some strong poison had been received into his system. To add to the misery of this prostration, the Queen, to whom he was devotedly attached, escaped but narrowly from an illness with which she was attacked in 1493. It was not till 1494 that the King began to show symptoms of some return to convalescence. The joy that pervaded the kingdom was universal, but was soon clouded by the presence of famine, accompanied by an epidemic which spread death and ruin among the people. The King's most earnest attention was directed to the remedy of these evils, when his own malady took the form of dropsy, and he was required to dismiss from his mind all thoughts of public business, and attend solely to the re-establishment of his health.

Meanwhile ever since the death of Prince Affonso, who

had married Isabella, Princess of Castile, the dominant anxiety of the King had been to establish the succession in his illegitimate son George, the child of Anna de Mendoza, whom he had made Duke of Coimbra and Grand Master of the Orders of Santiago and Aviz. He sent ambassadors to Rome to solicit his legitimization, but to this every possible objection was interposed by the King of Castile. The Queen and the people moreover declared themselves in favour of the King's cousin, Dom Manoel, the Duke of Beja, and to their influence, as well as to the claims of legitimacy, the King felt himself at length compelled to succumb. It will thus have been seen that the condition of the King's health and the personal anxieties accruing from the state of his kingdom, together with his domestic troubles, were of a nature to present serious obstacles to the development of those grander schemes which had been so vividly opened up to his ambition with respect to India. He died on the 25th of October, 1495, in the fortieth year of his age and the fourteenth of his reign; and it is hoped that enough has been said to explain how, as stated at the close of the preceding chapter, the momentous voyage of Bartholomeu Dias should be the last that distinguished the reign of King João II. His successor, King Manoel, received the name of "The Fortunate," from his good fortune in succeeding to the throne of a sovereign who had won for himself the designation of "The Perfect Prince." The first thought of the new King was to resume the distant maritime explorations which had already reflected so much honour on the far-sighted intelligence of their initiator, Prince Henry.

At length an experienced navigator of noble family was selected, in 1496, to attempt the passage to India by the newly-discovered southern cape of Africa. If we may trust an historian of good repute, and the holder of an important post in the Royal archives, this selection was the result of a mere whim on the part of King Manoel. We are told by Pedro de Mariz, in his "Dialogos de Varia Historia," that the King was one evening at one of the windows of his palace, meditating on the possibility of realizing the grand projects of his predecessor, João II., when Vasco da Gama happened to come alone into the court beneath the King's balcony. Without hesitation the King mentally resolved that he should be the chief in command of the fleet of the Indies.

The preparations for the enterprise were made by the King with the greatest forethought. Four vessels, purposely made small for the sake of easy and rapid movement, the largest not exceeding a hundred and twenty tons, were built expressly in the most solid manner, of the best-selected wood, well fastened with iron. Each ship was provided with a triple supply of sails and spars and rope. Every kind of needful store was laid in in superfluity, and the most skilful pilots and sailors that the country could furnish were sent out with Da Gama. The largest vessel, the *Sam Gabriel*, he of course took under his own command. The captaincy of the *Sam Raphael*, of one hundred tons, was given to his brother, Paolo da Gama; the *Berrio*, a caravel of fifty tons, was commanded by Nicolas Coelho; and a small craft laden with munitions was given to the charge of Pedro Nuñez, a servant of Da Gama. It had been intended that Bartholomeu Dias should accompany the expedition, but he was subsequently ordered to sail for San Jorge el Mina, perhaps for politic reasons, on a more profitable but less glorious mission. His pilot, however, Pero de Alemquer, who had carried him beyond the Stormy Cape, was sent out on board Vasco da Gama's ship, and the other two pilots were João de Coimbra and Pero Escolar.

It was on Saturday, the 8th of July, 1497, that Vasco da Gama started from Restello, an *ermida* or chapel, which had been built by Prince Henry about a league from Lisbon, and in which he had placed certain Friars of the Order of Christ, that they might receive confessions and administer the Communion to outward-bound or weather-bound sailors. Dom Manoel, who succeeded his uncle as Grand Master of the Order, subsequently built on the spot the splendid

Temple of Belem, or Bethlehem. As the first-fruits of the
success of that important voyage, on which Da Gama was
now starting, he transferred it to the Order of the Monks of
St. Jerome. The whole building is erected on piles of pine
wood. It is entered on the south side under a rich porch,
which contains more than thirty statues. The doorway is
double. Above the central shaft is a statue of Prince Henry
in armour.* (*See Engraving.*)

Without dwelling on such details of Da Gama's outward
voyage as present no important novelty, we shall pass over
four months, and on the 4th of November we shall find the
little fleet anchored in the Bay of St. Helena, on the west
coast of Africa, where for the first time they became acquainted with the Bosjesmans or Bushmen, that peculiar
race allied to the Hottentots, but so different from the
Caffirs. Here they landed in order to take in water, as well
as to take astronomical observations with the astrolabe,
newly invented by Behaim, for Da Gama mistrusted the
observations taken on board, on account of the rolling of the
vessel.† While he was thus occupied, they perceived two
negroes, one of whom they captured with very little difficulty, but were unable to make him understand them. They
therefore sent him back to his people laden with presents,
which had the effect of bringing them in crowds to beg for
similar gifts. These people were yellowish in colour, small
in stature, ill-formed, ugly, stupid, and stammering in their
speech. They proved, however, so friendly, that one of the
officers, Fernam Veloso, obtained permission to accompany
them to their home to make himself acquainted with their

* The late learned ecclesiologist, Dr. Mason Neile, says, "Belem is the last
struggle of Christian against Pagan art in Portugal. The visitor will be much
enchanted with the exquisite beauty of the details, more especially if he have
not previously seen the Capella Imperfeita at Batalha, with which Belem is not
for one moment to be compared."

† The astrolabe he used was of wood, three hands-breadth in diameter,
formed of three pieces like a triangle. They afterwards took out smaller ones
of latten. So humbly began the art which has since produced such mighty
results in navigation.

country. But his errand, as it happened, proved fruitless, for after journeying with them for some time, he was seized with a panic, and returned to the ships without having gained any information.

The ships remained for several days longer in St. Helena Bay, but nothing more was seen of the inhabitants, and Da Gama was balked in his hope of learning something about the country and the distance from the Cape of Good Hope. His pilot, Pero de Alemquer, who had been with Bartholomeu Dias in the first discovery of the Cape, was unable to inform him on that point, for, as we have already seen, in that voyage they had sailed southward too far from the land, and on the return voyage had sailed past this portion of the coast by night. Besides this, the stormy weather which Dias had encountered in the neighbourhood of the Cape had prevented him from making such observations as would have helped Da Gama in determining the distance of the southern point. On a rough estimate, however, Pero de Alemquer calculated the distance at about thirty leagues.

On the 16th of November they proceeded south. At length they came upon the open sea, but on the 19th made their course for the desired point. On Wednesday, the 22nd of November, at noon, Da Gama sailed before a wind past the formidable cape, to which King João II. had given the undying name of Good Hope, in anticipation of the achievement which was now about to be accomplished.

On Saturday, the 25th of November, he entered the bay which Bartholomeu Dias had named San Bras, and where the Portuguese had had a disagreement with the natives. The latter were now amiable enough, and exchanged with their visitors ivory bracelets for scarlet caps and other articles. Their cattle were remarkable for their size and beauty. A misunderstanding unhappily arose through unfounded suspicions on the part of the natives, but Da Gama prudently withdrew his men without bloodshed, and frightened the Hottentots by firing his guns from the ships.

In this bay Da Gama set up a padrão and cross, but they were thrown down before his eyes by the natives.

They left the bay of San Bras on Friday, December 8th. On Friday the 15th they sighted the Ilheos Chaos, or Flat Islets, five leagues beyond the Ilheo da Cruz (the Bird Islands in Algoa Bay), where Dias had left a padrão. On the night of Sunday, the 17th, they passed the Rio do Iffante, the extreme point of Dias's discovery, and here Da Gama became seriously alarmed at the force of the current that he encountered. Fortunately the wind was in his favour, and on Christmas Day he gained sight of land, to which, on that account, he gave the name of Natal.

On Wednesday, the 10th of January, 1498, they came to a small river, and on the next day landed in the country of the Caffirs, where an entirely new race of men from those they had hitherto seen met their eyes. With these, formidable as they were with their large bows and iron-tipped azagays, Da Gama established such friendly relations that he called the country the Terra da Boa Gente, or Country of the Good People, and the river he called the Rio do Cobre, on account of the copper which the natives brought in exchange for linen shirts. Barros confounds the Rio do Cobre, which appears to be the Inhambane, or Limpopo, with the Rio dos Reis, which the early maps make to debouch in Delagoa Bay, and is probably the river Manice.

On Monday, the 22nd of January, Da Gama reached a large river, where, to his great joy, he met with two richly dressed Mahometan merchants, who trafficked with the Caffirs, and from whom he gathered valuable information as to the route to India. Here he erected a pillar, which he named the padrão of Sam Rafael, and he called the river the Rio dos Bōos Signaes, or River of Good Signs (the Quilimane River). In an inferior sense the name was inappropriate, for here the scurvy broke out amongst the crew.

They set sail on Saturday, the 24th of January, and on the 10th of March anchored off the island of Mozambique. The people of the country told them that Prester John had

many cities along that coast, whose inhabitants were great
merchants, and had large ships, but that Prester John
himself lived a great way inland, and could only be reached
by travelling on camels.   This information filled the Portuguese with delight, for it was one of the great objects of
these explorations to find out the country of Prester John,
and they prayed God to spare them to see what they all
so earnestly desired.   The ships of this country were large
and without decks, not fastened with nails, but with leather.
Their sails were made of matting of palm leaves, and the
sailors had Genoese compasses to steer with, as well as
quadrants and sea-charts.   The viceroy of the island, whose
name was Colytam,* came very confidingly on board the
vessel with his suite, and the friendliest intercourse ensued ;
but it was afterwards discovered that treachery underlay
this seeming goodwill.   In fact the new comers had at first
been supposed to be Mahometans, but the mistake was soon
discovered.   A pilot whom the viceroy had given to the
Portuguese misled them, and conducted them to a place for
taking in water, where they found armed men hidden behind
palisades, who endeavoured with slings to drive them
from the water.   These, however, were soon dispersed by the
Portuguese guns,

Da Gama left this coast on the 29th of March, and on
Sunday, the 1st of April, came to some islands very near
the mainland, to the first of which he gave the name of
Ilha do Açoutado, or Whipping Island, because on the
Saturday afternoon the pilot they had taken in at Mozambique told the captain that these islands were mainland, and
for this falsehood he ordered him to be whipped.   These
islands were numerous, and so close to each other, that it
was difficult to distinguish them.   These were the Querimba
islands, of which the Ilha do Açoutado would be the
southernmost.   On Monday they saw other islands five
leagues out at sea, the more northern islands of the
Querimba group.

* Probably Çolytam or Sultan.

On Friday, the 6th of April, the *Sam Rafael* stranded on some reefs two leagues from the shore, and opposite a range of lofty and handsome hills, to which they gave the name of Serras de Sam Rafael, and they gave the same name to the reef.*

The day following, Saturday the 7th, they reached Mombaza, and were treated with great kindness by the King, who sent them presents, and offered to supply them with all that they might require. But having discovered a plot between the Moors of Mombaza and the pilots which he had brought from Mozambique, and being besides attacked by them in the night, Da Gama thought it wisest to continue his voyage, and on the 12th of April he set sail, though with little wind. The following morning, being about eight leagues distance from Mombaza, they saw two barks at sea about three leagues to leeward of them, and made for them, wishing to find pilots. By the evening they came upon one of them, and took it, but the other made for the shore. In the one they took were seventeen men, and gold and silver, and a quantity of maize and provisions, and a girl, the wife of an old man of rank, who was a passenger. On the Portuguese boarding, all in the vessel threw themselves into the water, and the former proceeded to pick them up in the boats. On Easter Day, the 15th of April, they reached Melinda, and their captives informed them that they would there find four ships belonging to Indian Christians, from whom they might procure Christian pilots, and every necessary in the way of meat and water, and wood, &c. On the Monday morning Da Gama sent the old man whom he had captured to the King, to tell him how happy he should be to enter into peaceful relations with him. After dinner the old man

---

* These appear to be the Waseen reefs, which make the coast inside of Pemba island unsafe of approach. "Although the coast is low, there is a range of hills in the background, and occasionally in the distance may be seen curiously isolated mountains, which present a remarkable contrast to the general flatness of the country. One of them, called Waseen Peak, is about two thousand five hundred feet high."—See "African Pilot," 1864, p. 206.

returned, attended by one of the King's household, and an officer, with three sheep from the King and a message that it would give the King great pleasure to enter into peaceful relations with the captain, and that he would be happy to supply him with pilots or anything that his country might afford. Da Gama sent word that he would enter the harbour on the following day, and immediately sent to the King an overcoat, two sprigs of coral, three copper basins, a hat, some bells, and two pieces of striped cloth. On Tuesday the King sent Da Gama six sheep, and a good quantity of cloves, and cummin seeds, and ginger, and nutmeg, and pepper, and also sent word that he would come to see him on the following day. After dinner on Wednesday the King came out in his boat to the ships, and Da Gama in his boat went to meet him. The King proposed that they should interchange visits, but Da Gama replied that he was not permitted by his sovereign to land. The King asked the name of Da Gama's King, and ordered it to be written down, and said that if Da Gama would return that way he would send an embassy, or would write to his sovereign. The King then went round the ships, and was delighted with seeing the guns fired. He spent three hours on board, and when he departed left one of his sons and an officer in the ship, and took with him two of the Portuguese to show them his palaces, and told Da Gama that since he would not come on shore that he should go along the coast the next day to see his horsemen ride. The King brought with him a close-fitting damask robe, lined with green satin, and a very rich head-dress, two chairs of bronze with their cushions, a round sunshade of crimson satin fastened to a pole, a sword in a silver scabbard, several trumpets, and two of a peculiar form made of elaborately carved ivory as high as a man, to be played at a hole in the middle. There were four ships here belonging to Indian Christians, who, when they came on board the first time, were shown by Da Gama an altar-picture, in which was the Virgin and child at the foot of the cross with the Apostles. The Indians immediately threw themselves on the ground in an attitude of prayer. These Indians warned Da Gama not to go on shore, nor place any faith in the joyous demonstrations that were made in his favour, for that they were not sincere. On Sunday, the 22nd of April, the King came on board, and Da Gama begged of him the pilots that he had promised. The King accordingly sent him a Christian pilot, and Da Gama gave up the hostage that he had retained. On the 24th of April they made sail for Calicut, under the guidance of their pilot, whose name was Malemo Canaca.

On Thursday, the 17th of May, 1498, Da Gama first sighted, at eight leagues distance, the high land of India, the object of so many anxieties and of so many years of persevering effort. On Sunday, the 20th of May, he anchored before Calicut. On the following day some boats came out to them, and Da Gama sent one of the " degradados," or condemned criminals, on shore with them, and they took the man to two Moors of Tunis, who spoke both Spanish and Genoese, and the first salutation they gave him was as follows : " The devil take you for coming here. What brought you here from such a distance?" He replied, " We come in search of Christians and spices." They said, " Why does not the King of Spain, and the King of France, and the Signoria of Venice send hither?" He replied that the King of Portugal would not consent that they should do so, and they said he was right. Then they welcomed him, and gave him wheatened bread with honey, and after he had eaten, one of the two Moors went back with him to the ships, and when he came on board said, " Happy venture ! happy venture ! abundance of rubies ! abundance of emeralds ! You ought to give many thanks to God for bringing you to a country in which there is such wealth." The Portuguese were utterly astounded at hearing a man at that distance from Portugal speak their own language. This Moor, whom Barros calls Monçaide and Castanheda Boutaibo, most probably Bou-said, proved very useful to Vasco da Gama, and went home with him to Portugal, where he died a Christian.

Calicut, the wealthy capital of that part of the Malabar coast, was governed at that time by a Hindoo sovereign, named Samoudri-Rajah (the King of the Coast), a name which the Portuguese afterwards converted into Zamorin. Gama had the good fortune to gain an audience of this prince, by whom he was favourably received, but with very little ultimate success, in consequence of his not being provided with presents suitable for an Eastern sovereign. This unlucky circumstance, combined with the hatred of the Arab merchants, whose ships crowded the harbour and who regarded with apprehension any rivals in the rich trade of spices, was near producing fatal results.

Da Gama thought it his duty to establish a factory, at the head of which he placed Diogo Dias, the brother of the first discoverer of the Cape. At the instigation of the Arabs, Dias and his men were taken prisoners. By way of reprisal, Da Gama kept as hostages twelve Hindoos who had visited his vessels; but when Dias and his comrades were allowed to return, he sent back only six of the Hindoos and retained the other six. When he set sail on Wednesday, the 29th of August, several vessels came to recover their countrymen. This Da Gama refused, and warned them to keep their distance, believing that their motives were treacherous. He told them at the same time that he meant to return as soon as possible, when they would know whether the Portuguese were thieves or not, as the Moors had represented them to be. Whatever might have been the danger of Da Gama, and doubtless it was great from the hostility of the Arabs, this conduct was indefensible, for there appears no reason to doubt either the integrity or the good-will of the Zamorin, inasmuch as the detention of Diogo Dias and his companions had been without his knowledge, and he himself not only discharged him, but sent by him a letter to Da Gama for the King of Portugal, written in Dias's own hand, to the following effect: " Vasco da Gama, a nobleman of your household, has visited my kingdom, which has given me great pleasure. In my kingdom there is abundance of cin-

namon, cloves, ginger, pepper, and precious stones in great
quantities. What I seek from thy country is gold, silver,
coral, and scarlet." The only shadow of an excuse for Da
Gama's retention of the six Hindoos was that he hoped to
take them to Portugal, and bring them back again, when
they might prove of the greatest assistance in establishing
friendly relations between the two countries. That it was a
genuine motive there can be little doubt, however harsh in
its first conception, but, alas! he was ignorant that the caste
of the poor captives would make them prefer death to their
present position, and it can only be supposed that they
speedily perished. They were becalmed about a league
below Calicut, and at noon, on Thursday the 30th, they were
beset by seventy boats crowded with people, whom they
kept at bay with their artillery. The contest continued for
an hour and a half, when fortunately a storm arose which
carried them out to sea, and the boats finding themselves
powerless returned, and Da Gama pursued his course. On
Monday, the 10th of September, as they had but little wind,
Da Gama put on shore one of the captives with letters to
the Zamorin, written in Arabic by a Moor who had come with
them. On the 15th they reached some islets about two
leagues from the shore, and on one of them they erected a
pillar, to which they gave the name of Santa Maria, for the
King had ordered Da Gama to erect three columns, which
he should name respectively Sam Rafael, Sam Gabriel, and
Santa Maria. That of Sam Rafael had been erected at the
Rio dos Bôos Signaes, that of Sam Gabriel at Calicut, and
now the last, that of Santa Maria, was placed on this islet,
and the group has since received the name of Santa Maria
from the pillar erected there. The group extends from lat.
37° 27' to 13°, 19¾ N.; Durrea or Deriah Bahauder Ghur in
lat. 13° 20' N. long., 70° 40¾, E., six leagues southward
from Cundapore River is the largest of the range, and probably that on which the padrão was erected. The inhabitants were pleased at the idea of the pillar with its cross
being set upon their island, as they were Christians, and were

happy to meet with those of the same creed.   Da Gama
then continued his course northward, and putting in for
water at a point of the coast opposite six little islands near
Hog Island, he became aware of the proximity of two barks
of unusual size.   He hastened his men on board, and found
from the look-out at the mast-head, that eight more such
were becalmed at about six leagues distance.   When the
wind arose he sailed straight for them, and they put in for
shore.   One of them, however, broke its rudder, and the crew
landed in their boat, leaving the ship at the mercy of the
Portuguese.   The other seven were run aground, and received the Portuguese fire as they pulled ashore in their boats.
These they found were vessels come in pursuit from Calicut.
    Thence Da Gama still proceeded north, till on Sunday,
the 23rd, he reached the little island of Anchediva,
where they drove the *Berrie* and the *Sam Gabriel* ashore to
caulk them, but the *Sam Rafael* remained afloat.   One day,
while they were on board the *Berrio*, two large row-boats
approached laden with men with trumpets and drums and
banners.   Da Gama found on inquiry that these were armed
pirates, who introduced themselves on board vessels under
the show of friendship, and once on board took possession if
they found themselves strong enough.   When, therefore,
they came within gun-shot the *Sam Rafael* fired at them.
They called out that they were Christians, but finding that Da
Gama was not to be duped, they put in for shore, and were
pursued for some time by Nicolao Coelho.   On the following
day came several with presents, asking to see the ships, but
they were coldly received.   Among them, however, came
one man of forty years of age, who spoke Venetian perfectly,
was well dressed in linen, with a handsome turban on his
head and a cutlass at his side.   He said that he came
originally from the west when he was a boy, that he lived
with a Moor who commanded forty thousand horsemen (in
fact the Rajah of Goa), and hearing that Franks, or people
from the west, were come, he had begged permission to come
to pay them a visit, and his master, sent word by him, that

he would be happy to offer them ships or provisions, or anything else in his dominions which might be of service to
them, or if they would take up their abode in his country
he would be very pleased. Meanwhile Paolo da Gama made
inquiries as to who the man was, and was informed that he
was the owner of the vessels that had come out to attack
him. When Da Gama learned this he had him flogged for
the purpose of extracting the truth from him. He confessed
that he knew that all the country was hostile, and that he
had come on board to ascertain the state of the Portuguese
defences. This man proved to be a Polish Jew, a native of
Posen, whence a cruel persecution had driven his family in
1456 to Palestine. They afterwards migrated to Egypt,
and he himself was born in Alexandria, whence he passed
by the Red Sea to India. He joined his fortunes to the Portuguese, and as he was an experienced and intelligent man,
Da Gama took him with him to Lisbon, where he embraced
Christianity, and at his baptism received the name of Gasparo da Gama. He proved of great use to Da Gama on the
homeward voyage, especially at Melinda, and was subsequently employed by King Manoel in different negotiations
with India, was made a knight of the king's household, and
received pensions and emoluments which afforded him an
honourable livelihood.

Da Gama remained twelve days in the island of Anchediva,
and after that he had repaired his vessels and taken in water,
set sail westwards on Friday, the 5th of October. When
they were some two hundred leagues away from land, this
same man said that he thought the time was come for him
to dissemble no longer, and confessed that while he was with
the Rajah his master, news was brought that the Portuguese
were wandering along the coast at a loss to find their way
back, and that a number of flotillas were trying to capture
them ; that his master then desired that an attempt should
be made not only to learn what strength the Portuguese
had for defence, but if possible to induce them to land, and
that once landed he would capture them, and as they were

courageous men, employ them in battle against his enemies
in the neighbourhood, but he reckoned without his host.

The passage across to Africa lasted for three months all
but three days, in consequence of the frequent calms and
contrary winds.   During this time the crews were attacked
so severely with scurvy that thirty men died, so that there
were only left seven or eight men to work each vessel, and
if the voyage had lasted a fortnight longer there would not
have been a soul left.   The commanders were even thinking
of putting back to India, but happily a favourable wind
arose which brought them in six days in sight of land, which
was almost as welcome to them as if it had been Portugal.
This was on Wednesday, the 2nd of January, 1499.   The
next day they found themselves off Magadoxo, but they
were in quest of Melinda, and did not know how far they
were from it.   On Monday, the 7th of January, they anchored
off that town.   The King sent to welcome them, and to say
that he had been long hoping to see them.   They spent here
five happy days of rest and relief from disease and the peril
of death, receiving princely proofs of kindness and hospitality from the King, who, at Da Gama's request, gave him
an ivory trumpet to convey to the King his master, as also
a young Moor, with a particular recommendation of him to
the King of Portugal, to whom he specially sent him to
show how much he desired his friendship.

On Friday, the 11th of January, they set sail; on Saturday
the 12th, passed Mombaza, and on Sunday anchored on the
Sam Rafael shoals, where they set fire to the *Sam Rafael*
herself, because they were too short of hands to work the
three vessels.   The people of the village off which they were,
and which was named Tamugata, brought an abundance of
fowls to barter for shirts and bracelets.   They sailed thence on
Sunday, the 27th of January, passed Zanzibar, called in the
narrative Jamgiber, and on the evening of Friday, the 1st of
February,* they anchored off the Ilhas de Sam Jorge (St.

---

* It stands February in the text, proving that the references to the preceding
month had been made to February erroneously instead of January.

George's Islands), in Mozambique, and on the following morning raised a pillar on the island, in which they had first heard mass on their outward voyage, though it rained so heavily that they were unable to light a fire to melt the lead that was needed for fixing the cross, so that the pillar was left without it.

On Sunday, the 3rd of March, 1499, they reached the bay of San Bras, where they took a quantity of anchovies and salted down penguins and sea-wolves for their home-ward voyage, and the wind being fair they doubled the Cape of Good Hope on Wednesday, the 20th of March. The survivors had recovered their health and strength, but were half numbed with the cold, which they attributed less to the actual cold of the climate than to their having come from a hot country. For twenty-seven days they sailed before a wind to within, as they reckoned by their charts, a hundred leagues of the island of Santiago, in the Cape Verdes. On Thursday, the 25th of April, they found ground in thirty-five fathoms varying to twenty fathoms, and the pilots said they were on the shoals of the Rio Grande. Shortly afterwards the caravel of Nicolao Coelho was sepa-rated from that of Da Gama, but whether the separation was the effect of a storm, or whether Coelho, who was aware of the superior sailing qualities of his vessel, availed himself of it to be the first to carry to Lisbon the news of the dis-covery of the Indies, has never been satisfactorily decided. However that may have been, Nicolao Coelho reached the bar of Lisbon on the 10th of July, 1499. When Vasco da Gama reached the island of Santiago, where his brother Paolo da Gama was seriously ill, he delegated the command of the vessel to his secretary, João de Sá. He then freighted a swifter caravel with the view of shortening the passage to Portugal. Meanwhile his brother died, and he put in at the island of Terceira and buried him there.

He reached Lisbon at the end of August or beginning of September, and was received with great pomp by the Court. His return from a voyage in which so mighty a discovery had been made was hailed with magnificent fêtes and public rejoicings, which by the King's order were repeated in all the principal cities throughout the kingdom.   In that important voyage he had lost his brother, more than half of his crew, and half his vessels, but he brought back the solution of a great problem which was destined to raise his country to the very acme of prosperity.

It has been seen in a former chapter what unsuccessful efforts have been made in later times by the French to establish a claim to discoveries on the coast of Guinea before the time of Prince Henry.   In like manner, it has been asserted, that Vasco da Gama was anticipated by a Frenchman in the discovery of the Cape of Good Hope.   In the "Mémoires Chronologiques pour servir à l'Histoire de Dieppe, par J. A. Desmarquets," 1785, tom. i. p. 92, it is asserted that a navigator named—

" Cousin sailed from Dieppe in the beginning of the year 1488. He was the first man in the universe who had been able to take the elevation in the midst of the ocean.   This he had done in pursuance of the lessons of Descaliers, so that he no longer hugged the coast as his predecessors had done.   After two months he reached an unknown land, where he found the mouth of a large river, which he named the Maragnon.   By the elevation which he there took, he perceived that in order to reach the coast of Adra, he must sail southwards, but bearing to the east.   By doing this he first made the discovery of the point of Africa, and gave the name of ' Aiguilles ' to a bank which he there observed.   This young captain having taken note of the places and their position, returned to the coasts of Congo and Adra, where he bartered his goods and arrived at Dieppe in the course of 1489.   The shipowners of this city agreed for their own interest to keep this discovery secret, for believing that they were the only ones who could reach India by this route, they reckoned upon deriving therefrom an immense revenue.   The French Government was occupied with intestine wars, and the Dieppese knew but too well how little attention the Government would give to maritime commerce.   They resolved therefore to profit by their discovery to the exclusion of all other nations, and accordingly equipped several ships for the

Indies, of access to which they were assured by Descaliers, from the facility now discovered of turning the south point of Africa."

At page 98 M. Desmarquets proceeds thus :—

" In order to turn to account the possibility of reaching India, the merchants gave Cousin the command of three well-armed ships laden with merchandise. Descaliers assured the captain of success, if he attended to the observations with which he supplied him in writing, and to the true position of India which he described to him. Cousin had learnt his lesson too well not to conform to it. He sailed midway between Africa and America, which he had discovered, turned the Cap des Aiguilles, reached India, where he exchanged his merchandise to very great profit, and returned to Dieppe, about two years after his departure."

The race begins to be exciting, and one longs to make a more intimate acquaintance with this able hydrographer Descaliers, to whose scientific acumen these great results were due. M. Desmarquets speaks of him as the *Abbé Descaliers, a priest of Arques, and the best mathematician and astronomer of his time.* Now I happen to have in my charge at the British Museum a most superb map of the world, on vellum, the execution of which might fairly warrant a compatriot in complimenting its author as " the best mathematician and astronomer of his time." The map records the name of its author and its date thus : " Faicte a Arques par Pierres Desceliers, Pbre, l'an 1550." " Done at Arques, by *Pierres (sic) Desceliers, priest,*" who with his own hand tells us that its *date is* " 1550."

Now that there should have been a Descaliers and a Desceliers, both priests at Arques, and both *super-excellent as mathematicians and hydrographers,* one in 1488, and the other in 1550, seems so improbable, that only remarkable accuracy in M. Desmarquets' statements in general would induce us to give credence to it. A few pages on, when I come to speak of the discovery of China by the sea, I shall have a valuable opportunity of showing what reliance is to be placed on his assertions, when he ventures on another claim to Dieppese discovery in that direction. But it may be suggested that

Desceliers and Descaliers were one and the selfsame person.
So I believe them to be. M. Desmarquets, however, who is
always remarkably circumstantial, tells us that Descaliers
was born in 1440, which would make him in that case the
constructor of the beautiful mappe-monde in the British
Museum, at the age of one hundred and ten. This is inad-
missible, and we have only the almost impossible alternative
that there were two such prodigies in scientific excellence of
the same name, place, and priestly office, and one of them
flourishing at a period when we find not a single evidence
of hydrographic skill existing at Arques. Moreover, the fact
of there having been two such marvellous persons would call
for especial mention by M. Desmarquets, whereas he speaks
only of one, although he mentions by name the successors
of his " Descaliers " in the school of hydrography at Arques
even beyond the period of the indubitable "Desceliers" of
the Museum map. But as I pledge myself to show further
on that M. Desmarquets could commit himself to assertions
of great moment which are demonstrably false, it may fairly
be concluded that the unquestionable Pierre Desceliers of
1550 has been carried back in his existence more than half
a century to give an appearance of reality to a discovery
which is not found recorded elsewhere.

In the year after Da Gama's return, at his recommendation, Pedro Alvarez Cabral, a scion of a noble house of
Portugal, was charged with the command of an expedition
to Calicut, with the view of establishing commercial intercourse with the Rajah of that country. The expedition was
a magnificent one. It consisted of thirteen ships formidably
armed with artillery, but at the same time sumptuously
provided with presents for the Rajah, and although sent out
with a purely commercial object, the boldest and most
famous seamen of the period were placed under the orders of
Cabral. Among these were Bartholemeu Dias, who fourteen
years before had rounded the Stormy Cape, Nicolao Coelho,
the able companion of Da Gama in 1497, and the talented interpreter Gasparo, whom Da Gama had brought home with

him from India. To these were added men of administrative intelligence, who might be able to treat with prudence on matters of commercial policy, it being intended to establish a factory on the coast of Malabar. Great as the importance of this object was, it was the fate of the expedition to make a discovery, before which even the results thus contemplated shrunk into insignificance. The expedition sailed on the 9th of March, 1500. After thirteen days, when off the Cape Verde Islands, one of the vessels, which was commanded by Pedro Dias, lost convoy, and after a short delay the fleet proceeded without her. Various have been the reasons assigned for the westerly course which the expedition now took. According to Barros the object was to avoid the calms off the coast of Guinea, while others have asserted that the fleet was driven westward by a storm. If, however, we take into consideration the intensity of the curiosity excited by the recent discoveries in the New World, and the noble emulation which such discoveries, made in the service of a rival nation, would inspire in the minds of men, who in another direction had gained so many laurels in the career of maritime enterprize, we may fairly doubt whether this south-westerly course was not pursued by Cabral in the hope of lighting on some part of the new-found western world. But whatever the inducement or the cause, the result was such as to satisfy both hope and curiosity. On Wednesday, the 22nd of April, Cabral perceived the rounded top of a mountain, on what he at first supposed to be an island, and as they were then in Holy Week or in the octave of Easter he gave the mountain the name of Monte Pascoal. It forms part of the chain of the Aymores, in Brazil.* To the country he gave the name of Vera Cruz, or, as it was afterwards called, Santa Cruz, which name it retained till the importation from it into Europe of the valuable dye-wood of the

* Fortunate as Cabral was in this discovery, he had been anticipated, as we have already seen, in landing on the coast of Brazil, although at a widely different part of that coast. On the 20th of January of the same year, viz., forty-eight days before the departure of Cabral, Pinzon had discovered Cape St. Augustine.

ibirapitanga, caused it to be called Brazil, from the name
which for centuries had been given to similar dye-woods
imported from India. On the 23rd, Nicolao Coelho was
despatched to examine the coast. On the 24th they anchored
in the bay afterwards named Porto Seguro. On the 1st of
May formal possession was taken of the country for Portugal,
and a large cross was set up on the coast in commemoration
of the event. The luxuriance of the vegetation, as well as
the sociable demeanour of the natives, and their respectful
bearing when witnessing the solemn celebration of mass,
were matters of surprise and gratification to the discoverers.
Cabral forthwith despatched Gaspar de Lemos to the King
with the important news, which was described most admirably
in a letter drawn up by Pedro Vaz de Caminha, the second
secretary of the Calicut Factory, accompanied by an astronomical diagram by Mestre João, the King's physician, who
had accompanied the expedition as doctor. By this means
the first information of the discovery of Brazil was brought
to Europe. Before the departure of the fleet an incident of
importance occurred. One of the natives who had come on
board the *Admiral,* was struck with the brightness of a brass
candlestick, and made signs to the effect that a similar metal
was found in that country. Cabral accordingly left behind him
two young *degradados,* or banished criminals, with orders to
make themselves acquainted with the products and habits of
the country, thus giving them the double chance of serving
their nation and retrieving their own position. One of these
subsequently became an able and respected agent of the
colony which King Manoel lost no time in establishing.
The fleet set sail on the 22nd of May, but the joy which had
been awakened by their success was soon to be turned into
mourning. The appearance of an immense comet produced
an alarm which was only too unhappily realized. A fearful
typhoon sunk four vessels, and the brave Bartholomeu Dias,
whose great achievement had converted his Stormy Cape
into a Cape of *Good Hope,* perished off that very cape which
for him was still to be a Cape of Storms.

Cabral, notwithstanding, pushed on, and reached Quiloa on the 20th of July, whence proceeding to Melinda, he renewed with the sovereign of that country the alliance which had been based upon his friendly treatment of Da Gama. Thence he crossed to India, and anchored before Calicut on the 13th of September. Through the medium of his intelligent interpreter, Gasparo da Gama, he succeeded in laying before the Zamorin or Rajah the objects of the embassy, which were favourably received. The splendid presents which he brought, and the formidable artillery with which he was protected, doubtless served to extinguish the recollection of the misunderstanding with Da Gama. Permission to establish a factory on the coast was readily granted, and the Rajah solemnly pledged himself to the terms of this new treaty of commerce, in which the future interests of Europe were so largely involved. The factory was peacefully established at Calicut, under the direction of Ayres Correa, but within a short time the treachery of the Mohammedans showed itself, and Correa and more than fifty of the Christians were massacred. Cabral took ample revenge for this unprovoked injury, and forthwith betook himself to the King of Cochin, the enemy of the Rajah of Calicut, with whom, as well as the King of Cananor, he succeeded in establishing peaceful relations. Having laden his remaining vessels with a most valuable cargo, he set sail for Portugal. Near Melinda, however, one of the most richly freighted of the ships, commanded by Sancho de Tovar, foundered on a reef. The vessel was of two hundred tons burthen, and laden with spices. The crew escaped with their lives, and they burnt the ship; but the King of Mombaza succeeded in recovering the guns, which he afterwards turned to account against the Portuguese. When they reached Cape Verde at the beginning of June, they fell in with a Portuguese flotilla of three ships, which had sailed from Lisbon on the 13th of May, for the purpose of making discoveries on the coast of Brazil, on board of which was Amerigo Vespucci.

In the letter addressed to Lorenzo di Pier Francesco de'

Medici, dated from that cape on the 4th of June, and recently discovered by Count Baldelli Boni, Vespucci relates the story of Cabral's discoveries as communicated to him by the interpreter Gasparo. He further mentions how, by a curious coincidence, on that very day one of Cabral's ships, that of Pedro Dias, which had lost convoy thirteen days after the expedition had set sail from Portugal, in March, 1500 (see page 409), again joined the squadron to which it belonged. It had wandered as far as the mouth of the Red Sea, and worked its way back through incredible hardships. Before it made its appearance two vessels alone remained with Cabral out of the thirteen with which he had set sail. The three returned to Lisbon in company. Of the wealth brought back Vespucci gives the following account. He says there was an immense quantity of cinnamon, green and dry ginger, pepper, cloves, nutmegs, mace, musk, civet, storax, benzoin, porcelain, cassia, mastic, incense, myrrh, red and white sandalwood, aloes, camphor, amber, canne (Indian shot, *Canna Indica*), lac, mummy,* anib,† and tuzzia (or Thuja, Indian cypress), opium, Indian aloes, and many other drugs too numerous to detail. Of jewels he knew that he saw many diamonds, rubies, and pearls, and one ruby of a most beautiful colour weighed seven carats and a half, but he did not see all.

They reached Lisbon on the 23rd of July, 1501, where, although Portuguese historians are silent on the subject, it may be inferred from the rewards subsequently conferred on his family that Cabral met with the reception due to one who had secured such important benefits to his country. Immense, however, as had been the successes of Cabral in some respects, it will have been seen that he had not been so fortunate as he had wished in establishing a

---

* Portions of mummy that had been prepared with bitumen were in those days used as a drug.

† The Aniba is an aromatic wood from Guyana, with which Vespucci may have made acquaintance in the West, and perhaps without sufficient precision have mentioned among these eastern products.

factory at Calicut, although he had left some agents behind at Cochin. Nevertheless he had paved the way for effecting the object he had in view, which was not long in being carried into execution.

Before Cabral's return King Manoel had sent out a noble Galician named Juan de Nova with four vessels. He set sail from Belem on the 5th of March, 1501. In his voyage out he discovered the island of Ascension, but which he called the island of Conception. It appears first to have received its name of Ascension from Alfonso d'Albuquerque, who saw it again in May, 1503, and mentioned it in his journal, probably by mistake, under the latter name, which it has ever since retained. On the 7th of July, De Nova anchored at the watering-place of San Bras, beyond the Cape of Good Hope. Here Pedro de Ataide, who had been separated from Cabral in the great storm already described, had left in a shoe, so as to be sheltered from the winds, a letter announcing his having passed that way, and with what object, and urging all captains bound for India to go by way of Mombaza, where they would find other letters in charge of one Antonio Fernandes. By this means De Nova, who of course possessed no further information of those parts than what had been gathered from Vasco da Gama, became aware of the existence of two friendly and safe ports in India where he could take in a cargo, namely, Cochin and Cananor. At Quiloa he fell in with Antonio Fernandes, who delivered him Cabral's letter. He then proceeded to Cananor, where he was well received by the Rajah, who pressed him to freight his ship with spices from that port. From this De Nova courteously excused himself, stating that he had orders from the King to take a cargo first from the place where his agents had been left. He however desired that while he went to Cochin, a certain quantity of ginger, cinnamon, and other drugs, should be got in readiness, which quantity he would deduct from the cargo he would take in at Cochin. On the way he encountered the fleet of the King of Calicut, and with his artillery sunk

five large vessels and nine proas. At Cochin he was received with great warmth on account of the victory he had
gained over the Rajah of Calicut, and the King of Cochin
readily met the wishes of De Nova. The latter added six
or seven men to the number of agents already settled there,
returned to Cananor, completed the freighting of his ships
with a rich cargo, and set sail for Portugal. On his homeward voyage another piece of good fortune awaited him in
the discovery of the island of St. Helena, which seemed to
be providentially placed by the Almighty as a watering
station for vessels returning from India. De Nova reached
Portugal on the 11th of September, 1502, and was received
by the King with distinguished honour for the valuable
services which he had rendered to the country.

In the next year Antonio de Saldanha, on his way out
to India, gave his name to the Agoada de Saldanha near
the Cape of Good Hope, a fact to which we shall presently
have occasion to refer; and in this year the two Albu-
querques, Francisco and Affonso, sailed for India. The
former restored to the King of Cochin his territory, from
which he had been driven by the King of Calicut, and
founded the first Portuguese fort in India at Cochin, leaving
the famous Duarte Pacheco Pereira defender of the kingdom.
Affonso de Albuquerque, after touching on the coast of the
Terra de Santa Cruz discovered by Cabral, reached Coulam,
now Quilon, in Travancore, as yet unknown to the Portuguese, made terms of friendship with its King, and established a factory there.

In 1504, Diogo Fernandes Pereira wintered at Socotra,
which had not previously been reached by the Portuguese.

In 1505, King Manoel sent out a great expedition of two-
and-twenty ships and fifteen thousand men, which sailed
from Lisbon on March 25th, 1505, under Dom Francisco de
Almeida, the first Viceroy of the Indies, with instructions
to build fortresses at Sofala and Quiloa, and to free the
Portuguese commerce in India from the difficulties with
which it was oppressed. Juan de Nova sailed in this expe-

dition. As a proof of his success Almeida sent back, in the beginning of the following year, eight ships loaded with spices to Portugal, under the command of Fernam Soares. On their way they discovered, on the 1st of February, 1506, the east coast of the island of Madagascar, to which was subsequently given the name of Ilha de San Lourenço. In his outward passage Almeida conquered Quiloa, and dethroned the King, who refused to pay the stipulated tribute, and who had showed himself an enemy to the Portuguese. He set a new King on the throne, and himself crowned him with great solemnity. He also founded a fort there, which he named Santiago. On his arrival in India he founded the forts of Anchediva and Cananor. He solemnly crowned the King of Cochin, to whom King Manoel sent a rich crown of gold. Almeida also received ambassadors from the King of Narsinga and other princes, with whom he had entered on terms of alliance and friendship.

In 1505, Francisco de Almeida's son, Lourenço, discovered Ceylon, already known by overland accounts. He entered the Porto de Galle, and made its King an annual tributary to Portugal of four hundred bahars (about 300 pounds each) of cinnamon.

In this year also Pedro de Anhaya made the King of Sofala tributary to Portugal, and laid the foundations of a fort there on the 21st of September.

The high command which had been given to Almeida had been intended by the King for Tristam da Cunha, who was prevented from accepting it by a malady in the eyes, but, that obstacle being now removed, he was sent out on the 6th of April, 1506, with the command of sixteen vessels and thirteen hundred men to strengthen the dominion of Portugal in Africa and India. Affonso d'Albuquerque went out under his orders. It was in this voyage that the three islands bearing the name of Tristam da Cunha were discovered. In consequence of information brought to the King by Diogo Fernandes Pereira, the discoverer of the island of Socotra, to the effect that the Moors had a fortress therein, and held the Christians in subjection, Tristam da Cunha and Albuquerque were commissioned to take the fortress, which they succeeded in doing.

In this year João Gomez d'Abreu discovered the west coast of Madagascar on the 10th of August, St. Laurence's Day, from which circumstance the island received the name of San Lourenço. He gave the name of Bahia Formosa to the bay which he first entered (apparently the bay between Point Barrow and Point Croker). Tristam da Cunha, hearing of this discovery, visited various points of the same coast, and reached the end of the island on Christmas Day, and accordingly gave it the name of Cape Natal (now Cape Amber). The ship of Gomez d'Abreu doubled this cape, and running along the east coast reached the mouth of a river in the province of Matatana, where he landed, and left some Portuguese on shore. In a letter to King Manoel from Affonso de Albuquerque, dated Mozambique, 8th of February, 1507, he speaks of the discovery of the island of San Lourenço.

In 1506 Affonso de Albuquerque returned to India to succeed Francisco de Almeida so soon as the term of his governorship should expire, and on his way explored the strait of Bab-el-Mandeb. In this year the first elephant was sent to Portugal from India by Francisco de Almeida. In 1507 Lourenço de Almeida discovered the Maldives. In this year Duarte de Mello founded the fort of Mozambique. Affonso de Albuquerque explored the coasts of Arabia and Persia, made the King of Ormuz tributary to Portugal, and on the 24th of October laid the foundations of the fort there, which he named Nossa Senhora da Vittoria. In 1507 Tristam da Cunha sent on shore at Melinda three envoys charged with letters from the King of Portugal to the Emperor of Abyssinia. One was a Portuguese named Fernam Gomez of Sardo; another a Tunisian Moor named Sidi Mohammed; and the third a Christian Morisco named João Sanchez. The friendly King of Melinda undertook to further them on the way, but found himself unable to afford them such security as he wished, and the expedition for that time failed. The next year Albuquerque, having charged them with letters from himself to the King of Abyssinia, landed these same men at a point three leagues from Cape Guardafui, and they succeeded in reaching the court of Abyssinia, which was then governed by Helena, the grandmother of King David, who was in his minority. The result of this embassy was that an Armenian named Matthew was, some years after, sent as envoy from Abyssinia to the King of Portugal. He first proceeded to Goa, and thence to Lisbon, where he met with a gratifying reception from the King in the month of February, 1514. In 1515, Matthew returned with Duarte Galvão as ambassador from King Manoel to Abyssinia, but the latter died in 1517, in the island of Camaran in the Red Sea, and was unable to reach the court. Indeed it was not till 1520 that Matthew himself gained admission into Abyssinia by the Port of Massowah, when he was accompanied by Rodrigo de Lima, sent by the Viceroy as ambassador in lieu of Duarte Galvão.

In 1508 Diogo Lopez de Sequeira was commissioned by the King to examine the coasts of Madagascar and to discover Malacca. He discovered the islands which he named Santa Clara; he thence passed to Matatane, and coasted the island till he came to a bay, which he named San Sebastian, because he discovered it on the 20th of September, 1509. In August of that year he sailed for Malacca. Passing the islands of Nicobar, he went to Pedir and Pacem, about twenty leagues south-west of Pedir, in Sumatra, and raised pillars in both places, after having made terms of peace with their respective sovereigns. On the 11th of September he anchored at Malacca, the great emporium of the east, to which were brought cloves from the Moluccas, nutmegs from Banda, sandal-wood from Timor, camphor from Borneo, gold from Sumatra and Loo Choo, and gums, spices, and other precious commodities from China, Japan, Siam, Pegu, &c. There he established a factory. Fernam de Magalhaens was in this expedition.

In the year 1510 the illustrious Francisco de Almeida, on his way home to Portugal, was slain on the 1st of March in an encounter with the natives in the Agoada de Saldanha, near the Cape of Good Hope, which had been, as already stated, discovered by Antonio de Saldanha in 1503.

On the 25th of November of this year Affonso de Albuquerque conquered Goa, where he built a fort, and organized a municipal government, adopting measures of wise administration which paved the way for this city becoming the capital of the eastern empire of Portugal. In 1511 he conquered the city of Malacca, the King of which had treacherously plotted the death of Sequeira, with whom he had made terms of friendly intercourse. He then sent out expeditions to Siam, Birmah, and the East India Islands, and in this year and 1512, Antonio de Abreu discovered the island of Amboyna, and Francisco Serrão went to Ternate in the Moluccas. In 1512 or 1513 the Mascarenhas islands are supposed to have been discovered by Pedro de Mascarenhas, who appears to have sailed for India in 1511, and to have remained a considerable time during 1512 at Mozambique, but nothing certain is known of this discovery. In 1517 Fernam Peres de Andrade sailed to China, and entered into commercial relations with the Governor of Canton. He also sent to Nankin as ambassador Thomé Pires, who, however, was cast into prison and died after a captivity of many years, in consequence of his commission to the Emperor not being worded in conformity with the rules of Chinese etiquette, that sovereign being addressed by the Governor of the Indies in the same style as he was accustomed to address the Indian rajahs who were tributary to Portugal. Andrade returned to India in 1519.

I will here fulfil my promise to show the facility with which the author of the "Mémoires Chronologiques de Dieppe" could attribute to Dieppese the honour of greater discoveries than they could rightfully claim. He tells us that the Dieppese, Jean Parmentier, "had conjectured, from what some Indians had told him, that beyond the Indies there must be some great islands in which grew nutmegs, pepper, and cloves, and that these islands separated the Indian from the China Sea. He made an offer to Ango [a wealthy and enterprising Dieppese shipowner], who was then in his prosperity, to go out to explore them. The latter entertained the project, and entrusted two of his ships to the charge of Parmentier, who made a successful voyage, in which he visited these islands and reached the coasts of China. After a navigation of two years and a half he returned to Dieppe in 1529, with his two ships laden with nutmegs, cloves, and other spices."

Now, so far is this from the truth, that we know to an indisputable certainty that in the year 1529 this Jean Parmentier made a voyage to Sumatra, and in that voyage he died. He was accompanied therein by his intimate friend, the poet Pierre Crignon, who on his return to France published in 1531 the poems of Parmentier with a prologue containing his eulogium, in which in his endeavour to express the highest praise of him that he could, he says that he was " the first Frenchman who undertook to be a pilot to guide ships to the American land called Brazil, and apparently the first Frenchman who had discovered the Indies *so far as Sumatra, and if death had not prevented him, I believe he would have gone as far as the Moluccas.*" * This from a man of education, a shipmate and bosom friend of Parmentier in his last voyage. What then becomes of the veracity of M. Desmarquets ?

* " Le premier François qui a entrepris à estre pilotte pour mener navires à la terre Amérique qu'on dit Brésil, et semblablement le premier François qui a descouvert les Indes *jusqu' à l'Isle de Taprobane (Sumatra) et si mort ni l'eust pas prevenu, je crois qu'il eust été jusques aux Moloques.*" (See Léon Guérin. "Les Navigateurs Français," p. 157.)

CHAPTER XXI.

THE discovery of the New World was a fertile source of misconstruction and misnomer. Columbus to his dying day believed that Cuba was a part of Asia. Three years after his death the vast continent which his genius and perseverance had disclosed received a name which was other than his, and when at length the great ocean which bathed the western shores of that continent was revealed, the very points of the compass were dislodged from their natural position in the process of providing it a name. A glance at the map of the world is enough to show that the Atlantic and Pacific Oceans in their general extent lie east and west of America, yet from the simple accident that the Pacific lay south of that part of America (the Isthmus of Darien) from which it was first discovered, it received the inappropriate name of the South Sea. This designation was applied to it even in its most northern part, and, by way of antithesis, the Atlantic has occasionally been called the North Sea, even in its most southern part.

From 1505 to 1507 the court of Spain was earnestly engaged in the project of finding a direct route to the Spice Islands by the west, and on the 29th of June, 1508, Vicente Yañez Pinzon and Juan Diaz de Solis, reputed to be the ablest navigator and pilot of his day, sailed from San Lucar and explored the coasts of South America from Cape St. Augustine to the fortieth degree of south latitude, and yet missed the mouth of the La Plata. It was not till 1513, on the 25th of September, that Vasco Nuñez de Balboa, who had in 1510 been placed in command of a small colony at Santa Maria on the Gulf of Darien, perceived the Pacific from the ridge of the Sierra de Quarequa. Kneeling on the scarped summit from which he gazed on this vast and unknown ocean, he raised his hands to heaven in wonder and gratitude at the immensity of the revelation that had been made to him. But he had to encounter the resistance of the natives before he reached the shore. These he managed without much difficulty to subdue. Meanwhile he despatched Francisco Pizarro, Juan de Escaray, and Alonzo Martin de Don Benito in search of the shortest pathway to the sea. Alonzo Martin on reaching the shore threw himself into a canoe which happened to be lying alongside, and was the first European who can be said to have navigated the Pacific. On the evening of the 29th of September, Balboa, with twenty-six of his companions, reached the strand, and walking into the water knee-deep, with his sword in one hand and the flag of Spain in the other, took formal pos-session of the newly-found ocean on behalf of his sovereign the King of Spain, and vowed to defend it against all his enemies. In token of possession he erected piles of stones on the shore. When the King of Spain heard of this dis-covery, he sent out Pedro Arias de Avila as Governor of Darien. Avila sailed from San Lucar with fifteen vessels and fifteen hundred men, and by his tyranny and exactions after his arrival spread desolation over the whole country from the Gulf of Darien to the Lake of Nicaragua. A dissension arose between him and Balboa, and in 1517 the latter, charged with calumny against the Government, was put in chains, tried, condemned, and beheaded.

In October, 1515, Juan Diaz de Solis was sent out with the express purpose of discovering a passage to the Moluccas by the west, and in January, 1516, he entered the Rio de la Plata, to which was originally given the name of Rio de Solis. According to Herrera the name of La Plata, which means "silver," was not given it till 1527, when Diego

Garcia found some plates of that metal, probably from the mines of Potosi, in the hands of the Guarani Indians. The expedition was fatal to De Solis. Having anchored in the mouth of the river, he attempted a descent in the country, and he and eight of his men were massacred by the natives, and their bodies were cut in pieces, roasted, and devoured in sight of the ships. This was probably in August, 1516. The survivors had no heart to proceed further, but returned to Europe, and thus King Ferdinand died without seeing the accomplishment of the great object of his anxiety.

It was not till 1517 that Magalhaens laid before Charles V., at Valladolid, his proposals for effecting the great discovery, but here we have to deal with a character and an achievement of colossal proportions, which demand especial description in a work devoted to the "Results" of the life of Prince Henry the Navigator.

Fernam de Magalhaens, better known by the Spanish form of his name, Magellan, was of noble Portuguese parentage, but we know little for certain of his early youth, except that he was brought up in the household of Queen Leonora, the wife of Dom João II. The instruction in mathematics and geography which he would there receive would be of an advanced kind, as at that time these sciences, which had received large development in Portugal under the auspices of Prince Henry, were taught by the two eminent Jews, named Josef and Rodrigo, of whom mention has been already made. He afterwards entered the service of Dom Manoel. In March, 1505, when little more than twenty years old, he joined the expedition of Francisco de Almeida, first Viceroy of the Indies, to Quiloa, in which were João de Nova, the constant rival of Albuquerque, already known to the reader as the discoverer of St. Helena, Diego Correa, and Magalhaens' own bosom friend João Serrão. We have also seen that he was at the discovery of Malacca by Diogo Lopez de Sequeira in 1509. His sojourn in India and his campaigns in the extreme East, enabled him to gather information on which he afterwards based his memorable enterprise.

One of his cousins, Francisco Serrão, who in 1511 first went to Ternate, married a woman of that island, and settled there, having contrived to secure the good-will of the Malay sovereign. He thence communicated to Magalhaens the great commercial advantages which might be secured by foreigners from intercourse with his adopted country.

Duarte Barbosa, also, the future brother-in-law of Magalhaens, contributed by his explorations, the account of which he completed in 1516, to that information which influenced the subsequent movements of Magalhaens. After his return from the East, Magalhaens served in Africa, and during a razzia at Azamor, was wounded in the knee, from which wound he remained lame all the rest of his life. In the distribution of some cattle then captured some disagreement arose, which led to complaints against him at court, and to much dissatisfaction. Conceiving himself unjustly treated by the king in the matter of these complaints and the mode of their reception, Magalhaens resolved to renounce his nationality, and to leave Portugal. His experience in navigation, and his acquaintance with the geography of the Moluccas, made him an acceptable visitor to Charles V., who was then but just returned from Flanders. Magalhaens arrived in Seville on the 20th of October, 1517, accompanied by two other malcontents, Rui Faleiro, a learned cosmographer, and Christovam de Haro, a wealthy merchant, who already possessed immense commercial relations with India. The Papal Bull of Alexander VI., which had determined that a line drawn from pole to pole a hundred leagues west of the Azores should be the boundary between the claims of Spain and Portugal, was practically indecisive on account of the difficulty of measuring longitudes. Nor were matters improved by the Convention of 1494, in which the line of demarcation was removed to three hundred and seventy leagues west of the Azores, for though Portugal thereby gained in South America, Spain became also a considerable gainer in the East, the sea way to which had been first opened up by Portugal. The Moluccas formed, moreover, the very garden

of those spices, the commerce of which was so eagerly coveted.
Magalhaens gave it as his opinion that the Moluccas fell within
the Spanish boundary, and undertook to take a fleet thither
by the south of the American continent. The position of
Magalhaens at Seville was strengthened by his marriage, in
January, 1518, with the daughter of his relative, Diogo Barbosa, with whom he had taken up his quarters, and who had
sailed to the Indies in 1501 under the order of Juan de Nova.
He was now commander of the Order of Santiago, and
lieutenant to the Alcaide of the Castle of Seville. Magalhaens had further the good fortune to secure the friendship
and aid of Juan de Ovando, the principal factor of the
Contratacion or chamber of commerce. To the latter was
mainly owing the arrangement with the Emperor for that
great expedition which was afterwards to hold so distinguished a position in the history of nautical discovery.

In August, 1519, Charles V. gave Magalhaens five ships,
with the rank of Captain General, and it is remarkable that
every one of the vessels was accompanied by a Portuguese pilot.
The *Trinidad*, of one hundred and twenty tons,* on board of
which Magalhaens hoisted his flag, had Stevam Gomez for
pilot ; the *San Antonio*, also of one hundred and twenty
tons, commanded by Juan de Cartagena, had indeed a
Spaniard, Andres de San Martin, for pilot, but he was
accompanied by the Portuguese pilot, João Rodrigues de la
Mafra ; the *Concepcion*, of ninety tons, commanded by
Gaspar de Quesada, had for pilot the Portuguese, João
Lopez de Caraballo ; the *Vittoria*, of eighty-five tons, under
the command of Luis de Mendoza, was piloted by the Portuguese Vasco Gallego ; and the *Santiago*, of seventy-five tons,
was commanded by João Serrão, a Portuguese pilot, on
whose skill and knowledge of the East, especially of the
Moluccas, of which they were in search, Magalhaens placed
great reliance.

* To produce a correct impression on our minds of the size of these vessels,
one-fifth may be added to the recorded tonnage to make the equivalent of the
measurement of the present day ; *e. g.*, the *Trinidad*, recorded as of one hundred
and twenty tons, may be estimated at one hundred and forty-four tons.

The fleet, which consisted of two hundred and sixty-five persons, set sail from San Lucar de Barrameda on the 21st of September, 1519, and reached what is now called Rio de Janeiro on the 13th of December. Magalhaens named it Porto de Santa Lucia. Thence they came to the Rio de la Plata, where at first they supposed they had found a channel to the Pacific; but giving up this hope, they proceeded south, and on the 31st of March, 1520, entered Port St. Julian, where Magalhaens stayed five months. The voyage, if destined to be a great one in the world's history, was a most unhappy one. It is not improbable that national jealousy had much to do with the insubordination exhibited by some of the Spanish captains on the one side, and the extreme severity resorted to by Magalhaens by way of re-pression. The revolt was initiated off the coast of Africa by Juan de Cartagena, Captain of the *San Antonio*. Discontent had arisen from Magalhaens having deviated from the course previously settled in a consultation with the principal officers, and by which deviation, unfortunately, much time was lost. Juan de Cartagena took upon himself to remon-strate with Magalhaens, who simply replied that it was his duty to follow his commander, and not call him to account. On a later occasion, this was followed by conduct so muti-nous that Cartagena was not only deprived of his command, but made a prisoner, and the command of his vessel—the *San Antonio*—was given to a kinsman of Magalhaens, Alvaro de Mezquita. This led to worse. On the morrow of their arrival at Port St. Julian, which was Easter Day, the whole fleet was summoned to attend mass on shore; but Luis de Mendoza and Gaspar de Quesada, the captains of the *Vittoria* and *Concepcion*, absented themselves. This looked strongly like disaffection, and so it proved. Magalhaens having decided to winter at Port San Julian, and finding fish abundant, judged it expedient to retrench the allowance of provisions. This, with the cold and barrenness of the country, caused great murmuring, and the crews desired that their commander would either issue the usual allowance, or return, for they had already gone further than any other, and it was impossible to say what dangers lay before them. To this Magalhaens replied that what he had undertaken he intended to perform : that the King had ordered the voyage, and that it was his duty to go on till he found a termination to the land, or a strait. He reminded them that the place where they were to winter abounded in wood and water and fish and fowl, and he engaged that they should have no lack of bread and wine. He further held out to them the confident hope that they should discover a world as yet unknown. But they contended that all the spices of the Moluccas were not worth so long a voyage, in which they had to cross the line and coast the whole of Brazil, spending seven or eight months in passing through so many climates, and to reach a point so much easier of attainment in the opposite direction. But besides these reasonings, no doubt, the being commanded by a Portuguese was hateful to them, and they mistrusted his loyalty to their country. The result was, that one night Gaspar de Quesada boarded and took possession of the *San Antonio*, made Alvaro de Mezquita, the captain, and Mafra, the pilot, prisoners, and released Juan de Cartagena.

Magalhaens now saw plainly that summary measures were more prudent than leniency. He first secured the *Vittoria* by sending thirty men on board her under Gonzalo Gomez de Espinosa, who by the commander's orders poniarded Mendoza. At midnight it happened that the ebb-tide caused the *San Antonio* to drag her anchors, and to float down the river towards Magalhaens' ship, who supposing her to be come with the intention of fighting, fired into her a ball, which made its way into the cabin, and passed between the legs of Mafra, the pilot, who was confined there. The ship was presently boarded, and Quesada, with the rest of the mutineers, was captured. The *Concepcion* after this surrendered at discretion. After a long inquiry, Quesada was condemned to be strangled, and a servant of his, who would otherwise have been hanged, was reprieved on condition of his being the executioner. Juan de Cartagena was sentenced to be put on shore with a French priest who had shared in the mutiny. The rest by a wise policy were pardoned, and the regulations respecting the provisions were modified.

When May set in, João Serrão was sent southward to examine the coast, and at twenty leagues distance discovered on the 3rd of May a river, which in honour of the day he named Santa Cruz, but he had only passed three leagues beyond it, when his ship, the *Santiago*, was driven violently ashore by a gust of wind from the east, and became a total wreck. The crew, who happily were all saved, contrived in a few days, during which they had to live on herbs and shell-fish, to make a small boat of some planks which were driven on shore, and by this means two men crossed the Santa Cruz and managed to reach Port San Julian, though in a most exhausted state. Assistance and provisions were immediately sent overland, but the weather was so severe that they had to thaw ice for drink. The crew were rescued and distributed among the other ships, Serrão being appointed to the command of the *Concepcion*, and Duarte Barbosa to the *Vittoria*.

It was not till they had lain two months in Port St. Julian that any sign of a native appeared. At length a man of gigantic size was seen on the beach, singing and dancing and sprinkling dust upon his head. As this was supposed to imply friendliness, a sailor was sent on shore to imitate his movements, which he did so well that the giant accompanied him on board. He first pointed to the sky, by way of asking whether they had come down from that region. When he saw his reflection in a looking-glass, he started back with such sudden surprise that he overturned four Spaniards who were behind him. Other natives soon came, the smallest of whom was taller and stouter than the largest Spaniard. They had bows and arrows, and coats made of skins. A kettle full of pottage with biscuit was served to them, enough for twenty Spaniards, but six of these men ate it all up. They then went on shore.

Two of them the next day brought some of the meat of

the animals whose skins they wore, and were much pleased with the present of a red jacket to each in return.   One of them came often afterwards, and having been taught the Lord's Prayer, was baptized and received the name of Juan Gigante (John Giant).   Observing that the mice were thrown overboard, he begged to have them to eat, and took on shore all that they could give him, but after six days they saw no more of him.   In all they saw only eighteen of the natives.   They wore shoes made of the skins of the guanaco, which gave their feet the appearance of paws, for which reason Magalhaens gave them the name of Patagones, *pata*, in Spanish, meaning a paw.   After a lapse of twenty days, four of them re-appeared, and a most treacherous plan was resorted to in order to capture them, the object being to carry away the two youngest and to exchange the other two for their wives, with the view of importing this gigantic race into Europe.   After the two youngest had had their hands filled with presents of different kinds, bright iron rings were offered them, but as, much as they wished for them, they could not take them in their hands, it was proposed to put them on their legs, and thus unsuspectingly they were chained.   As soon as they perceived the ungenerous trick, they struggled furiously, imploring Setebos, their demon, to come to their help.   Nine Spaniards seized the other two and with difficulty bound them, but one broke loose at the time and the other afterwards escaped.   The next day seven of the Spaniards had an encounter with nine of the natives, in which one of the former was shot; to avenge whose death, Magalhaens sent out twenty men to take or slay all they might meet, but happily, though they were eight days absent, they encountered none.   The natives were wanderers, and carried with them their huts, which were made of light framework covered with skins.   The men were about seven feet six inches high, and remarkably swift of foot; the women not so tall, but stouter.

After taking possession of the country for the King of Spain, by erecting a cross on a hill which they named

Monte Christo, the ships set sail on the 24th August, leaving Juan de Cartagena and Sanchez de Reino on shore, with a supply of bread and wine.

Finding in the river of Santa Cruz a great abundance of fish, with wood and water, the fleet put in there till the 18th of October, when they proceeded southward, and on the 21st reached a cape, from which the coast turned directly *due west.* In honour of the day, which was the feast of St. Ursula, they named the cape Cabo de las Virgenes. Magalhaens then sent on two small ships to explore the inlet, but not to be absent more than five days. At the end of that time they returned with the report that while one of them had only found some bays containing many shoals, the other had sailed three days westward without finding an end to the strait, and that the tide was stronger when it flowed westward than when it ran to the east. This news was so encouraging that the whole fleet entered the channel. The *San Antonio* was sent on to explore, and after sailing fifty leagues brought back the same promising account as the others. There now remained only provisions enough for three months, and Magalhaens wisely called a council of the officers, at which the majority agreed with him in the desirableness of proceeding, but Stevam Gomez was for returning to Spain, lest they might be caught in calms and perish of starvation. Magalhaens, however, declared that "even if they were to be reduced to eating the leather on the ship's yards, he would fulfil his promise to the Emperor, and, by the help of God, he hoped to succeed." He accordingly set sail, forbidding any on pain of death to say a word about returning to Spain, or about shortness of provisions.

In consequence of many fires being seen on the southern shore of the strait, Magalhaens named that country Tierra del Fuego—(The Land of Fire).

As they proceeded westward another arm of the sea towards the south-east made its appearance and invited examination, and the *San Antonio* was sent to explore it, with orders to return in three days. As six days passed

without her re-appearance, the *Vittoria* went in search of her,
and subsequently the whole fleet: but as no sign appeared
of her, it was concluded that she had sailed for Spain, as
afterwards proved to be the case. The fleet now resumed its
course westward, and on the 27th of November, 1520, thirty-
seven days after the discovery of the eastern entrance,
emerged from the strait into an open sea. The Cape which
terminated the strait at the westward on Tierra del Fuego
was named Cabo Deseado (The Desired Cape), and that in-
flexible man, whom neither danger could deter nor death
intimidate, is said to have shed tears of gratitude as he
beheld this realization of his hopes. His illustrious name,
as was only just, was subsequently given to the strait which
had thus been traversed, although at first it was named after
the *Vittoria*, which had first sighted the eastern entrance.

Now that the great discovery was effected, it was desirable
to make for the warm latitudes, and Magalhaens with the three
remaining ships, the *Trinidad*, the *Vittoria*, and the *Concepcion*, steered north-west. On their way they discovered on
the 24th of January, 1521, an uninhabited island in 16° 15′ S.,
which they named San Pablo (Saint Paul), probably from
the remaining Patagonian who, dying on the passage, had
been baptized in that name. His fellow-captive had been
carried off in the *San Antonio*, on board of which he also
perished as he neared the warm latitudes. Two days the
ships remained off San Pablo in the hope of recruiting their
diminished stores with fish, but without success. At two
hundred leagues distance they discovered on the 4th of
February, another equally profitless, which from the number
of sharks near it, they named Tiburones. In their disappointment they named the two islands, though so wide apart,
Las Desventuradas (The Unfortunate Islands), for their distress was becoming so great that they even ate sawdust
and the leather on the rigging. To save the fresh water,
they mixed one-third of salt water with the fresh to boil
their rice, which brought on the scurvy, and twenty of the
men died of that disease.

They crossed the line on the 13th of February, and on the 6th of March, they had the happiness of reaching some beautiful islands, the natives of which came out to meet them in canoes bringing cocoa nuts, yams, and rice. They were a sturdy race with olive complexions and long hair. They dyed their teeth red and black, and were naked with the exception of an apron of bark.

Magalhaens would gladly have stayed here, but the pilfering habits of the people made it impossible. After some contentions on this account, at length they stole a skiff, which act Magalhaens punished by landing ninety men, and firing their houses. Several natives were killed, and all the provisions that could be found were carried off. The skiff was soon set adrift and re-captured, but the event made Magalhaens decide on leaving these islands, which from the propensity of their inhabitants received the name of The Ladrones (The Thieves).

From the Ladrones, Magalhaens steered W. and by W.S.W., and on the 16th of March reached a group of islands to which he gave the name of Archipielago de San Lazaro, a name which was afterwards replaced by that of the Philippines. He anchored off the island of Humunu (Jumonhol). On the 18th nine natives came out to him in a canoe from the island of Zuluan. He received them cordially, and they gave him fruits and palm wine. They afterwards brought provisions in exchange for trinkets, and the ships remained there nine days. These people were tattooed and went nearly naked, but some of the chiefs wore earrings and bracelets of gold, and a light covering of cotton, embroidered with silk, on the head. Their weapons also were sometimes ornamented with gold. On the 28th of March, Magalhaens anchored off Mazagua, with the chief of which island he entered into very friendly relations. On the 31st of March, being Easter-day, mass was celebrated on shore with great solemnity. The rajah, who was named Colambu, and his brother were present, and when the Spaniards knelt in adoration, they followed their example. On inquiry Magalhaens found that they worshipped a supreme being, whom they named Abba.

On the 5th of April Magalhaens sailed under the guidance of Colambu himself to the large island of Zebu, the King of which was Colambu's relation. They reached the town of Zebu on the east side of the island on the 7th. Their arrival at first occasioned great alarm, which was allayed by Colambu, who represented the new-comers as a peaceable people who wished to barter goods with the islanders. At length all difficulties were removed and presents were interchanged. Here Magalhaens obtained supplies of provisions in great abundance.

Being anxious to introduce the Christian religion, for which the people seemed favourably inclined, with the King's consent he erected a stone chapel on the shore, and it having been duly consecrated, and also ornamented with tapestry and palm branches, he landed on Sunday, the 14th April, with many of his people to hear mass. The procession was headed by the royal ensigns and two men in complete armour. The King and a large number of natives came to observe the service, and behaved with the greatest decorum. By means of the interpreter, a native of Sumatra, who had accompanied the expedition, the priest endeavoured to instruct them in the Christian faith, and soon the King and the chief of Mazagua requested to be baptized. The former had given him the name of Carlos from the Emperor. Colambu was named Juan, and in a short time the Queen, the Princesses, and the residents of the town followed their example. Unfortunately, while explaining the duties required by Christianity, the foremost of which was the destruction of their idols, Magalhaens held out an inducement to conversion which savoured much more of worldliness than of the religion he was advocating. He assured the King of Zebu that one of the benefits of Christianity would be the power of easily subduing his enemies. Now the island contained several little independent sovereignties which were often at war with each other, and a temptation like this was a great

provocation to his zeal. Whether resting on so insecure a
foundation it was likely to be permanent, we shall presently
see. To show the King of Zebu the value of Spanish protection, Magalhaens called a meeting of the other chiefs,
four of whom attended. These he required on pain of death
to pledge themselves to obey the new Christian king. They
yielded, but one of them afterwards slighting the command,
was attacked in the night by Magalhaens with forty men,
who ransacked and burnt one of his villages and set up a cross
on the spot.

Near Zebu was a small island named Matan, to the
chief of which, who was named Cilapulapo, he sent a
similar requirement that he would submit to the Christian
King of Zebu, on pain of having his town, named also Matan,
similarly destroyed. The gallant chief replied that he wished
to be on good terms with the Spaniards, and to prove his
words sent them a present of provisions, but absolutely
refused to obey strangers of whom he knew nothing, or to
submit to those whom he had long been accustomed to command. Against the advice of the King of Zebu as well as
of João Serrão, Magalhaens determined to punish the chief
of Matan for his contumacy. At midnight, on the 26th of
April, Magalhaens sailed for Matan with three boats and
sixty men, accompanied by the King of Zebu and a thousand
natives. It wanted two hours of daylight when they arrived,
but it was low water, and while waiting for the morning,
Magalhaens sent a messenger to the chief, proposing that if
he would then make submission, all would be forgotten.
The only answer was a defiance. The King of Zebu would
have led the attack with his thousand men, but his offer
was declined, and he himself ordered to remain quiet with
his men in their canoes and see how the Spaniards would
fight. Eleven men were left to guard the boats, and forty-
nine, including Magalhaens, landed. They first set fire to
some houses, when a strong body of Indians appeared in one
direction, and as soon as the Spaniards had prepared to
attack them, another body of Indians made their appearance

from another quarter. Magalhaens was thus obliged to
divide his little band into two. The battle was kept up with
projectiles during the greater part of the day, the Indians
using stones, lances, and arrows, and the Spaniards their
muskets and cross-bows. After a time it was perceived that
the fire of the Spanish musketry was not so deadly as had
been apprehended, and the islanders had further noticed
that the legs of their enemies could be assailed with greater
effect than their heads and bodies, which were covered with
armour. Moreover, Magalhaens had detached a small party
to set fire to some houses, more than twenty of which were
burnt, but two of the party were killed by the Indians. The
latter became now bolder, and approached nearer with a view
to taking the life of Magalhaens himself. His ammunition
now began to fail, and he ordered a retreat, but immediately
after received a wound from an arrow in the leg. It
was some distance to the boats, and the Spaniards lost all
order in their retreat, but Magalhaens himself bravely con-
fronted the Indians, looking back from time to time to see
if the men had reached the boats. He had just slain an
Indian by hurling back at him his own lance, when in
attempting to draw his sword, he found that a wound in
the right arm prevented him from doing so more than half-
way. The islanders seeing this, attacked him boldly. A
severe wound in the right leg caused him to fall forward
on his face, and he was speedily despatched. In obedience to the unfortunate order which he had received,
the King of Zebu and his people had remained in their
canoes, looking quietly on, but seeing the failing condition
of the Spaniards at the close, came to their relief and saved
many of them. Eight Spaniards died with Magalhaens, and
twenty-two were wounded.

Thus fell this great navigator, second only to Columbus
in the history of nautical exploration. Midway in the execution of a feat such as the world had never witnessed, the
very hardihood which already had rendered that achievement
possible, had now, by degenerating into presumption, deprived him of the glory of its fulfilment.

The Spaniards who escaped elected Duarte Barbosa and the pilot João Serrão, as joint Commanders-in-chief. We have now to witness the hollowness of that rapid conversion to Christianity professed by the King of Zebu, which very naturally betrayed itself so soon as the false hope on which he had depended was proved to be unfounded. The defeat of Magalhaens was the most conclusive proof that Christianity did not insure victory in battle. The friendly co-operation of allies was now replaced by the basest treachery. On Wednesday, May the 1st, this Christian King invited the commanders and officers to an entertainment on shore, in order that he might deliver to them the presents intended for the Emperor, which were now ready. But he plotted their death, a threat from the King of Matan rendering it necessary for him to prove himself their enemy. Against the advice of Serrão, who had his misgivings, Barbosa accepted the invitation, and by means of an unmerited taunt, induced Serrão to accompany him. Twenty-six Spaniards went on shore, and in the midst of the feast were attacked by a body of armed men, who had been concealed. All of them were murdered with the exception of the interpreter and Serrão, who had been a favourite amongst them and was spared. The Spaniards on board hearing the cry of the victims, whose bodies were presently dragged to the shore and cast into the sea, weighed anchor and fired upon the town. The Indians now brought forward Serrão, naked and in chains, who implored his countrymen to desist from firing and to ransom him; but Caraballo, the principal officer then in command, knowing that Serrão's detention left him without a superior, threw aside every feeling of humanity and made sail, basely abandoning Serrão to the mercy of the natives, who, doubtless, put him to death.

About one hundred and fifteen only now survived of the armada, too small a number to manage the three remaining ships. They sailed for the island of Bohol, S.E. of Zebu, where they burned the *Concepcion*, which was the oldest of their three vessels. Caraballo was elected Commander-in-chief, and Goncalo Gomez de Espinosa was appointed to the command of the *Vittoria*.

From Bohol they sailed S. S. W., and after touching on the west coast of Mindanao, passed by Cagayan to the island of Palawan, and thence to Borneo.

On the 8th of July, 1521, they anchored at about three leagues distance from the city of Borneo, and the next morning were visited by the King's Secretary, inquiring their business and whence they came. They sent a present to the King, whose name was Siripada, and received from him an invitation for two of the Spaniards to visit the city. Espinosa, Captain of the *Vittoria*, accordingly landed with six men, and was conveyed on an elephant to the King, who when he had satisfied his curiosity, dismissed him with a present of Chinese damask. This city was inhabited by Mohammedans, but there was another larger one inhabited by the earlier natives, who worshipped the sun and moon. Both cities were built on piles over the water. As the people continued friendly, and the Spaniards had occasion to caulk the ships, five men were sent to the city of Borneo to procure wax to serve instead of pitch for that purpose, but as three days elapsed without their returning, and some large junks were seen to enter the port and anchor near them, while a host of smaller craft put off from the shore, the Spaniards tripped their anchors, attacked and captured two of the junks, and with their artillery dispersed the smaller vessels. Two days after, the 31st of July, they captured another junk, containing one hundred men, and among them, one of the King of Borneo's generals, said to be the son of the King of Luzon. This man was sent on shore with several others to the King, to tell him that if the five Spaniards were not restored, every vessel coming into the port should be destroyed. Two of them only were returned, but the Spaniards being anxious to proceed on their voyage, set sail a few days after without the other three ; an inexplicable proceeding, as they had so large a number of natives to give in exchange. It afterwards appeared that Caraballo had privately released the son of the King of Luzon for a sum of money, a circumstance which caused his removal from the command by general consent when the ships were ready to sail. Espinosa was made Commander-in-chief, and Juan Sebastian del Cano, a Biscayan, was appointed to the command of the *Vittoria*.

Off the coast of Mindanao, they captured a vessel of the country containing some of the chiefs of Mindanao, by whose instructions they altered their course to south-east with the view of finding the Moluccas. In the passage their Mindanao prisoners leaped overboard in the night-time, and made their escape. It was on Wednesday, November 6th, 1521, that four islands were descried, which, as they learned from one of the pilots, who being wounded could not escape, were the long-sought-for Molucca Islands, the object for the discovery of which by a western route this most wearisome voyage was undertaken.

On Friday, the 8th, they anchored at Tidor, and the next day the King, whose name was Almanzor, richly dressed in embroidered linen and silks, came on board the *Capitana*, where he met with a cordial welcome, and on taking leave embraced the Captain, expressing himself much gratified with his reception. As at first the demand for spices was not promptly responded to, the Spaniards thought of seeking a cargo at one of the other islands, but when the King heard this, he made a solemn promise to them on the Korán, to provide them with all the spices they desired if they would engage not to seek them elsewhere. To this Espinosa consented, and to show his consideration for the King, at his request liberated his prisoners and killed all the hogs on board, the dislike of the Mohammedans to these animals being intense. By way of compensation, the King made the Spaniards an ample present of goats and poultry.

On the 13th of November, a Portuguese named Pedro Affonso de Lorosa, who had come to the Moluccas with the first discoverers, came from Ternate and informed the Spaniards that the news of their voyage had reached the

Moluccas, nearly twelve months before. Francisco Serrão, the discoverer, had died at Ternate eight months previously. Lorosa begged permission to return with his wife to Europe in the Spanish ships, which was granted.

On the 25th, Almanzor having procured a large quantity of cloves from the neighbouring islands, invited the Spaniards to a banquet on shore, saying that it was customary to entertain merchants on the occasion of their first taking in a cargo. But the Spaniards having a wholesome recollection of the fatal feast at Zebu, prudently declined the invitation with thanks.

Having laid in their stock of spices and provisions, they prepared to sail on Wednesday, December 18th, and the Kings of the islands came to pay their farewell visit, but while the *Trinidad* was weighing, it was found that she had sprung a leak. As several days were spent in vain endeavours to remedy the mischief, or to find out where the water entered, it was resolved that the *Vittoria* should sail forthwith to Europe by the Cape of Good Hope, and that the *Trinidad*, after repairing, should sail eastward for Panama. The *Vittoria* had in her forty-seven Europeans, and thirteen Indians. The *Trinidad* fifty-three Europeans. The King of Tidor sent two pilots to steer the *Vittoria* clear of the neighbouring islands, and she proceeded on her homeward voyage.

On the 10th of January they reached Mallua or Ombay, where they remained fifteen days to repair, and on the 26th they came to the island of Timor, where they took in white sandal-wood, cinnamon, and wax. Here a mutiny broke out, in which several lives were lost. Some of the mutineers were executed, and others left the ship.

On February 11th they sailed from Timor, and in order to avoid Portuguese ships in rounding the Cape, sailed as far south as 42° S., but with all their caution, when they passed the Cape on the 6th of May, they approached it within five leagues. Their sufferings must have been great, for the ship was leaky and provisions scarce, and at all hazards, when they reached the Cape Verde Islands, they were compelled to put in at Santiago, for their provisions were now exhausted. This was on the 9th of July.

To prevent the Portuguese from suspecting whence they came, they said that they had come from America, and thus they procured some rice from the shore. It was here they discovered that in sailing round the world they had lost a day in reckoning their time, for whereas at Santiago it was Thursday, the 10th of July, the *Vittoria's* account made it Wednesday, the 9th.

Through the imprudence of one of the sailors who offered spices in payment for what he wanted to purchase, the remnant of this extraordinary expedition narrowly escaped even at this late period from a ruinous disaster. The boat was stopped, and the Portuguese made preparations to attack the ship, but fortunately the movement was perceived in time, and Del Cano weighed anchor and left the island.

On Saturday, the 6th of September, 1522, the *Vittoria* arrived at San Lucar, with eighteen survivors only of the noble fleet which had sailed from the same port on the 20th of September, 1519. Thus three years, all but fourteen days, had been expended in this most eventful and wonderful voyage—a miracle of resolute perseverance under inconceivable hardships. It was appropriate that the only ship which had effected this great achievement should have borne the name of *Vittoria*, for a victory had been gained such as the world had never witnessed. On his arrival, Del Cano, the fortunate recipient of the honours which had been toiled for and deserved by the talents and indomitable resolution of his great commander, Magalhaens, was summoned by the Emperor to Valladolid, and received with great distinction. A life pension of five hundred ducats was conferred on him, with a patent of nobility. The coat of arms granted him by the Emperor bore branches of the clove, cinnamon, and nutmeg trees, with a globe for a crest, and the motto, " *Primus circumdedisti me.*"

Thus far we have had great things to record as links in one continuous chain of magnificent discoveries. Eastward and westward the bold hand of man had torn away the veil from the bosom of ocean, and at length he has encircled her waist with his grasp. One little century has transpired since the budding manhood of Prince Henry had seen him bent upon the investigation of the unknown paths of the Atlantic, and now the world has been encompassed by men whose daring was fostered by the example of his perseverance. A century shall not have transpired since his first feeble but persistent efforts had succeeded in rounding the then formidable Cape Boyador, before another gem shall be added to the crown of that glorious little nation, to whose courageous efforts his genius and constancy had given the first impulse.

The great discoveries of Francisco Pizarro and Diego de Almagro on the western coast of South America soon followed.

It was not long after the completion of the ever memorable voyage which has been just related, before Australia, that vast insular continent, with whose discovery we are accustomed generally to connect the name of our illustrious Cook, or at furthest that of Dampier, was explored both on its eastern and western sides by the Portuguese. True, the knowledge of this fact lay dormant till the close of the last century, hidden in the testimony of some valuable old French manuscript maps, whose worth and importance, nay, even whose existence, appears to have been unrecognised till after the gallant Cook had completed his unrivalled series of explorations.* Long previously, no doubt, the great fact had

* It is greatly to be regretted that Alexander Dalrymple, at that time Hydrographer to the Admiralty and East India Company, to whom England is largely indebted for its commercial prosperity, and who panted for the glory of discovering a great southern continent, should have allowed his jealousy of Captain Cook's appointment to the *Endeavour* to lead him into an injurious insinuation that the great Captain's discoveries on the coast of New Holland were the result of his acquaintance with one of these pre-existent maps. This he bases on the resemblance of the names of bays and coasts on the map to those given by Captain Cook to parts of New Holland which he had himself discovered. The

been faintly indicated on engraved maps, but in a fashion far less definite than on these yet older manuscript maps, which, drawn before copper plates were used for cartography, tell forth unequivocally the story of the important discovery. The earliest engraved indication of Australia occurs on a mappe-monde, in the third volume of the Polyglot Bible of Arias Montanus, dated 1572, and is the more striking that it stands unconnected with any other land whatever, and bears no kind of description. It is simply a curved line indicating the north part of an unexplored land exactly in the position of the north of Australia, distinctly implying an imperfect discovery. In the other *engraved* maps of the sixteenth century, we find the Terra Australis occupying the whole of the southern part of the globe, that portion which lay in the real longitude of Australia being brought up to its right position, far more to the north than elsewhere; thus recognising the genuine discovery of the north part of that continent. The vast remainder was but a fancied continuation of the Tierra del Fuego, the southern point of which not having been yet discovered, was supposed to form a portion of a great southern land which from remote ages had been supposed to be in existence, and as Magalhaens had been the discoverer of this Tierra del Fuego in passing through the strait which bears his name, this supposed great southern continent received from its reputed discoverer the name of Terra Magellanica. In some of these early engraved maps, New Guinea and the Terra Australis are united, but no greater proof can be adduced of the fact, that Australia was then known to be discovered, although as yet no *authenticated* discovery by any given ship or navigator had been recorded, than the fact that on other maps of this period is found the legend, "Nova Guinea, quæ an sit insula aut pars continentis adhuc ignotum est." These facts respecting the early *engraved* maps are interesting, because

unworthy insinuation met with a sensible and generous refutation from the pen of a Frenchman, M. Frederic Metz, in a paper printed at p. 261, Vol. 47 of *La Revue, ou Décade Philosophique, Littéraire et Politique*, Nov., 1805.

while utterly distinct from the earlier and more valuable manuscript documents, they precede the period of what, for want of a better word, I must call the *authenticated* discovery of Australia, *i.e.*, made *by a known ship or navigator*. The earliest *authenticated* discovery, in this sense of the word, was, till recently, supposed to have been made on the 18th of November, 1605, by the Dutch yacht, the *Duyfhen*, which had been despatched from Bantam to explore the island of New Guinea, and which sailed along what was thought to be the west side of that country to $19\frac{3}{4}°$ of south latitude. In the year 1861, however, it was my good fortune to light upon a manuscript mappemonde in the British Museum which enabled me to carry back that first *authenticated* discovery to the year 1601, and to transfer the honour thereof from the Dutch to the Portuguese, by whom the discovery, although unauthenticated by the known name of the ship or navigator, had been made, as I shall presently show, some seventy years before.

On the mappe-monde in question was delineated the northwest portion of Australia, and on the extreme north was the following legend : " Nuça Antara foi descuberta o anno 1601 por mano (*sic*) el godhino de Evedia (*sic*) por mandado de (*sic*) Vico Rey Aives (*sic*) de Saldaha (*sic*) :" which translated is : "Nuça Antara (the local name) was discovered in the year 1601 by Manoel Godinho de Eredia, by command of the Viceroy Ayres de Saldanha." Unfortunately, the map is only a copy, apparently, as I have since discovered from a letter addressed to Navarrete by the Vicomte de Santarem in 1835, of a map in a MS. Atlas, made in the 17th century by one Teixeira, and hence the blunders marked by " sic." For this reason I could only have adduced it for *quantum valeret*, were it not that I am able to corroborate it by independent evidence to which the blunders themselves are but a further corroboration. The name Nuça Antara is shown in Sir Stamford Raffles' " Java " to apply also to the island of Madura, north-east of Java, but, as that island is distinctly given in this very mappe-monde, it is clear that no mistake is involved

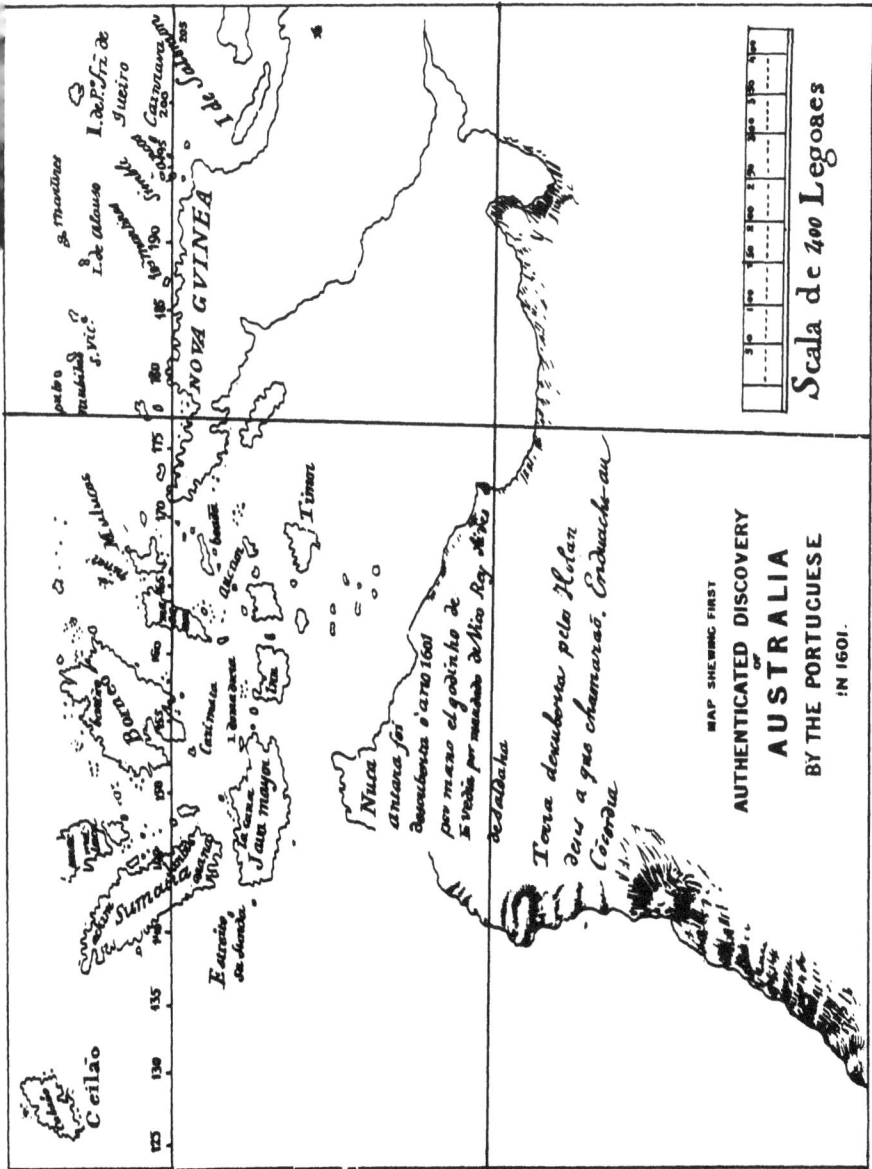

Ceilão

C. Molucos

Borneo

Sumatra

J. aux mayor

Timor

NOVA GVINEA

S. Martinez
I. de alasio
I. de P.ͤ Fr.ᵈᵉ Jueiro
I de Salvaron

Nua antara for descuberta i anno 1601 por meno el godinho de Eredia por mandado de Vico Rey N Sᵉ

I de Daha

Terra descuberta pllo Holan dez a qᵘᵉ chamaraõ Cornha

Scala de 400 Legoaes

MAP SHEWING FIRST
or
AUTHENTICATED DISCOVERY
or
AUSTRALIA
BY THE PORTUCUESE
IN 1601.

on that account; and that the country delineated is really Australia is proved by a second legend below the first, thus : " Terra descuberta pelos Holandeses a que chamarao Enduacht (*sic*) ou Concordia," (land discovered by the Dutch, which they called Endracht or Concord). Eendraghtsland, as we all know, was the name given to a large tract on the west coast of Australia, discovered by the Dutch ship, the *Eendraght*, in 1616. The map being a copy, it may be objected that the important legend declaring the discovery in 1601 may have been fraudulently inserted. But to give such a suggestion weight a motive must be shown, the most reasonable one being that of assigning the honour of the first authenticated discovery to Portugal instead of to Holland. For this purpose we must suppose the falsifier to have been a Portuguese. To this I reply, that while all the writing of the map is in Portuguese, the copy was made by a person who was not only not a Portuguese himself, but who was ignorant of the Portuguese language. For example, the very legend in question, short as it is, contains no less than five blunders, all showing ignorance of the language : thus, the words " por Manoel " are written " por mano el," " Eredia " is written " Evedia," " do " is written " de," " Ayres " is written " Aives," " Saldanha " is written " Saldaha," without the circumflex to imply an abbreviation.

But further, if we attribute to such supposed falsification the ulterior object of claiming for the Portuguese the honour of a prior discovery, whence comes it that that object has never been carried out ? It was not till 1861 that the fact was made known by me, and those most interested in the ancient glory of the Portuguese nation were ignorant of the discovery which this map declares to have been made. That it never became matter of history may be explained by the comparatively little importance which would at the time be attached to such a discovery, and also by the fact that the Portuguese, being then no longer in the fulness of their prosperity, were not keeping the subject before their attention by repeated expeditions to that country, as the Dutch shortly afterwards really began to do.

Again, the speculation might be hazarded that, as this map is a copy, the date of the discovery may have been carelessly transcribed, as, for example, 1601 may easily have been written in the original 1610 and erroneously copied. Fortunately the correctness of the date can be proved beyond dispute. It is distinctly stated that the voyage was made by order of the Viceroy Ayres de Saldanha, the period of whose viceroyalty extended only from 1600 to 1604, thus precluding the possibility of the error suggested, and terminating before the period of the earliest of the Dutch discoveries.

Moreover, if the legend in question were not a genuine copy from a genuine ancient map, how came the modern falsifier to be acquainted with the name of a real cosmographer who lived at Goa, at a period which tallies with the state of geographical discovery represented on the map, but none of whose manuscript productions had been put into print at the time when the supposed fictitious map was made, or the legend fictitiously inserted?

I think these arguments are conclusive in establishing the legitimacy of the modern copy from the ancient map. As regards the discoverer, Manoel Godinho de Eredia (or rather Heredia, as written by Barbosa Machado and by Figaniere), I find the following work by him : "Historia do Martyrio de Luiz Monteiro Coutinho que padeceo por ordem do Rey Achem Raiamancor no anno de 1588, e dedicada ao illustrissimo D. Aleixo de Menezes, Arcebispo de Braga;" which dedication was dated Goa, 11th of November, 1615, fol. MS. with various illustrations.

Barbosa Machado calls him a distinguished mathematician; and Figaniere, a cosmographer resident at Goa. It follows as a most likely consequence that the original map was made by himself. The copy came from Madrid, and was purchased by the British Museum in 1848, from the Señor de Michelena y Roxas. It will be matter of interest to discover at some future day the existence of the original map, but whether that be in the library at Madrid, or elsewhere, must be a subject for future inquiry.

In a scarce pamphlet entitled " Informacão da Aurea Chersoneso, ou Peninsula e das Ilhas Auriferas, Carbunculas e Aromaticas, ordenada por Manoel Godinho de Eredia, Cosmographo," translated from an ancient MS., and edited by Antonio Lourenço Caminha, in a reprint of the " Ordenacões da India do Senhor Rei D. Manoel," Lisbon, Royal Press, 1802, 8vo., occurs a passage which may be translated as follows : —

" *Island of Gold.* While the fishermen of Lamakera, in the island of Solor,* were engaged in their fishing, there arose so great a tempest that they were utterly unable to return to the shore, and thus they yielded to the force of the storm, which was such that in five days it took them to the Island of Gold, which lies in the sea on the opposite coast, or coast outside of Timor, which properly is called the Southern Coast. When the fishermen reached the Land of Gold, not having eaten during those days of the tempest, they set about seeking for provisions. Such happy and successful good fortune had they, that, while they were searching the country for yams and *batatas*, they lighted on so much gold, that they loaded their boat so that they could carry no more. After taking in water and the necessary supplies for returning to their native country, they experienced another storm, which took them to the island of Great Ende ;† there they landed all their gold, which excited great jealousy amongst the Endes. These same Endes therefore proposed, like the Lamacheres fishermen, to repeat the voyage ; and when they were all ready to start, both the Endes and Lama-

---

* The inhabitants of the coast of Solor are specially mentioned as fishermen by Crawfurd, in his " Dictionary of the Indian Islands."

† This is the island of Flores. In a " List of the principal Gold-mines obtained by the explorations (curiosidade) of Manoel Godinho de Heredia, Indian cosmographer, resident in Malacca for twenty years and more," also published with the "Ordenações da India," Lisbon, 1807, the same story is told, but the island Ende is there called Ilha do Conde.

cheres, there came upon them so great a trepidation that
they did not dare, on account of their ignorance, to cross
that Sea of Gold.

"Indeed, it seems to be a providential act of Almighty
God, that Manoel Godinho de Eredia, the cosmographer,
has received commission from the Lord Count-Admiral, the
Viceroy of India within and beyond the Ganges, that the
said Eredia may be a means of adding new patrimonies to
the Crown of Portugal, and of enriching the said Lord Count
and the Portuguese nation. And therefore all, and especially the said Lord, ought to recognise with gratitude this
signal service, which, if successful, will deserve to be regarded as one of the most happy and fortunate events in the
world for the glory of Portugal. In any case, therefore, the
discoverer ought for many reasons to be well provided for the
gold enterprise. First, On account of the first possession
of the gold by the crown of Portugal. Secondly, For the
facility of discovering the gold. Thirdly, Because of the
gold mines being the greatest in the world. Fourthly,
Because the discoverer is a learned cosmographer. Fifthly,
That he may at the same time verify the descriptions of
the Southern Islands. Sixthly, On account of the new
Christianity. Seventhly, Because the discoverer is a skilful
captain, who proposes to render very great services to the
King of Portugal, and to the most happy Dom Francisco da
Gama, Count of Vidigueira, Admiral and Viceroy of the
Indies within and beyond the Ganges, and possessor of the
gold, carbuncle, and spices of the Eastern Sea belonging to
Portugal."

Short of an actual narrative of the voyage in which the
discovery, thus newly announced by this map, was made,
we could scarcely ask for fuller confirmation of the truth
of that discovery, than that which is supplied by the above
extract. Manoel Godinho de Eredia is there described as a
learned cosmographer and skilful captain, who had received
a special commission to make explorations for gold mines,
and at the same time to verify the descriptions of the

Southern Islands. The Island of Gold itself is described " as on the opposite coast, or coast outside of Timor, which properly is called the Southern Coast." It is highly probable from this description that it is the very Nuça Antara of our MS. map, which does lie on the southern coast opposite to Timor. It is still further most remarkable that, by the mere force of facts, the period of the commission here given to Eredia is brought into proximity with the date of his asserted discovery of Australia. The viceroy, Francisco de Gama, who gave that commission, was the immediate predecessor of Ayres de Saldanha. His viceroyalty extended only from 1597 to 1600, and the asserted discovery was made in 1601, though we know not in what month. A more happy confirmation of a discovery, unrecorded except in a probably unique map, could scarcely have been hoped for.

But all this, though important and far from irrelevant to our subject, is a digression from the more important consideration of the story told by the invaluable manuscript maps of the beginning of the sixteenth century. It occurs in similar form on seven maps, five of them in England and two in France, on which, immediately below Java, and separated from that island only by a narrow strait, is drawn a large country stretching southward to the verge of the several maps. The first of these maps to which a fixed date is attached, are two in one atlas, which was made in 1542 by a Frenchman named Jean Rotz or Roty, who came to England and dedicated his work to Henry VIII. Another, probably older but without a date, is anonymous, but as it bears the arms of the Dauphin it appears to have been executed in the time of Francis I. for his son the Dauphin, afterwards Henry II. Both this map and the atlas by Rotz are in the British Museum. As the other maps all tell the same story, it will be needless here to make any further reference to them.*

· The two selected will suffice for our consideration as to

* For a more detailed account of these maps see my "Early Voyages to Terra Australis," printed for the Hakluyt Society, 1859.

what the large country is which is thus delineated to the
south of Java, who were the discoverers, and what was the
approximate period of the discovery.   On the Dauphin map
this great country is called " Jave la Grande ; " on Rotz's
map " The Londe of Java," and is distinguished from the
smaller island of Java itself by the latter being called "The
lytil Java."

1. The first question that naturally arises, is, how far does
this large country correspond in latitude, longitude, and outline with our present surveys of Australia?

And first as respects latitude.   In all of these maps the
latitude of the north of Java, which is the first certain starting point, is correct.   The south coast of " the lytil Java,"
though separated from the " Londe of Java " by a narrow
channel, has no names indicating any pretension to a survey.

The maps show plainly that it was unknown whether the
two countries were connected or not.   Happily, Rotz's map
is an exception to the rest in one important respect.   Whereas the rest connect this great country with a vast continent
occupying the whole south part of the globe, including and,
as it were, springing from the Tierra del Fuego, from whose
discoverer it is called Magellanica, Rotz's map exhibits the
degrees of latitude in which both the western and eastern
coasts were supposed to cease, and by their accuracy, prove an
amount of exploration which fairly throws back the original
discovery to a period very much earlier than 1542, the date
of this map.   On the western side the coast line ceases
altogether at 35°, the real south-western point of Australia.
On the eastern coast, for the Portuguese the remotest for
investigation, there is greater inaccuracy, the line terminating
in the sixtieth degree, a parallel far exceeding in its southing
even the southernmost point of Tasmania, which is in 43°
35', but there is strong reason for supposing that the eastern
side of Tasmania was included in this line.   With respect
to longitude, it may be stated that while there is no other
country but Australia lying between the same parallels and
of the same extent, between the east coast of Africa and the

IAVE LA GRANDE

AUSTRALIA.

FROM THE

DAUPHIN MAP

[1530?]

west coast of America, so Australia does in reality lie between the same meridians as the great mass of the country
here laid down.

There are also many points of correctness in contour. As
regards the west coast, a single glance of the eye is sufficient to detect the general resemblance. On the eastern
side, as might be expected, the discrepancies are much greater,
but nothing can be more remarkable than the great number
of islands and reefs laid down along the north-east coast as
coinciding with the Great Barrier Reefs, and with the Cum-
berland and Northumberland Islands, and a host of others,
which skirt this part of the shores of Australia.

2. It being, therefore, indubitable that the extensive
country thus laid down is Australia, we proceed to examine
who were the discoverers. All these maps are French, and
that they are all repetitions with slight variations from one
source is shown by the fact that the inaccuracies are alike in
all of them. But although the maps are in French, there
are indications of Portuguese in some of the names, such as,
" Terre ennegade," a Gallicized form of Tierra Anegada, *i.e.*
" land under water " or " sunken shoal," " Graçal " and
" Cap de Fromose." The question then arises, Were the
French or the Portuguese the discoverers? In reply, I present the following statement.

In limine, 1 have to recall to the reader's memory, that
in the year 1529 a voyage was made to Sumatra by Jean
Parmentier of Dieppe, and in this voyage he died. Parmentier was a poet and a classical scholar, as well as a navigator and good hydrographer. He was accompanied in this
voyage by his intimate friend, the poet Pierre Crignon, who, on
his return to France, published in 1531 the poems of Parmentier, with a prologue containing his eulogium, in which
he says of him, that he was " semblablement le premier
François qui a descouvert les Indes jusqu'à l'Isle de Taprobane, et, si mort ne l'eust pas prévenu, je crois qu'il eust
été jusques aux Moluques." This is high authority upon this
point, coming as it does from a man of education and a

shipmate and intimate of Parmentier himself. The French, then, were not in the South Seas beyond Sumatra before 1529. The date of the earliest of our quoted maps is not earlier than 1535, as it contains the discovery of the St. Lawrence by Jacques Cartier in that year; but even let us suppose it no earlier than that of Rotz, which bears the date of 1542, yet we find no voyages of the French in the South Seas between the years of 1529 and 1542. Neither the Abbé Raynal, nor any modern French writer, nor even antiquaries, who have entered most closely into the history of early French explorations, as for example, M. Léon Guérin, the author of the *Histoire Maritime de France*, Paris, 1843, 8vo., and of *Les Navigateurs Français*, 8vo., Paris, 1847, offer the slightest pretension that the French made voyages to those parts in the early part or middle of the sixteenth century. Indeed, France was at that time too poor and too much embroiled in political anxieties, to busy herself with extensive nautical explorations. Had she so done the whole of North America and Brazil might now have belonged to her. Now we do know from Barros and Galvão that, at the close of 1511, Albuquerque sent from Malacca Antonio de Abreu and Francisco Serrão, with three ships to Banda and the Moluccas; they passed along the east side of Sumatra to Java, and thence by Madura, Bali, Sumbava, Solor, &c., to Papua or New Guinea. From thence they went to the Moluccas and to Amboyna. See Barros, D. 3, i. 5, c. 6, p. 131, and Galvão, translated by Hakluyt, p. 378. Now of these islands we have some which are incorporated into the northern portion of the Grande Jave, but that which is totally wanting between 1511 and 1529, is the account of the various Portuguese explorations of the eastern and western coasts of the vast country described under that name. That this obscurity is mainly due to a jealous apprehension lest lands of large extent and great importance in the southern seas might fall into the hands of rival powers to their own displacement or prejudice, may not only be suspected, but is affirmable from historical evidence.

It is stated by Humboldt (*Histoire de la Géographie du Nouveau Continent*, tom. iv. p. 70), upon the authority of the letters of Angelo Trevigiano, secretary to Domenico Pisani, ambassador from Venice to Spain, that the Kings of Portugal forbad upon pain of death the exportation of any marine chart which showed the course to Calicut. We find also in Ramusio (*Discorso sopra il libro di Odoardo Barbosa*, and the *Sommario delle Indie Orientali*, tom. i. p. 287 *b*), a similar prohibition implied. He says that these books "were for many years concealed, and not allowed to be published, for convenient reasons that I must not now describe." He also speaks of the great difficulty he himself had in procuring a copy, and even that an imperfect one, from Lisbon. "Tanto possono," he says, "gli interessi del principe." Again, in tom. iii. of the same collection, in the passage which I have already had occasion to quote from the "Discorso d'un gran Capitano del Mare Francese del luogo di Dieppa," &c., now known to be the voyage of Jean Parmentier to Sumatra in 1529, and in all probability written by Pierre Crignon, the covetousness and exclusiveness of the Portuguese are inveighed against. "They seem," he says, "to have drunk of the dust of the heart of King Alexander, for that they seem to think that God made the sea and the land only for them, and that if they could have locked up the sea from Finisterre to Iceland, it would have been done long ago," &c.

At the same time, however, we know that the Portuguese had establishments before 1529 in the East Indian Islands, and the existence of Portuguese names on the countries of which we speak as thus delineated on these French maps, is in itself an acknowledgment of their discovery by the Portuguese, as assuredly the jealousy implied in the sentence just quoted from Pierre Crignon's Prologue, would not only have made the French most ready to lay claim to all they could in the shape of discovery, but would have prevented any gratuitous insertion of Portuguese names on such remote countries had they themselves discovered them.

But though we have no evidence to show that the French made any original discoveries in the South Seas in the first half of the sixteenth century, we have the evidence that they were good hydrographers. Crignon describes Parmentier as " bon cosmographe et géographe," and says, " par luy ont esté composez plusieurs mape-mondes en globe et en plat, et maintes cartes marines sur les quelles plusieurs ont navigué seurement." It is dangerous to draw conclusions from negatives ; but it is both legitimate and desirable that we should give due weight to evidence of high probability when such fall within our notice. If all the French maps we have quoted are, as we have shown, derived from one source, since they all contain the same errors ; and if Parmentier, who was a good hydrographer, was the only French navigator whom we find mentioned as having gone so far as Sumatra before the period of the earliest of these maps ; and further, if these maps exhibit Portuguese names laid down on a country beyond Parmentier's furthest point of exploration, we think the inference not unreasonable that Parmentier may have laid down, from Portuguese maps, the information which has been copied into those we have quoted, and that the descriptions round the coast, which are all (as may be plainly seen), with the exception of those which bear the stamp of Portuguese, convertible into French, have been naturally written by French map-makers, in that language. We can but throw out this suggestion for *quantum valeat*. All positive evidence, in spite of laborious research, is wanting. The Portuguese names are but few, but there they are, and bear their stubborn evidence.

Our surmises, therefore, lead us to regard it as highly probable that Australia was discovered by the Portuguese between the years 1511 and 1529, and to a demonstrable certainty, that it was discovered before the year 1542.

We have now seen how, within the small compass of a

single century from the date of the rounding of Cape Boyador, more than one-half of the world was opened up to man's knowledge, and brought within his reach by an unbroken chain of discovery which originated in the genius and the efforts of one whose name is all but unknown.

The Coasts of Africa visited; the Cape of Good Hope rounded; the New World disclosed; the sea-way to India, the Moluccas, and China, laid open; the globe circumnavigated, and Australia discovered: such were the stupendous results of a great thought and of indomitable perseverance in spite of twelve years of costly failure and disheartening ridicule. Had that failure and that ridicule produced on Prince Henry the effect which they ordinarily produce on other men, it is impossible to say what delays would have occurred before these mighty events would have been realized; for it must be borne in mind that the ardour not only of his own sailors, but of surrounding nations, owed its impulse to this pertinacity of purpose in him. True it is, that the great majority of these vast results were effected after his death; and it was not granted to him to affix his quaint signature to charters and grants of territory in those Eastern

I[FFANTE] D[OM] A[NRIQUE].

and Western Empires which at length were won by means of the explorations he had fostered. True, he lived not to see the proof, in his own case unparalleled, that the courageous pursuit of a grand idea may produce consequences even greater than that idea had comprehended. No doubt that from Sagres no beam of light brought to his mental vision the prospect of an America to brighten the horizon of the Sea of Darkness; yet enough has, I trust, been said in the preceding pages to establish the correctness of the statement with which I set out, that "if, from the pinnacle of our present knowledge, we mark on the world of waters those bright tracks which have led to the discovery of mighty continents, we shall find them all lead us back to that same inhospitable point of Sagres, and to the motive which gave to it a royal inhabitant."

INDEX.

Covilham (Pedro de), his letter to King
João proves him to be the *theoretical*
discoverer of the Cape of Good Hope,
339, 340.

Crignon (Pierre), poet and friend of
Parmentier, quoted on the subject
of his voyage, 419, 449.

Cross fixed by Diogo Affonso at Cape
Branco, Azurara's remarks; M.
d'Avezac's deductions therefrom
erroneous, 194.

Cunha (Pedro Correa de), son-in-law
of Perestrello, Captain of Graciosa,
240.

Cunha (Pedro Vaz de), in a fit of anger
basely kills Bemoi, Prince of the
Jaloffs, 342.

Cunha (Tristam da), 1506, Affonso
d'Albuquerque goes with him; discovers the three islands that bear
his name; they take the fortress of
Socotra, 415, 416.

——, 1507, at Melinda sends envoys
to the Emperor of Abyssinia, but
the mission failed, 416, 417.

Da Gama (Gasparo), Polish Jew
taken prisoner by Da Gama in
India as a pirate; brought by him
to Lisbon, and baptized; afterwards
employed by King Manoel in negotiations with India, 402, 403.

—— goes out with Cabral's expedition, 1500, 408.

Da Gama (Paolo) sails in his brother's
expedition, in the *San Raphael*,
392.

—— dies on the way home, 405.

Da Gama (Vasco) made commander
of the fleet of the Indus; his ship,
the *San Gabriel;* his pilot, Pero
de Alemquer, who had been with
Bartholomeu Dias, sails 8th July,
1497, 392; the fleet anchors in the
bay of St. Helena, where they
become acquainted with the Bosjesmans, 393.

—— passes the Cape of Good Hope,
394.

——passes the extreme point of Dias's
discovery, and reaches Natal, which
he names Terra da Boa Gente; Rio
do Cobre; goes on to a river, which
he names Rio dos Bôos Signaes,
where he erects a pillar, which he
calls padrão of San Rafael; 10th of
March, anchors off Mozambique,
395.

Da Gama (Vasco) hears of Prester
John; comes to the Quebrima Islands, 396; (the *San Rafael* stranded on reefs) on Easter-day reaches
Melinda, 397.

—— receives a visit from the King,
398; who gives him a Christian
pilot, 399.

—— 20th of May anchors off Calicut,
399.

—— sends on shore one of the " degradados;" curious salutation, 399.

—— audience with the Samondri Rajah; establishes a factory, where he
places Diogo Dias; his indefensible
conduct with regard to exchange
of prisoners, 400.

—— erects a pillar, which he calls
Santa Maria, 401.

—— his adventure with pirates, 402,
403.

—— at Melinda has an interview
with the King, 404.

—— raises a pillar on one of the
Ilhas de Sam Jorge, 405; in the
return, Nicolas Coelho reaches
Lisbon first, 405.

—— his brother, Paolo da Gama,
dies on the way, 405.

—— he arrives at Lisbon, 405.

—— his reception, 406.

Dalmeida (Diego Lopez) sent by
Baldaya to reconnoitre at the Rio
d'Ouro, 84.

Dalrymple (Alexander), his injurious
insinuation against Captain Cook
refuted by M. Frederic Metz, 440,
441.

Dapper (Olivier) adduces a supposed,
but defective, date in the "Batterie
Françoise," to show that Fort
Mina was built by the French
in the 14th century, 123, 124;
M. d'Avezac's comments thereon
refuted, 124—127.

Dati (Giuliano), his narrative of the
discovery by Columbus in ottava
rima, 357.

Denis (M. Ferdinand) discovers the
MS. of Azurara, 1837, vii.; published, Paris, 1841; information
about Sagres, 52.

De Nova (Juan), sent out by King
Manoel, 1501; discovers the island
of Ascension; at San Bras, beyond
the Cape of Good Hope, he finds
a letter left by Pedro de Ataide,
directing all captains bound for

India to go by way of Mombasa ; gets further information from Antonio Fernandes, and proceeds to Cananor ; encounters the fleet of the King of Calicut, 413.

De Nova (Juan) is well received at Cochin, and having freighted his ships, sets sail for Portugal; discovers St. Helena on the way home, and on his arrival is received with distinguished honour by the King, 414.

—— sails in Almeida's expedition, 414.

Desceliers (Pierre), priest at Arques, author of the Mappemonde (1550) in the British Museum ; erroneous mention of him by Desmarquets, 406—408.

Desert of Sahara, 252.

Desmarquets (J. A.), in his "Memoires Chronologiques pour servir à l'Histoire de Dieppe," asserts that a navigator named Cousin sailed from Dieppe in 1488, and sailed round the Cape of Good Hope, 406 ; this assertion disproved, 407, 408.

—— his assertion, that Parmentier reached the coast of China, disproved, 418, 419.

Dias (Bartholomeu), 1486, with his brother, Pedro Dias, and João Infante, goes out in search of the country of Prester John, 338.

—— erects a pillar at Angra dos Ilheos, now Dias Point, 343.

—— passes Cape Voltas, and is driven south, 343.

—— finding no land when he steers eastward, sails north, and finds Angra dos Vaqueiros (Flesh Bay), 343.

—— going east, he reaches Algoa Bay ; sets up a pillar on a small island there, 344 ; the first land beyond the Cape trodden by Europeans, 344.

—— finds a river, which he names Rio do Infante, from João Infante, 345.

—— is obliged by his crew to return, 345.

—— names the Cape, Cabo Tormentoso (Stormy Cape), 345.

—— on his return to Portugal, the king named it Cape of Good Hope, 345.

Dias (Bartholomeu) was to have accompanied the expedition of Da Gama, but is subsequently ordered to sail for San Jorge el Mina, 392.

—— goes out with Cabral's expedition, 1500, 408.

—— perishes in a storm off the Cape of Good Hope, 410.

Dias (Dinis), called by Barros, Diniz Fernandez, 191 ; obtains permission to make explorations in the service of Prince Henry ; sails past the Senegal to the land of the Jaloffs, the first real blacks, 192.

—— reaches Cape Verde, to which he gives its name, 194.

—— from Lisbon, joins Lançarote's expedition, in a caravel of Dom Alvaro de Castro, 200.

—— in company with Pallenço, makes a capture, 207 ; then with Rodrigueannes proceeds to Cape Verde and the Madeleine Islands, 207.

Dias (João) with Lançarote in his first expedition, 178.

Dias (Lourenço), with Lançarote's expedition, is the first to reach the island of Arguin, 200.

—— goes further south with Lançarote, 203.

—— goes out with Gil Eannes, in 1446, 219.

Dias (Pedro) accompanies his brother Bartholomeu in the expedition in search of Prester John's country, 338.

—— goes out with Cabral's expedition, 1500, but loses convoy off Cape Verde Islands, 409.

—— again joins the squadron, after having been to the mouth of the Red Sea, 412.

Dias (Vicente) the outfitter, 202.

Dias (Vicente), the merchant, proceeds southward with Lançarote, 203.

—— his encounter with the African at the mouth of the Senegal, 205.

—— after reaching Cape Verde, he returns to Portugal, 207.

—— sailing captain of Cadamosto's caravel, 247.

Dieppese claims to prior discovery set up, 117—120 ; disproved, 120—130.

Doria (Jacopo) writes the account of Tedisio Doria's voyage, 100.

Doria (Tedisio), his voyage, 99.

466

INDEX.

Dornellas (Alvaro) joins Lançarote's expedition (from Madeira), 200.
—— his adventures at the Canary Islands, 211.
Dornellas (João) goes to the assistance of his cousin, 211.
Dragon's-blood, 139, 248.
Duarte (Dom), eldest brother of the Prince, 30.
—— present at the Queen's death, 30.
—— lands with Prince Henry at Ceuta, 34.
—— knighted by the King, 39.
—— takes great interest in meteorology, 60.
—— his Leal Conselheiro, xv., xvi., 81.
—— testifies his satisfaction with Prince Henry's efforts by the charters of the 26th of September, 1433, and of the 26th of October, 1434, giving him the islands of Madeira, Porto Santo, and the Desertas, 81.
—— his personal qualities, 153.
—— prediction of his astrologer, 153.
—— new title given to his heir, 153.
—— gives a reluctant consent to the attack on Tangier, 154.
—— applies to the Pope, but makes preparations before receiving an answer, and sends out the expedition, 154.
—— his grief at the disastrous result and the fate of his brother Fernando, 163.
—— attempts his brother's rescue, but in vain, 164.
—— his grief for his brother's sufferings undermines his health, 166.
—— his death, 167.
—— his character, 167.
—— the Lei Mental, 168.
—— his anxiety to replace the royal revenues, 168.

Edrisi, introduced the name of Niger, 192.
Edward III. of England the Prince's great-grandfather, 4.
Elephant hunting described by Cadamosto, 282, 283.
Elvas (Lourenço d') with Gil Eannes, 1446, 219.
English esquires at the battle of Aljubarrota, 16.
English ships which joined in the expedition against Ceuta, 42.
Equatorial Nile Lakes first mentioned by Pigafetta from Duarte Lopes, 334.

Eratosthenes speaks of the Phœnician colonies on the west coast of Africa as numerous, 90.
Eredia (Manoel Godinho de), or Heredia, the discoverer of Australia, 442—447.
Escobar (Pedro de), commander selected by Fernam Gomez for the expedition beyond Sierra Leona, 321.
—— discovers La Mina, and goes thirty-seven leagues beyond Cape Lopo Gonsalves, 322.
"Esmeraldo de Situ Orbis," MS. by Duarte Pacheco, xiii.
Estancelin, his summary of Bellefond's narrative of the Dieppese, 117—120.
—— of the claims re-asserted by Père Labat, 120.
Esteves (Alvaro), of Lagos, pilot in Fernam Gomez' expedition; reputed the best navigator in Portugal, 321.
—— discovers Principe, Annabon, and S. Thomé, 328.
Eudoxus of Cyzicus, his voyages, 97.
Evora (Bishop of) at the siege of Tangier, 157.

Faleiro (Rui) accompanies Magalhaens to Seville, 423.
Fayal, first donatary Jobst Van Heurter, 240.
—— question of grant discussed, 240—243.
Fernandes (Alvaro) joins Lançarote's expedition, 200.
—— is the first to go to the Madeleine Islands, 208.
—— reaches a cape, which he names Cabo dos Mastos, 209.
—— in 1446 passes beyond Cape Verde; is wounded with a poisoned arrow, but recovers by the use of an antidote, 217, 218.
—— reaches 110 leagues south of Cape Verde, 219.
—— meets with Ahude Maimom at the Cabo do Resgate, 219.
—— is rewarded by the Regent and Prince Henry for having gone further south than any of his predecessors, 219.
Fernandes (Antonio) gives information of Juan de Nova, 413.
Fernandes (João) goes with Gonsalves to the Rio d'Ouro, and remains seven months in the interior, 190.

Moluccas (the), Francisco Serrão went to Ternate, 1511, 418.
—— after Magalhaens' death, the remainder of his ships arrive there, 437.
Monomoezi, first mentioned by Pigafetta, from Duarte Lopes, 335.
Montucla mistaken in attributing the invention of hydrographic plane charts to Prince Henry, 54.
Morales (Juan de), as fellow captive, hears from Machin's people of the discovery of Madeira; imparts this knowledge to Zarco, and goes with him to the re-discovery, 68—74.
Mungo Park mistaken in ascribing to the Joliba, or Quorra, the name of Niger, 192.
Murr (M.) the chief among those who have sought to exalt Behaim at the expense of Columbus, 326.

Natal, named by Da Gama, 395.
Neale, (Dr. J. Mason), his description of Batalha, 305.
—— his observations on Belem, 393.
Neckam (Alexander) makes the earliest allusion to the use of the compass in the Middle Ages, 57.
Negro fair, 269.
Negus (the), Prince of Abyssinia, detains Covilham in his dominions, and treats him with honour, 340.
Nile of the Negroes (the Senegal), 203.
Nile (equatorial) lakes first mentioned by Pigafetta, from Duarte Lopes, 334.
Nile, sources of, referred to by Pigafetta, from Duarte Lopes, 1591, 334.
Nolli (Antonio de), his letter, 102—105.
—— is joined by Cadamosto, 271.
—— it falls to his lot to set an interpreter on shore at the Joombas, who is killed by the natives, 273.
—— goes with Cadamosto on his second voyage, 278.
—— joins Diogo Gomez, 297.
—— arrives in Portugal before him, 298.
Nombre de Dios founded by Diego de Nicuesa 1510, 369.
Nomimansa, King of the Barbacins, makes peace with Diogo Gomez, 293.
—— wishes to be baptized, 294.

Nomimansa, the Prince sends the Abbot of Soto de Cassa to instruct him and his people in the faith, 296.
Nova (Juan de) with Almeida's expedition, 1505, 422.
Nuñez (Pedro), the mathematician, his statement respecting the early navigations of the Portuguese, 55.
Nuñez (Pedro), servant of Da Gama, who accompanied the expedition in a small craft laden with munitions, 392.
Nurembergers, their claims on behalf of Behaim; claims refuted, 327.
Nyanza (Lake) apparently referred to by Pigafetta, from Duarte Lopes, 1591, 334.

Odjein, or Ougein, sacred city, whence the Indians reckoned their first meridian, 100.
Ojeda (Alonzo de) with Columbus on his second voyage, 368.
—— quarrels with Columbus, and receives from the Bishop Fonseca, 1498, a fragment of a map showing the admiral's last discoveries, 368.
—— he sets sail for South America, May, 1499, with the pilot Juan de la Cosa and Amerigo Vespucci, 368.
—— his discoveries, 368.
—— raises a revolt in Hispaniola against the authority of Columbus, 369.
—— returns to Cadiz, 369.
Orchil, for dyeing, 138, 139, 250.
Ottmar (Johann) published at Augsburg the "Mundus Novus," containing the description of Vespucci's third voyage, in a letter from himself to Lorenzo di Pier Francesco de' Medici, 367.
Ouro (Rio d'), whence it received its name, 176.
Ovando (Juan de), his friendship for Magalhaens, 424.

Pacheco (Gonsalo) goes out with Diniz Eannes da Grãa, Alvaro Gil, and Mafaldo, 198—201.
Paes (Ruy) goes out with Zarco's expedition, 73.
—— finds the tomb of Machin, 74.
Pajola, or Palola, 112, 113.
Pallenço with Lançarote's expedition, 200.
—— his adventures in company with Diniz Dias, 207.

Panso Aquitimo, younger son of the King of Congo, rejects the faith, 333.

Parmentier (Jean), of Dieppe, his voyage to Sumatra ; the statement of M. Desmarquets, that he went to China, disproved, 419.

—— his voyage referred to, 449, 450.

Patagonia discovered by Magalhaens, 427.

Payva (Affonso de) sent out with Covilham by João II., but after parting with him at Aden, dies on the journey, 339.

Pedro the Severe, Prince Henry's grandfather, 5.

—— his sons, of whom King João was the youngest, and illegitimate, 5.

Pedro (Dom), second brother of the Prince, present at the Queen's death, 30.

—— at the siege of Ceuta, 37.

—— knighted by the King, 39.

—— receives the titles of Duke of Coimbra, &c., 42.

—— his travels, 61.

—— receives from the Venetians a copy of Marco Polo's travels, and a map, 61.

—— visits England, and is made Knight of the Garter, 62.

—— shares Prince Henry's studies, 62.

—— remonstrates with Dom Duarte on the measures taken for the attack on Tangier, 154.

—— the arrangements for his government of the kingdom after Dom Duarte's death, and during the minority of his son, 169.

—— grants a charter to Prince Henry in consequence of the discoveries of Gonsalves and Tristam, 174.

—— sends out Gomes Pires, in 1445, with Antam Gonsalves, to the Rio d'Ouro, 190.

—— summons Prince Henry to Coimbra to invest with knighthood his eldest son, Pedro, 199.

—— grants the charter of the Canary Islands to Prince Henry, 214.

—— rewards Alvaro Fernandes for going further south than any of his predecessors, 219.

—— his history taken up from the time of his return to Portugal, after his travels in 1428, 226.

Pedro (Dom), his marriage, 226.

—— is appointed guardian to the Infant Affonso, 226.

—— incurs the ill-will of the Queen, 226.

—— is nominated Regent, 227.

—— his disagreement with the Duke of Braganza, 227.

—— his daughter married to the King, 228.

—— the continued enmity of the Duke of Braganza, formerly Count de Barcellos, his bastard brother, 228.

—— retires to Coimbra, 229.

—— the accusations of his enemies and the apathy of his friends, with the exception of Alvaro Vaz d'Almado, Count d'Avranches, 229.

—— provoked to battle by the Duke of Braganza, 230.

—— goes to meet the King, 232.

—— his daughter's intercessions for him, 232.

—— his death ; the injustice of his enemies, 232.

—— his funeral, 233.

—— honour to his memory, 234.

Pegolotti mentions Malaguette pepper as imported into Nismes and Montpellier in the first half of the 14th century, 114.

Père Labat re-asserts the claims of the Dieppese, 120.

Pereira (Diogo Fernandes), 1504, wintered at Socotra, then first reached by Portuguese, 414.

Pereira (Duarte Pacheco) left by Francisco Albuquerque at Cochin as defender of the kingdom, 414.

Pereira (Martim) distinguishes himself in an encounter with the natives at the Madeleine Islands, 270.

Pereira (Nuño Alvarez), called the Holy Constable, his victory over the Castilians, 11.

—— his important help to the Grand Master after the siege of Lisbon, 13.

—— his high character, 14.

—— at the battle of Aljubarrota, 16,17.

—— his victory at Valverde, 21.

—— retires to the convent of Carmo, 78.

Perestrello (Bartollomeu) accompanies Zarco and Vaz in their expedition to Porto Santo, 77.

—— receives the governorship of the island, 77.

LIST OF PRINCIPAL BOOKS CONSULTED.

ABREU DE GALINDO. History of the Canary Islands. Glas. London, 1764.
ABU ÁL-FIDA ISMAIL BNU ALI. Géographie. Reinaud. Paris, 1848.
ALVAREZ (Francisco). Verdadera informaçam das Terras do Preste Joam das Indias. Lisboa, 1540.
ANNAES Maritimos e Coloniaes. Serie 1 to 6. Lisboa, 1840, &c.
ANNALES des Voyages, &c. Tome 7—8. Paris, 1809, &c.
ARTHYS GOTARDO. Historia Indiæ Orientalis, &c. Coloniæ, 1608.
AYALA (Pedro Lopez de). Colleccion de las Cronicas de los Reyes de Castilla desde el ano 1340 hasta el de 1396, con las emiendas de Geronimo Zurita, por E. Llaguno Amirola. Madrid, 1779. 8vo.

BACON (Roger). Opus Majus. Londini, 1733.
BARTH. Travels in North and Central Africa. London, 1858.
BARKER—WEBB et BERTHELOT. Histoire Naturelle des Iles Canaries. Paris, 1842, &c.
BARROS Y SOUZA, Visconde de Santarem. Quadro Elementar das Relações Politicas e Diplomaticas de Portugal. Pariz, 1842, &c.
BARROS (João de). Asia. Lisbon, 1777, &c.
BEAUPRE (J. N.). Recherches Historiques et Bibliographiques sur les Commencements de l'Imprimerie en Lorraine. St. Nicolas de Port, 1845. 8vo.
—— Nouvelles Recherches de Bibliographie Lorraine. Nancy, 1856.
BOILAT. Esquisses Sénégalaises. Paris, 1853.
BOSSI. I negri della Nigrizia. Torino, 1838.
BULLETIN de la Société de Géographie. IIIme Série, 3—6, 1835-6, 1844, 1846; IIIme Série, 1857—60; IVme Série; vols. 14—16, 18, 19.
BURNEY. Discoveries in the South Sea. Vol. i. London, 1803.

CAÇEGAS (L.). Historia de S. Domingos. Lisboa, 1767.
CAETANO DE SOUZA (Ant.). Historia genealogica da casa real Portugueza desde a sua origem até o presento com as familias que procedem dos reys e dos serenissmos duques de Braganza. 1735-48. 20 vols. in 4to.
CAILLIÉ (René). Travels through Central Africa to Timbuctoo. London, 1830. 2 vols.
CANARY ISLANDS. Biblioteca Isleña. Santa Cruz. 1847-8.
CARDOSO (Jorge). Agiologio Lusitano. 4 tom. Lisboa, 1652—1744, fol.
CARLI. Opere. Milano. 1785.
CASTANHEDA (Lopez de). Conquista da India pelos Portugueses. Lisboa, 1833.
CATALOGO dos Manuscriptos da Bibliotheca Publica Eborense, Lisboa, 1850.
CHRONICA do Descobrimento e Conquiste de Guiné, escripta por mandado d'El Rei D. Affonso V., pelo chronista Gomes Eannes de Azurara, fielmente trasladada do manuscripto original contemporaneo, que se acha na Bibliotheca real de Pariz, e dada pela primeira vez à luz, por diligencia do visconde da Carreira, precedida de una introduçao, e illustrada com notas do Visconde de Santarem, &c. Paris, 1841.

CLADERA (C.) Investigaciones Historicas sobre los Descubrimientos de los Españoles. Madrid, 1794.

COLLECÇÃO de noticias para a historia e geographia das naçoes ultramarinas que viven. nos dominios Portuguezes ou lhes são vizinhas, publicada pela Academia Real das Sciencias. Lisboa, 1836-41.

COLLECÇÃO dos documentos, estatutos e memorias da Academia Real de Historia Portugueza. Lisboa, 1727, in fol.

COLLINA (D. A.). Sopra la Bussola. Faenza, 1748.

COOLEY (W. D.). Negroland of the Arabs. London, 1841.

CORDEYRO. Historia Insulana, &c. Lisboa, 1717.

CORREA DA SERRA. Collecção de livros ineditos de historia Portogueza, publicados de ordem da Acad. R. das Sciencias de Lisboa. Lisboa, 1790—1824. 5 vols. fol.

COSTA (Antonio Carvalho da). Corografia Portugueza e descripção topografica do famoso reino de Portugal, &c. Lisboa, 1706, 1708, et 1712. 3 vols. in 4to.

COSTA DE MACEDO (J. J. de). Memoria sobre as verdadeiras epochas em que principiaram as nossas navigações e descobrimentos no Oceano Atlantico. Hist. e Mem. da Acad. R. das Sciencias, tom. ii. p. 2.

—— Memoria em que se pretende provar que os Arabes não conheceram as Canarias antes dos Portuguezes. Hist. e Mem. da Acad. R. das Sciencias. Serie 2, tom. i. pp. 37—268.

COSTA E SILVA (José Maria da). Ensaio Biographico-Critico sobre os melhores Poetas Portuguezes. Lisboa, 1850.

COSTA QUINTELLA. Annaes da Marinha Portugueza. Lisboa, 1839-40.

D'AVEZAC. Les Voyages de A. Vespuce. Paris, 1858.

DE BRY. Indiæ Orientalis, Pars vi. Anno 1604.

DENIS (Ferdinand). Portugal. Paris, 1846.

DEPPING (G. B.). Histoire du Commerce, &c. Paris, 1830. Vols. i. and ii.

DESMARQUETS. Mémoires Chronologiques pour servir à l'Histoire de Dieppe, &c. Paris, 1785.

DOS SANTOS (F. Manoel). Monarchia Lusitana. Parte 8. Lisboa, 1727.

DUARTE. Leal Conselheiro. Santarem. Paris, 1842.

EDRISI. Géographie. Trad. Par. Jaubert.

ESTANCELIN (L.). Recherches sur les Voyages et Découvertes des Navigateurs Normands en Afrique. Paris, 1832.

FARIA Y SOUSA (Manoel de). Historia de Portugal. Brusselas, 1730.

—— Asia Portuguesa. Lisboa, 1674.

—— Epitome de las historias Portuguesas dividido em quatro partes. Bruselas, 1677.

—— Historia del Reyno de Portugal. Brusselas, 1730.

—— Historias Portuguezas, &c. Lisboa, 1673.

FERRET. Notices sur Dieppe, Arques, &c. Paris, 1824.

FERNANDEZ DE NAVARRETE (Martin). Historia de la Nâutica. Madrid, 1846.

—— Colleccion de los Viages y Descubrimientos, &c. Madrid, 1825—1837.

FIGANIERE (F. F.). Catalogo de MSS. Portuguezes no Museo Britannico. Lisboa, 1853.

FRACAN. Itinerarium Portugallensium e Lusitania in Indiam, &c. 1508.

FREIRE (Franc. José). Vida do Infante D. Henrique por Candido Lusitano. Lisboa, 1758.

FOSCARINI (M.). Letteratura Veneziana. Padoua, 1752.

GALVÃO (Antonio). Descobrimentos. Lisboa, 1555.

GARCÃO STOCKLER (F. de B.). Ensaio Historico sobre a origem e progressos das mathematicas em Portugal. Paris, 1819.
GHILLANY (F. W.). Geschichte des Seefohrers Ritter Martim Behaim. Nurnberg, 1853.
GLAS. History of the Canary Islands. London, 1764.
GOES (Damião de). Historia geral de Portugal e suas conquistas. Lisboa.
—— Chronica do Rey Dom Emanuel. Lisboa, 1619.
GOMARA (F. L. de). Historia de las Indias. Medina del Campo, 1553, fol.
GRÄBERG DE HEMSÖ. Annali de Geografia, &c. Genova, 1802.

HERCULANO and PAIVA. Roteiro da Viagem de Vasco da Gama. Lisboa, 1861.
HERRERA. Historia de las Indias. Madrid, 1730.
HUMBOLDT (Alexander von). Examen critique de l'histoire du Nouveau Continent. Paris, 1839.

INDICE Chronologico das navegações, viagens, descobrimentos e conquistas dos Portuguezes, &c. Lisboa, 1841.
ILACOMILUS (M.). Cosmographiæ Introductio. Deodate, 1507.

JANSEN (J.). Histoire Diplomatique du Chevalier Portugais Martin Behaim. Strasb. et Paris, 1802.

KERHALLET (Philippe de). Manuel de la Navigation à la Côte Occidentale d'Afrique. 3 tom. Paris, 1851-52.
KLAPROTH (Jules). Lettre à M. le Baron de Humboldt sur l'invention de la Boussole. Paris, 1834. 8vo.
KÖNIGLICH Bayerische Akademie der Wissenchaften. Abhandlungen den Philosophisch-philolog. Classe. Vierter Band. München, 1847.

LA CLÈDE. Histoire Générale de Portugal. Paris, 1735. 2 vols. in 4to., or 8 vols. in 12mo.
LAMBINET sur l'Origine de l'Imprimerie. Brux., 1799.
LIÃO (Duarte Nunez de). Chronicas dos reis de Portugal, reformados pelo licenciado. Lisboa, 1600.
—— Descripcão do reino de Portugal. Lisboa, 1610.
LIBRI (Count Guglielmo). Histoire des Sciences Mathématiques en Italie. Paris, 1838-41.
LOPES DE LIMA (J. J.). Ensaios sobre a Statistica das Possessões Portuguezas no Ultramar. Lisboa, 1846.
LUDD (Gualtier). Speculi Orbis succinctiss. Declaratio. Argent., 1507.
LUDOLFF. Historia Æthiopica. 1681.

MAFFEI (J. F.). Historiarum indicarum Libri xvi. 1593.
MANUEL DE MELLO (Francisco). Epanaphoras de Historia Portugueza. Lisboa, 1676.
MARIZ (Pedro de). Dialogos de varia historia sobre os reys de Portugal. Lisboa, 1594. 2 vols. in 8vo.
MARMOL CARAVAJAL. Descripcion de Africa. 1573.
MEMORIAS de Litteratura Portugueza. Tom. 8. Lisboa, 1812.
MORELET (A.). Journal du Voyage de Vasco da Gama en 1497. Lyon, 1864.
MUÑOZ (J. B.). Historia del Nuevo Mundo. Tom. i. Madrid, 1793.
MURPHY. History of the Mahometan Empire in Spain. London, 1816.

NAVARRETE (M. Fernandez de). Colleccion de Viajes. Madrid, 1825-37.
NOUVELLES Annales des Voyages, &c. Tom. 7, 8, 41, 42, 109, 110. Paris, 1820, &c.

PASQUAL. Descubrimiento de la Aguja Nautica. Madrid, 1789.
PIGAFETTA (Antonio). Viaggio attorno il Mondo, in Ramusio. 1 vol. 1550 fol.
PIGAFETTA (Felippe). Relatione del Reame di Congo et delle Circonvicine contrade, &c. Roma, 1591.
PIMENTEL (M.). Arte de Navegar. Lisboa, 1762.

RACZYNSKI. Les Arts en Portugal. Paris, 1846.
RAMUSIO. Viaggi. Venetia, 1588, &c.
RESENDE (Garcia de). Livro das obras de Garcia de Resende, que tracta da vida e grandissimas virtudes e bondades, etc., do excellentissimo Don Joam ho segundo deste nome. Evora, 1554, in fol.
RETRATOS e Elogios dos Varoes e Donas que illustram a Nação Portugueza. Tom. i. Lisboa, 1817.
REVISTA Litteraria. Tom. i.—iv. Porto, 1838.

SAN LUIZ (D. Francisco de). Obras Completas. Lisboa, 1855.
SCHOEFFER (H.). Geschichte von Portugal. Hamburg, 1836-9.
SILVA LOPES. Corografia ou Memoria Economica, Estadistica do Reino do Algarve. Lisboa, 1841.
SLANE. Abd Al-Rahman. Historie des Berbères, &c. Alger, 1852-4.
SOARES DA SYLVA. Memorias para a Historia de Portugal. João I. Lisboa, 1732.
SPRENGEL. Geographischen Entdeckungen. Halle, 1792.
—— Ursprung des Negerhandels. Halle, 1779.
SYLVA (Soares de). Memorias para a historia de Portugal que comprehendem o Governo del Rey João I. Lisboa, 1730-4. 4 vols. gr. in 4.

TORRES AMAT (Felix). Diccionario Critico de los Escritores Catalanes. Barcelona, 1836. 4to.

VARNHAGEN (F. A. de). Amerigo Vespucci. Lima, 1865.
VIERA Y CLAVIJO. Historia de las Islas de Canaria. Madrid, 1772. Vol. i.
VILLANUEVA. Noticia del Viage Literario. Valen., 1820.
VINCENT. Commerce and Navigation of the Ancients in the Indian Ocean. London, 1807.
VITET. Anciennes villes de France. Paris, 1833.

WAPPÆUS (E. J.). Untersuchungen über die geographischen Entdeckungen der Portugiesen unter Heinrich dem Seefahrer. Gottingue, 1842, in 8vo.

ZURLA (Placido). Il Mappemondo di fra Mauro. Venezia, 1806.
—— Di Marco Polo e degli altri Viaggiatori Veneziani. 2 parts. Venezia. 1818.

For Product Safety Concerns and Information please contact our EU
representative GPSR@taylorandfrancis.com
Taylor & Francis Verlag GmbH, Kaufingerstraße 24, 80331 München, Germany